Problems and Theorems in Linear Algebra

TRANSLATIONS OF
MATHEMATICAL
MONOGRAPHS
VOLUME 134

V. V. Prasolov

Problems and Theorems in Linear Algebra

American Mathematical Society

Виктор Васильевич Прасолов

ЗАДАЧИ И ТЕОРЕМЫ ЛИНЕЙНОЙ АЛГЕБРЫ

Translated by D. A. Leites from an original Russian manuscript
Translation edited by Simeon Ivanov

1991 *Mathematics Subject Classification.* Primary 15–01.

ABSTRACT. This book contains the basics of linear algebra with an emphasis on nonstandard and neat proofs of known theorems. Many of the theorems of linear algebra obtained mainly during the past thirty years are usually ignored in textbooks but are quite accessible for students majoring or minoring in mathematics. These theorems are given with complete proofs. There are about 230 problems with solutions.

Library of Congress Cataloging-in-Publication Data

Prasolov, V. V. (Viktor Vasil′evich)
[Zadachi i teoremy lineĭnoĭ algebry. English]
Problems and theorems in linear algebra/V. V. Prasolov; [translated by D. A. Leites from an original Russian manuscript; translation edited by Simeon Ivanov]
p. cm. — (Translations of mathematical monographs, ISSN 0065-9282; v. 134)
Includes bibliographical references.
ISBN 0-8218-0236-4
1. Algebras, Linear. I. Ivanov, Simeon. II. Title. III. Series.
QA184.P7313 1994 94-13332
512′.5—dc20 CIP

∞ The paper used in this book is acid-free and falls within the guidelines
established to ensure permanence and durability.
♻ Printed on recycled paper.

10 9 8 7 6 5 4 3 2   01 00 99 98 97 96

CONTENTS

PREFACE

There are very many books on linear algebra, among them many really wonderful ones (see the list of recommended literature). One might think that no more books on this subject are necessary. Choosing the words more carefully, it is possible to deduce that these books contain all that one needs and in the best possible form, and therefore any new book will, at best, only repeat the old ones.

This opinion is manifestly wrong, but nevertheless almost ubiquitous.

New results in linear algebra appear constantly and so do new, simpler and neater proofs of the known theorems. Besides, more than a few interesting old results are ignored by textbooks.

In this book I tried to collect the most attractive problems and theorems of linear algebra still accessible to students majoring in mathematics.

The computational aspects of linear algebra were left somewhat aside. The major part of the book contains results known from journal publications only. I believe that those results will be of interest to many readers.

I assume that the reader is acquainted with the main notions of linear algebra: linear space, basis, linear map, and the determinant of a matrix. Apart from this, all the essential theorems of the standard course of linear algebra are given here with complete proofs, and some definitions from the above list of prerequisites are recollected. I placed the prime emphasis on nonstandard neat proofs of known theorems.

In this book I only consider finite dimensional linear spaces.

The exposition is mostly performed over the fields of real or complex numbers. The peculiarity of the fields of finite characteristics is mentioned when needed.

Cross-references inside the book are natural: 36.2 means subsection 2 of §36; Problem 36.2 is Problem 2 from §36; Theorem 36.2.2 stands for Theorem 2 from 36.2.

ACKNOWLEDGMENTS. This book is based on a course I read at the Independent University of Moscow, 1991/92. I am thankful to the participants for comments and to D. V. Beklemishev, D. B. Fuchs, A. I. Kostrikin, V. S. Retakh, A. N. Rudakov, and A. P. Veselov for fruitful discussions of the manuscript.

For the second printing (1996), the author added the proof of the Kronecker theorem for pairs of linear maps (Sec. 12.6) and corrected several small errors.

MAIN NOTATIONS AND CONVENTIONS

$A = \begin{pmatrix} a_{11} & \dots & a_{1n} \\ \dots & \dots & \dots \\ a_{m1} & \dots & a_{mn} \end{pmatrix}$ denotes a matrix of size $m \times n$; we say that a square $n \times n$ matrix is of *order* n;

a_{ij}, sometimes denoted by $a_{i,j}$ for clarity, is the element or the entry from the intersection of the ith row and the jth column;

(a_{ij}) is another notation for the matrix A;

$\left\|a_{ij}\right\|_p^n$ still another notation for the matrix (a_{ij}), where $p \leq i, j \leq n$;

$\det(A)$, $|A|$ and $\det(a_{ij})$ all denote the *determinant* of the matrix A;

$|a_{ij}|_p^n$ is the *determinant* of the matrix $\left\|a_{ij}\right\|_p^n$;

E_{ij} — the (i, j)th *matrix unit* — the matrix whose only nonzero element is equal to 1 and occupies the (i, j)th position;

AB — the product of a matrix A of size $p \times n$ by a matrix B of size $n \times q$ — is the matrix (c_{ij}) of size $p \times q$, where $c_{ik} = \sum\limits_{j=1}^{n} a_{ij}b_{jk}$, is the scalar product of the ith row of the matrix A by the kth column of the matrix B;

$\operatorname{diag}(\lambda_1, \dots, \lambda_n)$ is the *diagonal matrix* of size $n \times n$ with elements $a_{ii} = \lambda_i$ and zero offdiagonal elements;

$I = \operatorname{diag}(1, \dots, 1)$ is the *unit matrix*; when its size, $n \times n$, is needed explicitly we denote the matrix by I_n;

the matrix aI, where a is a number, is called a *scalar* matrix;

A^T is the transposed of A, $A^T = (a'_{ij})$, where $a'_{ij} = a_{ji}$;

$\overline{A} = (a'_{ij})$, where $a'_{ij} = \overline{a_{ij}}$;

$A^* = \overline{A}^T$;

$\sigma = \binom{1 \dots n}{k_1 \dots k_n}$ is a permutation: $\sigma(i) = k_i$; the permutation $\binom{1 \dots n}{k_1 \dots k_n}$ is often abbreviated to $(k_1 \dots k_n)$;

$$\operatorname{sign} \sigma = (-1)^\sigma = \begin{cases} 1 & \text{if } \sigma \text{ is even,} \\ -1 & \text{if } \sigma \text{ is odd;} \end{cases}$$

$\operatorname{Span}(e_1, \dots, e_n)$ is the linear space spanned by the vectors $e_1, \dots, e_n$.

Given bases $e_1, \dots, e_n$ and $\varepsilon_1, \dots, \varepsilon_m$ in spaces V^n and W^m, respectively, we assign to a matrix A the operator $A : V^n \longrightarrow W^m$ which sends the vector $\begin{pmatrix} x_1 \\ \vdots \\ x_n \end{pmatrix}$ into the vector $\begin{pmatrix} y_1 \\ \vdots \\ y_m \end{pmatrix} = \begin{pmatrix} a_{11} & \dots & a_{1n} \\ \vdots & \dots & \vdots \\ a_{m1} & \dots & a_{mn} \end{pmatrix} \begin{pmatrix} x_1 \\ \vdots \\ x_n \end{pmatrix}$.

Since $y_i = \sum\limits_{j=1}^{n} a_{ij}x_j$, we have

$$A\left(\sum_{j=1}^{n} x_j e_j\right) = \sum_{i=1}^{m}\sum_{j=1}^{n} a_{ij}x_j\varepsilon_i;$$

in particular, $Ae_j = \sum\limits_{i} a_{ij}\varepsilon_i$;

everywhere except for §37 the notations $A > 0$, $A \geq 0$, $A < 0$, or $A \leq 0$ mean that a real symmetric or Hermitian matrix A is positive definite, nonnegative definite, negative definite, or nonpositive definite, respectively; $A > B$ means that $A - B > 0$; *whereas in* §37 *they mean that* $a_{ij} > 0$ *for all* i, j, *etc.*

Card M is the cardinality of the set M, i.e, the number of elements of M;

$A|_W$ denotes the restriction of the operator $A : V \longrightarrow V$ onto the subspace $W \subset V$;

sup is the least upper bound (supremum);

$\mathbb{Z}, \mathbb{Q}, \mathbb{R}, \mathbb{C}, \mathbb{H}, \mathbb{O}$ denote, as usual, the sets of all integer, rational, real, complex, quaternion and octonion numbers, respectively;

$\mathbb{N}$ denotes the set of all positive integers (without 0);

$\delta_{ij} = \begin{cases} 1 & \text{if } i = j, \\ 0 & \text{otherwise.} \end{cases}$

CHAPTER I

DETERMINANTS

The notion of a determinant appeared at the end of the 17th century in works of **Leibniz** (1646–1716) and a Japanese mathematician, **Seki Kova**, also known as **Takakazu** (1642–1708). Leibniz did not publish the results of his studies related to determinants. The best known is his letter to **l'Hôspital** (1693) in which Leibniz writes down the determinant condition of compatibility for a system of three linear equations in two unknowns. Leibniz particularly emphasized the usefulness of two indices when expressing the coefficients of the equations. In modern terms he actually wrote about the indices i, j in the expression $x_i = \sum_j a_{ij} y_j$.

Seki arrived at the notion of a determinant while solving the problem of finding common roots of algebraic equations.

In Europe, the search for common roots of algebraic equations soon also became the main trend associated with determinants. Newton, Bezout, and Euler studied this problem.

Seki did not have the general notion of the derivative at his disposal, but he actually got an algebraic expression equivalent to the derivative of a polynomial. He searched for multiple roots of a polynomial $f(x)$ as common roots of $f(x)$ and $f'(x)$. To find common roots of polynomials $f(x)$ and $g(x)$ (for f and g of small degrees) Seki got determinant expressions. The main treatise by Seki was published in 1674; there applications of the method are published, rather than the method itself. He kept the main method secret confiding only in his closest pupils.

In Europe, the first publication related to determinants, due to **Cramer**, appeared in 1750. In this work Cramer gave a determinant expression for a solution of the problem of finding the conic through 5 fixed points (this problem reduces to a system of linear equations).

The general theorems on determinants were proved only *ad hoc* when they were needed to solve some other problem. Therefore, the theory of determinants developed slowly, left behind as compared with the general development of mathematics. A systematic presentation of the theory of determinants is mainly associated with the names of **Cauchy** (1789–1857) and **Jacobi** (1804–1851).

1. Basic properties of determinants

The *determinant* of a square matrix $A = \left\|a_{ij}\right\|_1^n$ is the alternated sum

$$\sum_\sigma (-1)^\sigma a_{1\sigma(1)} a_{2\sigma(2)} \dots a_{n\sigma(n)},$$

where the summation is over all permutations $\sigma \in S_n$. The determinant of the matrix $A = \left\|a_{ij}\right\|_1^n$ is denoted by $\det A$ or $|a_{ij}|_1^n$. If $\det A \neq 0$, then A is called *invertible* or *nonsingular*.

The following properties are often used to compute determinants. The reader can easily verify (or recall) them.

1. Under the permutation of two rows of a matrix A its determinant changes the sign. In particular, if two rows of the matrix are identical, $\det A = 0$.

2. If A and B are square matrices, $\det\begin{pmatrix} A & C \\ 0 & B \end{pmatrix} = \det A \cdot \det B$.

3. $|a_{ij}|_1^n = \sum_{j=1}^n (-1)^{i+j} a_{ij} M_{ij}$, where M_{ij} is the determinant of the matrix obtained from A by crossing out the ith row and the jth column of A (the *row (echelon) expansion* of the determinant or, more precisely, *the expansion with respect to the ith row*).

(To prove this formula one has to group the factors of a_{ij}, where $j = 1, \dots, n$, for a fixed i.)

4.

$$\begin{vmatrix} \lambda\alpha_1 + \mu\beta_1 & a_{12} & \dots & a_{1n} \\ \vdots & \vdots & \dots & \vdots \\ \lambda\alpha_n + \mu\beta_n & a_{n2} & \dots & a_{nn} \end{vmatrix} = \lambda \begin{vmatrix} \alpha_1 & a_{12} & \dots & a_{1n} \\ \vdots & \vdots & \dots & \vdots \\ \alpha_n & a_{n2} & \dots & a_{nn} \end{vmatrix} + \mu \begin{vmatrix} \beta_1 & a_{12} & \dots & a_{1n} \\ \vdots & \vdots & \dots & \vdots \\ \beta_n & a_{n2} & \dots & a_{nn} \end{vmatrix}.$$

5. $\det(AB) = \det A \det B$.
6. $\det(A^T) = \det A$.

1.1. Before we start computing determinants, let us prove *Cramer's rule*. It appeared already in the first published paper on determinants.

THEOREM (Cramer's rule). *Consider a system of linear equations*

$$x_1 a_{i1} + \dots + x_n a_{in} = b_i \quad (i = 1, \dots, n),$$

i.e.,

$$x_1 A_1 + \dots + x_n A_n = B,$$

where A_j is the jth column of the matrix $A = \left\|a_{ij}\right\|_1^n$. Then

$$x_i \det(A_1, \dots, A_n) = \det(A_1, \dots, B, \dots, A_n),$$

where column B is inserted instead of A_i.

PROOF. Since for $j \neq i$ the determinant of the matrix $\det(A_1, \dots, A_j, \dots, A_n)$, a matrix with two identical columns, vanishes,

$$\begin{aligned} \det(A_1, \dots, B, \dots, A_n) &= \det\left(A_1, \dots, \textstyle\sum x_j A_j, \dots, A_n\right) \\ &= \sum x_j \det(A_1, \dots, A_j, \dots, A_n) = x_i \det(A_1, \dots, A_n). \quad \square \end{aligned}$$

If $\det(A_1, \dots, A_n) \neq 0$ the formula obtained can be used to find solutions of a system of linear equations.

1.2. One of the most often encountered determinants is the *Vandermonde determinant*, i.e., the determinant of the *Vandermonde matrix*

$$V(x_1,\dots,x_n)=\begin{vmatrix}1 & x_1 & x_1^2 & \dots & x_1^{n-1}\\ \vdots & \vdots & \vdots & \dots & \vdots\\ 1 & x_n & x_n^2 & \dots & x_n^{n-1}\end{vmatrix}=\prod_{i>j}(x_i-x_j).$$

To compute this determinant, let us subtract the $(k-1)$st column multiplied by x_1 from the kth one for $k=n,n-1,\dots,2$. The first row takes the form $(1,0,0,\dots,0)$, i.e., the computation of the Vandermonde determinant of order n reduces to a determinant of order $n-1$. Factorizing each row of the new determinant by bringing out x_i-x_1 we get

$$V(x_1,\dots,x_n)=\prod_{i>1}(x_i-x_1)\begin{vmatrix}1 & x_2 & x_2^2 & \dots & x_1^{n-2}\\ \vdots & \vdots & \vdots & \dots & \vdots\\ 1 & x_n & x_n^2 & \dots & x_n^{n-2}\end{vmatrix}.$$

For $n=2$ the identity $V(x_1,x_2)=x_2-x_1$ is obvious, hence,

$$V(x_1,\dots,x_n)=\prod_{i>j}(x_i-x_j).$$

Many of the applications of the Vandermonde determinant are occasioned by the fact that $V(x_1,\dots,x_n)=0$ if and only if there are two equal numbers among $x_1,\dots,x_n$.

1.3. The *Cauchy determinant* $|a_{ij}|_1^n$, where $a_{ij}=(x_i+y_j)^{-1}$, is slightly more difficult to compute than the Vandermonde determinant.

Let us prove by induction that

$$|a_{ij}|_1^n=\frac{\prod_{i>j}(x_i-x_j)(y_i-y_j)}{\prod_{i,j}(x_i+y_j)}.$$

For a base of induction take $|a_{ij}|_1^1=(x_1+y_1)^{-1}$.

The step of induction will be performed in two stages.

First, let us subtract the last column from each of the preceding ones. We get

$$a'_{ij}=(x_i+y_j)^{-1}-(x_i+y_n)^{-1}=(y_n-y_j)(x_i+y_n)^{-1}(x_i+y_j)^{-1} \text{ for } j\neq n.$$

Let us take out of each row the factors $(x_i+y_n)^{-1}$ and take out of each column, except the last one, the factors y_n-y_j. As a result we get the determinant $|b_{ij}|_1^n$, where $b_{ij}=a_{ij}$ for $j\neq n$ and $b_{in}=1$.

To compute this determinant, let us subtract the last row from each of the preceding ones. Taking out of each row, except the last one, the factors x_n-x_i and out of each column, except the last one, the factors $(x_n+y_j)^{-1}$ we make it possible to pass to a Cauchy determinant of lesser size.

1.4. A matrix A of the form

$$\begin{pmatrix} 0 & 1 & 0 & \dots & 0 & 0 \\ 0 & 0 & 1 & \dots & 0 & 0 \\ \vdots & \vdots & \ddots & \ddots & \ddots & \vdots \\ 0 & 0 & 0 & \ddots & 1 & 0 \\ 0 & 0 & 0 & \dots & 0 & 1 \\ a_0 & a_1 & a_2 & \dots & a_{n-2} & a_{n-1} \end{pmatrix}$$

is called *Frobenius's matrix* or the *companion matrix* of the polynomial

$$p(\lambda) = \lambda^n - a_{n-1}\lambda^{n-1} - a_{n-2}\lambda^{n-2} - \dots - a_0.$$

With the help of the expansion with respect to the first row it is easy to verify by induction that

$$\det(\lambda I - A) = \lambda^n - a_{n-1}\lambda^{n-1} - a_{n-2}\lambda^{n-2} - \dots - a_0 = p(\lambda).$$

1.5. Let $b_i, i \in \mathbb{Z}$, such that $b_k = b_l$ if $k \equiv l \pmod{n}$ be given; the matrix $\|a_{ij}\|_1^n$, where $a_{ij} = b_{i-j}$, is called a *circulant matrix*.

Let $\varepsilon_1, \dots, \varepsilon_n$ be distinct nth roots of unity; let

$$f(x) = b_0 + b_1 x + \dots + b_{n-1}x^{n-1}.$$

Let us prove that the determinant of the circulant matrix $|a_{ij}|_1^n$ is equal to

$$f(\varepsilon_1)f(\varepsilon_2)\dots f(\varepsilon_n).$$

It is easy to verify that for $n = 3$ we have

$$\begin{pmatrix} 1 & 1 & 1 \\ 1 & \varepsilon_1 & \varepsilon_1^2 \\ 1 & \varepsilon_2 & \varepsilon_2^2 \end{pmatrix} \begin{pmatrix} b_0 & b_2 & b_1 \\ b_1 & b_0 & b_2 \\ b_2 & b_1 & b_0 \end{pmatrix} = \begin{pmatrix} f(1) & f(1) & f(1) \\ f(\varepsilon_1) & \varepsilon_1 f(\varepsilon_1) & \varepsilon_1^2 f(\varepsilon_1) \\ f(\varepsilon_2) & \varepsilon_2 f(\varepsilon_2) & \varepsilon_2^2 f(\varepsilon_2) \end{pmatrix}$$
$$= f(1)f(\varepsilon_1)f(\varepsilon_2) \begin{pmatrix} 1 & 1 & 1 \\ 1 & \varepsilon_1 & \varepsilon_1^2 \\ 1 & \varepsilon_2 & \varepsilon_2^2 \end{pmatrix}.$$

Therefore,

$$V(1, \varepsilon_1, \varepsilon_2)|a_{ij}|_1^3 = f(1)f(\varepsilon_1)f(\varepsilon_2)V(1, \varepsilon_1, \varepsilon_2).$$

Taking into account that the Vandermonde determinant $V(1, \varepsilon_1, \varepsilon_2)$ does not vanish, we have:

$$|a_{ij}|_1^3 = f(1)f(\varepsilon_1)f(\varepsilon_2).$$

The proof of the general case is similar.

1.6. A *tridiagonal* matrix is a square matrix $J = \|a_{ij}\|_1^n$, where $a_{ij} = 0$ for $|i-j| > 1$.

Let $a_i = a_{ii}$ for $i = 1, \dots, n$, let $b_i = a_{i,i+1}$ and $c_i = a_{i+1,i}$ for $i = 1, \dots, n-1$. Then the tridiagonal matrix takes the form

$$\begin{pmatrix} a_1 & b_1 & 0 & \dots & 0 & 0 & 0 \\ c_1 & a_2 & b_2 & \dots & 0 & 0 & 0 \\ 0 & c_2 & a_3 & \ddots & 0 & 0 & 0 \\ \vdots & \vdots & \vdots & \ddots & \ddots & \vdots & \vdots \\ 0 & 0 & 0 & \ddots & a_{n-2} & b_{n-2} & 0 \\ 0 & 0 & 0 & \dots & c_{n-2} & a_{n-1} & b_{n-1} \\ 0 & 0 & 0 & \dots & 0 & c_{n-1} & a_n \end{pmatrix}.$$

To compute the determinant of this matrix we can make use of the following recurrent relation. Let $\Delta_0 = 1$ and $\Delta_k = |a_{ij}|_1^k$ for $k \geq 1$.

Expanding $\|a_{ij}\|_1^k$ with respect to the kth row it is easy to verify that

$$\Delta_k = a_k \Delta_{k-1} - b_{k-1} c_{k-1} \Delta_{k-2} \text{ for } k \geq 2.$$

The recurrence relation obtained indicates, in particular, that Δ_n (the determinant of J) depends not on the numbers b_i, c_j themselves but on their products of the form $b_i c_i$.

The quantity

$$(a_1 \dots a_n) = \begin{vmatrix} a_1 & 1 & 0 & \dots & 0 & 0 & 0 \\ -1 & a_2 & 1 & \dots & 0 & 0 & 0 \\ 0 & -1 & a_3 & \ddots & 0 & 0 & 0 \\ \vdots & \vdots & \ddots & \ddots & \ddots & \vdots & \\ 0 & 0 & 0 & \ddots & a_{n-2} & 1 & 0 \\ 0 & 0 & 0 & \ddots & -1 & a_{n-1} & 1 \\ 0 & 0 & 0 & \dots & 0 & -1 & a_n \end{vmatrix}$$

is associated with continued fractions, namely:

$$a_1 + \cfrac{1}{a_2 + \cfrac{1}{a_3 + \ddots + \cfrac{1}{a_{n-1} + \cfrac{1}{a_n}}}} = \frac{(a_1 a_2 \dots a_n)}{(a_2 a_3 \dots a_n)}.$$

Let us prove this equality by induction. Clearly,

$$a_1 + \frac{1}{a_2} = \frac{(a_1 a_2)}{(a_2)}.$$

It remains to demonstrate that

$$a_1 + \cfrac{1}{\cfrac{(a_2a_3\dots a_n)}{(a_3a_4\dots a_n)}} = \frac{(a_1a_2\dots a_n)}{(a_2a_3\dots a_n)},$$

i.e., $a_1(a_2\dots a_n) + (a_3\dots a_n) = (a_1a_2\dots a_n)$. But this identity is a corollary of the above recurrence relation, since $(a_1a_2\dots a_n) = (a_n\dots a_2a_1)$.

1.7. Under multiplication of a row of a square matrix by a number λ the determinant of the matrix is multiplied by λ. The determinant of the matrix does not vary when we replace one of the rows of the given matrix with its sum with any other row of the matrix. These statements allow a natural generalization to simultaneous transformations of several rows.

Consider the matrix $\begin{pmatrix} A_{11} & A_{12} \\ A_{21} & A_{22} \end{pmatrix}$, where A_{11} and A_{22} are square matrices of order m and n, respectively.

Let D be a square matrix of order m and B a matrix of size $n \times m$.

THEOREM. $\begin{vmatrix} DA_{11} & DA_{12} \\ A_{21} & A_{22} \end{vmatrix} = |D| \cdot |A|$ *and* $\begin{vmatrix} A_{11} & A_{12} \\ A_{21} + BA_{11} & A_{22} + BA_{12} \end{vmatrix} = |A|$.

PROOF.

$$\begin{pmatrix} DA_{11} & DA_{12} \\ A_{21} & A_{22} \end{pmatrix} = \begin{pmatrix} D & 0 \\ 0 & I \end{pmatrix} \begin{pmatrix} A_{11} & A_{12} \\ A_{21} & A_{22} \end{pmatrix} \text{ and}$$

$$\begin{pmatrix} A_{11} & A_{12} \\ A_{21} + BA_{11} & A_{22} + BA_{12} \end{pmatrix} = \begin{pmatrix} I & 0 \\ B & I \end{pmatrix} \begin{pmatrix} A_{11} & A_{12} \\ A_{21} & A_{22} \end{pmatrix}. \quad \square$$

Problems

1.1. Let $A = \left\|a_{ij}\right\|_1^n$ be skew-symmetric, i.e., $a_{ij} = -a_{ji}$, and let n be odd. Prove that $|A| = 0$.

1.2. Prove that the determinant of a skew-symmetric matrix of even order does not change if to all its elements we add the same number.

1.3. Compute the determinant of a skew-symmetric matrix A_n of order $2n$ with each element above the main diagonal being equal to 1.

1.4. Prove that for $n \geq 3$ the terms in the expansion of a determinant of order n cannot be all positive.

1.5. Let $a_{ij} = a^{|i-j|}$. Compute $|a_{ij}|_1^n$.

1.6. Let $\Delta_3 = \begin{vmatrix} 1 & -1 & 0 & 0 \\ x & h & -1 & 0 \\ x^2 & hx & h & -1 \\ x^3 & hx^2 & hx & h \end{vmatrix}$ and define Δ_n accordingly. Prove that $\Delta_n = (x+h)^n$.

1.7. Compute $|c_{ij}|_1^n$, where $c_{ij} = a_ib_j$ for $i \neq j$ and $c_{ii} = x_i$.

1.8. Let $a_{i,i+1} = c_i$ for $i = 1, \dots, n$, the other matrix elements being zero. Prove that the determinant of the matrix $I + A + A^2 + \dots + A^{n-1}$ is equal to $(1-c)^{n-1}$, where $c = c_1 \dots c_n$.

1.9. Compute $|a_{ij}|_1^n$, where $a_{ij} = (1 - x_iy_j)^{-1}$.

1.10. Let $a_{ij} = \binom{n+i}{j}$. Prove that $|a_{ij}|_0^m = 1$.

1.11. Prove that for any real numbers a, b, c, d, e and f

$$\begin{vmatrix} (a+b)de-(d+e)ab & ab-de & a+b-d-e \\ (b+c)ef-(e+f)bc & bc-ef & b+c-e-f \\ (c+d)fa-(f+a)cd & cd-fa & c+d-f-a \end{vmatrix} = 0.$$

Vandermonde's determinant.

1.12. Compute

$$\begin{vmatrix} 1 & x_1 & \dots & x_1^{n-2} & (x_2+x_3+\dots+x_n)^{n-1} \\ \vdots & \vdots & \dots & \vdots & \vdots \\ 1 & x_n & \dots & x_n^{n-2} & (x_1+x_2+\dots+x_{n-1})^{n-1} \end{vmatrix}.$$

1.13. Compute

$$\begin{vmatrix} 1 & x_1 & \dots & x_1^{n-2} & x_2x_3\dots x_n \\ \vdots & \vdots & \dots & \vdots & \vdots \\ 1 & x_n & \dots & x_n^{n-2} & x_1x_2\dots x_{n-1} \end{vmatrix}.$$

1.14. Compute $|a_{ik}|_0^n$, where $a_{ik} = \lambda_i^{n-k}(1+\lambda_i^2)^k$.

1.15. Let $V = \left\|a_{ij}\right\|_0^n$, where $a_{ij} = x_i^{j-1}$, be a Vandermonde matrix; let V_k be the matrix obtained from V by deleting its $(k+1)$st column (which consists of the kth powers) and adding instead the nth column consisting of the nth powers. Prove that

$$\det V_k = \sigma_{n-k}(x_1, \dots, x_n) \det V.$$

1.16. Let $a_{ij} = \binom{in}{j}$. Prove that $|a_{ij}|_1^r = n^{r(r+1)/2}$ for $r \le n$.

1.17. Given $k_1, \dots, k_n \in \mathbb{Z}$, compute $|a_{ij}|_1^n$, where

$$a_{i,j} = \begin{cases} \dfrac{1}{(k_i+j-i)!} & \text{for } k_i+j-i \ge 0\,, \\ a_{ij} = 0 & \text{for } k_i+j-i < 0. \end{cases}$$

1.18. Let $s_k = p_1x_1^k + \dots + p_nx_n^k$, and $a_{i,j} = s_{i+j}$. Prove that

$$|a_{ij}|_0^{n-1} = p_1 \dots p_n \prod_{i>j}(x_i - x_j)^2.$$

1.19. Let $s_k = x_1^k + \dots + x_n^k$. Compute

$$\begin{vmatrix} s_0 & s_1 & \dots & s_{n-1} & 1 \\ s_1 & s_2 & \dots & s_n & y \\ \vdots & \vdots & \dots & \vdots & \vdots \\ s_n & s_{n+1} & \dots & s_{2n-1} & y^n \end{vmatrix}.$$

1.20. Let $a_{ij} = (x_i + y_j)^n$. Prove that

$$|a_{ij}|_0^n = \binom{n}{1} \dots \binom{n}{n} \cdot \prod_{i>k} (x_i - x_k)(y_k - y_i).$$

1.21. Find all solutions of the system

$$\begin{cases} \lambda_1 + \dots + \lambda_n = 0 \\ \dots\dots\dots\dots \\ \lambda_1^n + \dots + \lambda_n^n = 0 \end{cases}$$

in $\mathbb{C}$.

1.22. Let $\sigma_k(x_0, \dots, x_n)$ be the kth elementary symmetric function. Set: $\sigma_0 = 1$, $\sigma_k(\widehat{x}_i) = \sigma_k(x_0, \dots, x_{i-1}, x_{i+1}, \dots, x_n)$. Prove that if $a_{ij} = \sigma_i(\widehat{x}_j)$, then $|a_{ij}|_0^n = \prod_{i<j}(x_i - x_j)$.

Relations among determinants.

1.23. Let $b_{ij} = (-1)^{i+j} a_{ij}$. Prove that $|a_{ij}|_1^n = |b_{ij}|_1^n$.

1.24. Prove that

$$\begin{vmatrix} a_1c_1 & a_2d_1 & a_1c_2 & a_2d_2 \\ a_3c_1 & a_4d_1 & a_3c_2 & a_4d_2 \\ b_1c_3 & b_2d_3 & b_1c_4 & b_2d_4 \\ b_3c_3 & b_4d_3 & b_3c_4 & b_4d_4 \end{vmatrix} = \begin{vmatrix} a_1 & a_2 \\ a_3 & a_4 \end{vmatrix} \cdot \begin{vmatrix} b_1 & b_2 \\ b_3 & b_4 \end{vmatrix} \cdot \begin{vmatrix} c_1 & c_2 \\ c_3 & c_4 \end{vmatrix} \cdot \begin{vmatrix} d_1 & d_2 \\ d_3 & d_4 \end{vmatrix}.$$

1.25. Prove that

$$\begin{vmatrix} a_1 & 0 & 0 & b_1 & 0 & 0 \\ 0 & a_2 & 0 & 0 & b_2 & 0 \\ 0 & 0 & a_3 & 0 & 0 & b_3 \\ b_{11} & b_{12} & b_{13} & a_{11} & a_{12} & a_{13} \\ b_{21} & b_{22} & b_{23} & a_{21} & a_{22} & a_{23} \\ b_{31} & b_{32} & b_{33} & a_{31} & a_{32} & a_{33} \end{vmatrix} = \begin{vmatrix} a_1a_{11} - b_1b_{11} & a_2a_{12} - b_2b_{12} & a_3a_{13} - b_3b_{13} \\ a_1a_{21} - b_1b_{21} & a_2a_{22} - b_2b_{22} & a_3a_{23} - b_3b_{23} \\ a_1a_{31} - b_1b_{31} & a_2a_{32} - b_2b_{32} & a_3a_{33} - b_3b_{33} \end{vmatrix}.$$

1.26. Let $s_k = \sum_{i=1}^n a_{ki}$. Prove that

$$\begin{vmatrix} s_1 - a_{11} & \dots & s_1 - a_{1n} \\ \vdots & \dots & \vdots \\ s_n - a_{n1} & \dots & s_n - a_{nn} \end{vmatrix} = (-1)^{n-1}(n-1) \begin{vmatrix} a_{11} & \dots & a_{1n} \\ \vdots & \dots & \vdots \\ a_{n1} & \dots & a_{nn} \end{vmatrix}.$$

1.27. Prove that

$$\begin{vmatrix} \binom{n}{m_1} & \binom{n}{m_1-1} & \dots & \binom{n}{m_1-k} \\ \vdots & \vdots & \dots & \vdots \\ \binom{n}{m_k} & \binom{n}{m_k-1} & \dots & \binom{n}{m_k-k} \end{vmatrix} = \begin{vmatrix} \binom{n}{m_1} & \binom{n+1}{m_1} & \dots & \binom{n+k}{m_1} \\ \vdots & \vdots & \dots & \vdots \\ \binom{n}{m_k} & \binom{n+1}{m_k} & \dots & \binom{n+k}{m_k} \end{vmatrix}.$$

1.28. Let $\Delta_n(k) = |a_{ij}|_0^n$, where $a_{ij} = \binom{k+i}{2j}$. Prove that

$$\Delta_n(k) = \frac{k(k+1)\dots(k+n-1)}{1 \cdot 3 \dots (2n-1)} \Delta_{n-1}(k-1).$$

1.29. Let $D_n = |a_{ij}|_0^n$, where $a_{ij} = \binom{n+i}{2j-1}$. Prove that $D_n = 2^{n(n+1)/2}$.

1.30. Given numbers $a_0, a_1, ..., a_{2n}$, let $b_k = \sum_{i=0}^k (-1)^i \binom{k}{i} a_i$ $(k = 0, \dots, 2n)$; let $a_{ij} = a_{i+j}$, and $b_{ij} = b_{i+j}$. Prove that $|a_{ij}|_0^n = |b_{ij}|_0^n$.

1.31. Let $A = \begin{pmatrix} A_{11} & A_{12} \\ A_{21} & A_{22} \end{pmatrix}$ and $B = \begin{pmatrix} B_{11} & B_{12} \\ B_{21} & B_{22} \end{pmatrix}$, where A_{11} and B_{11}, and also A_{22} and B_{22}, are square matrices of the same size such that $\operatorname{rank} A_{11} = \operatorname{rank} A$ and $\operatorname{rank} B_{11} = \operatorname{rank} B$. Prove that

$$\begin{vmatrix} A_{11} & B_{12} \\ A_{21} & B_{22} \end{vmatrix} \cdot \begin{vmatrix} A_{11} & A_{12} \\ B_{21} & B_{22} \end{vmatrix} = |A + B| \cdot |A_{11}| \cdot |B_{22}| .$$

1.32. Let A and B be square matrices of order n. Prove that $|A| \cdot |B| = \sum_{k=1}^n |A_k| \cdot |B_k|$, where the matrices A_k and B_k are obtained from A and B, respectively, by interchanging the respective first and kth columns, i.e., the first column of A is replaced with the kth column of B and the kth column of B is replaced with the first column of A.

2. Minors and cofactors

2.1. There are many instances when it is convenient to consider the determinant of the matrix whose elements stand at the intersection of certain p rows and p columns of a given matrix A. Such a determinant is called a *pth order minor of A*. For convenience we introduce the following notation:

$$A\begin{pmatrix} i_1 & \dots i_p \\ k_1 & \dots k_p \end{pmatrix} = \begin{vmatrix} a_{i_1 k_1} & a_{i_1 k_2} & \dots & a_{i_1 k_p} \\ \vdots & \vdots & \dots & \vdots \\ a_{i_p k_1} & a_{i_p k_2} & \dots & a_{i_p k_p} \end{vmatrix} .$$

If $i_1 = k_1, \dots, i_p = k_p$, the minor is called a *principal* one.

2.2. A nonzero minor of the maximal order is called a *basic minor* and its order is called the *rank* of the matrix.

THEOREM. *If $A\binom{i_1 \dots i_p}{k_1 \dots k_p}$ is a basic minor of a matrix A, then the rows of A are linear combinations of rows numbered $i_1, \dots, i_p$ and these rows are linearly independent.*

PROOF. The linear independence of the rows numbered $i_1, \dots, i_p$ is obvious since the determinant of a matrix with linearly dependent rows vanishes.

The cases when the size of A is $m \times p$ or $p \times m$ are also clear.

It suffices to carry out the proof for the minor $A\binom{1 \dots p}{1 \dots p}$. The determinant

$$\begin{vmatrix} a_{11} & \dots & a_{1p} & a_{1j} \\ \vdots & \dots & \vdots & \vdots \\ a_{p1} & \dots & a_{pp} & a_{pj} \\ a_{i1} & \dots & a_{ip} & a_{ij} \end{vmatrix}$$

vanishes for $j \le p$ as well as for $j > p$. Its expansion with respect to the last column is a relation of the form

$$a_{1j}c_1 + a_{2j}c_2 + \dots + a_{pj}c_p + a_{ij}c = 0,$$

where the numbers $c_1, \dots, c_p, c$ do not depend on j (but depend on i) and $c = A\binom{1\ \dots p}{1\ \dots p} \neq 0$. Hence, the ith row is equal to the linear combination of the first p rows with the coefficients $\frac{-c_1}{c}, \dots, \frac{-c_p}{c}$, respectively. □

2.2.1. COROLLARY. *If $A\binom{i_1 \dots i_p}{k_1 \dots k_p}$ is a basic minor, then all rows of A belong to the linear space spanned by the rows numbered $i_1, \dots, i_p$; therefore, the rank of A is equal to the maximal number of its linearly independent rows.*

2.2.2. COROLLARY. *The rank of a matrix is also equal to the maximal number of its linearly independent columns.*

2.3. THEOREM (The Binet-Cauchy formula). *Let A and B be matrices of size $n \times m$ and $m \times n$, respectively, and $n \leq m$. Then*

$$\det AB = \sum_{1 \leq k_1 < k_2 < \dots < k_n \leq m} A_{k_1 \dots k_n} B^{k_1 \dots k_n},$$

where $A_{k_1 \dots k_n}$ is the minor obtained from the columns of A whose numbers are $k_1, \dots, k_n$ and $B^{k_1 \dots k_n}$ is the minor obtained from the rows of B whose numbers are $k_1, \dots, k_n$.

PROOF. Let $C = AB$, $c_{ij} = \sum_{k=1}^m a_{ik} b_{ki}$. Then

$$\begin{aligned}
\det C &= \sum_\sigma (-1)^\sigma \sum_{k_1} a_{1k_1} b_{k_1 \sigma(1)} \cdots \sum_{k_n} b_{k_n \sigma(n)} \\
&= \sum_{k_1, \dots, k_n = 1}^m a_{1k_1} \dots a_{nk_n} \sum_\sigma (-1)^\sigma b_{k_1 \sigma(1)} \dots b_{k_n \sigma(n)} \\
&= \sum_{k_1, \dots, k_n = 1}^m a_{1k_1} \dots a_{nk_n} B^{k_1 \dots k_n}.
\end{aligned}$$

The minor $B^{k_1 \dots k_n}$ is nonzero only if the numbers $k_1, \dots, k_n$ are distinct; therefore, the summation can be performed over distinct numbers $k_1, \dots, k_n$. Since $B^{\tau(k_1) \dots \tau(k_n)} = (-1)^\tau B^{k_1 \dots k_n}$ for any permutation τ of the numbers $k_1, \dots, k_n$, we have

$$\begin{aligned}
\sum_{k_1, \dots, k_n = 1}^m a_{1k_1} \dots a_{nk_n} B^{k_1 \dots k_n} &= \sum_{k_1 < k_2 < \dots < k_n} (-1)^\tau a_{1\tau(1)} \dots a_{n\tau(n)} B^{k_1 \dots k_n} \\
&= \sum_{1 \leq k_1 < k_2 < \dots < k_n \leq m} A_{k_1 \dots k_n} B^{k_1 \dots k_n}. \quad \square
\end{aligned}$$

REMARK. Another proof is given in the solution of Problem 28.7

2.4. Recall the formula for expansion of the determinant of a matrix with respect to its ith row:

$$|a_{ij}|_1^n = \sum_{j=1}^n (-1)^{i+j} a_{ij} M_{ij}, \tag{1}$$

where M_{ij} is the determinant of the matrix obtained from the matrix $A = \left\|a_{ij}\right\|_1^n$ by deleting its ith row and jth column. The number $A_{ij} = (-1)^{i+j} M_{ij}$ is called the *cofactor of the element a_{ij} in A.*

It is possible to expand a determinant not only with respect to one row, but also with respect to several rows simultaneously.

Fix rows numbered $i_1, \dots, i_p$, where $i_1 < i_2 < \dots < i_p$. In the expansion of the determinant of A there occur products of terms of the expansion of the minor $A\binom{i_1 \dots i_p}{j_1 \dots j_p}$ by terms of the expansion of the minor $A\binom{i_{p+1} \dots i_n}{j_{p+1} \dots j_n}$, where $j_1 < \dots < j_p$; $i_{p+1} < \dots < i_n$; $j_{p+1} < \dots < j_n$ and there are no other terms in the expansion of the determinant of A.

To compute the signs of these products let us shuffle the rows and the columns so as to place the minor $A\binom{i_1 \dots i_p}{j_1 \dots j_p}$ in the upper left corner. To this end we have to perform

$$(i_1 - 1) + \dots + (i_p - p) + (j_1 - 1) + \dots + (j_p - p) \equiv i + j \pmod 2$$

permutations, where $i = i_1 + \dots + i_p$, $j = j_1 + \dots + j_p$.

The number $(-1)^{i+j} A\binom{i_{p+1} \dots i_n}{j_{p+1} \dots j_n}$ is called the *cofactor of the minor* $A\binom{i_1 \dots i_p}{j_1 \dots j_p}$.

We have proved the following statement:

2.4.1. THEOREM (Laplace). *Fix p rows of the matrix A. Then the sum of products of the minors of order p that belong to these rows by their cofactors is equal to the determinant of A.*

The matrix $\operatorname{adj} A = (A_{ij})^T$ is called the (*classical*) *adjoint*[1] of A. Let us prove that $A \cdot (\operatorname{adj} A) = |A| \cdot I$. To this end let us verify that $\sum_{j=1}^n a_{ij} A_{kj} = \delta_{ki}|A|$.

For $k = i$ this formula coincides with (1). If $k \neq i$, replace the kth row of A with the ith one. The determinant of the resulting matrix vanishes; its expansion with respect to the kth row results in the desired identity:

$$0 = \sum_{j=1}^n a'_{kj} A_{kj} = \sum_{j=1}^n a_{ij} A_{kj}.$$

If A is invertible, then $A^{-1} = \dfrac{\operatorname{adj} A}{|A|}$.

2.4.2. THEOREM. *The operation* adj *has the following properties:*

a) $\operatorname{adj} AB = \operatorname{adj} B \cdot \operatorname{adj} A$;

b) $\operatorname{adj} XAX^{-1} = X(\operatorname{adj} A)X^{-1}$;

c) *if $AB = BA$, then* $(\operatorname{adj} A)B = B(\operatorname{adj} A)$.

PROOF. If A and B are invertible matrices, then $(AB)^{-1} = B^{-1}A^{-1}$. Since for an invertible matrix A we have $\operatorname{adj} A = A^{-1}|A|$, headings a) and b) are obvious. Let us consider heading c).

If $AB = BA$ and A is invertible, then

$$A^{-1}B = A^{-1}(BA)A^{-1} = A^{-1}(AB)A^{-1} = BA^{-1}.$$

[1]We will briefly write *adjoint* instead of the *classical adjoint*.

Therefore, for invertible matrices the theorem is obvious.

In each of the equations a) – c) both sides continuously depend on the elements of A and B. Any matrix A can be approximated by matrices of the form $A_\varepsilon = A + \varepsilon I$ which are invertible for sufficiently small nonzero ε. (Actually, if $a_1, \dots, a_r$ is the whole set of eigenvalues of A, then A_ε is invertible for all $\varepsilon \neq -a_i$.) Besides, if $AB = BA$, then $A_\varepsilon B = BA_\varepsilon$. $\square$

2.5. The relations between the minors of a matrix A and the complementary to them minors of the matrix $(\operatorname{adj} A)^T$ are rather simple.

2.5.1. THEOREM. *Let $A = \left\|a_{ij}\right\|_1^n$, $(\operatorname{adj} A)^T = |A_{ij}|_1^n$, $1 \le p < n$. Then*

$$\begin{vmatrix} A_{11} & \dots & A_{1p} \\ \vdots & \dots & \vdots \\ A_{p1} & \dots & A_{pp} \end{vmatrix} = |A|^{p-1} \begin{vmatrix} a_{p+1,p+1} & \dots & a_{p+1,n} \\ \vdots & \dots & \vdots \\ a_{n,p+1} & \dots & a_{nn} \end{vmatrix}.$$

PROOF. For $p = 1$ the statement coincides with the definition of the cofactor A_{11}. Let $p > 1$. Then the identity

$$\begin{pmatrix} A_{11} & \dots & A_{1p} & A_{1,p+1} & \dots & A_{1n} \\ \vdots & \dots & \vdots & \vdots & \dots & \vdots \\ A_{p1} & \dots & A_{pp} & A_{p,p+1} & \dots & A_{pn} \\ & 0 & & & I & \end{pmatrix} \begin{pmatrix} a_{11} & \dots & a_{n1} \\ \vdots & \dots & \vdots \\ a_{1n} & \dots & a_{nn} \end{pmatrix}$$

$$= \begin{vmatrix} |A| & & 0 & & \\ & \dots & & & 0 \\ 0 & & |A| & & \\ a_{1,p+1} & \dots & & \dots & a_{n,p+1} \\ \vdots & \dots & & \dots & \vdots \\ a_{1n} & \dots & & \dots & a_{nn} \end{vmatrix}$$

implies that

$$\begin{vmatrix} A_{11} & \dots & A_{1p} \\ \vdots & \dots & \vdots \\ A_{p1} & \dots & A_{pp} \end{vmatrix} \cdot |A| = |A|^p \cdot \begin{vmatrix} a_{p+1,p+1} & \dots & a_{p+1,n} \\ \vdots & \dots & \vdots \\ a_{n,p+1} & \dots & a_{nn} \end{vmatrix}.$$

If $|A| \neq 0$, then dividing by $|A|$ we get the desired conclusion. For $|A| = 0$ the statement follows from the continuity of the both parts of the desired identity with respect to a_{ij}. $\square$

COROLLARY. *If A is not invertible, then* $\operatorname{rank}(\operatorname{adj} A) \le 1$.

PROOF. For $p = 2$ we get

$$\begin{vmatrix} A_{11} & A_{12} \\ A_{21} & A_{22} \end{vmatrix} = |A| \cdot \begin{vmatrix} a_{33} & \dots & a_{3n} \\ \vdots & \dots & \vdots \\ a_{n3} & \dots & a_{nn} \end{vmatrix} = 0.$$

Besides, the transposition of any two rows of the matrix A induces the same transposition of the columns of the adjoint matrix and all elements of the adjoint matrix change sign (look what happens with the determinant of A and with the matrix A^{-1} for an invertible A under such a transposition). $\square$

Application of transpositions of rows and columns makes it possible for us to formulate Theorem 2.5.1 in the following more general form.

2.5.2. THEOREM (Jacobi). *Let* $A = \|a_{ij}\|_1^n$, $(\operatorname{adj} A)^T = \|A_{ij}\|_1^n$, $1 \leq p < n$, $\sigma = \begin{pmatrix} i_1 & \dots & i_n \\ j_1 & \dots & j_n \end{pmatrix}$ *an arbitrary permutation. Then*

$$\begin{vmatrix} A_{i_1j_1} & \dots & A_{i_1j_p} \\ \vdots & \dots & \vdots \\ A_{i_pj_1} & \dots & A_{i_pj_p} \end{vmatrix} = (-1)^\sigma \begin{vmatrix} a_{i_{p+1},j_{p+1}} & \dots & a_{i_{p+1},j_n} \\ \vdots & \dots & \vdots \\ a_{i_n,j_{p+1}} & \dots & a_{i_n,j_n} \end{vmatrix} \cdot |A|^{p-1}.$$

PROOF. Let us consider matrix $B = \|b_{kl}\|_1^n$, where $b_{kl} = a_{i_kj_l}$. It is clear that $|B| = (-1)^\sigma|A|$. Since a transposition of any two rows (resp. columns) of A induces the same transposition of the columns (resp. rows) of the adjoint matrix and all elements of the adjoint matrix change their sings, $B_{kl} = (-1)^\sigma A_{i_kj_l}$.

Applying Theorem 2.5.1 to matrix B we get

$$\begin{vmatrix} (-1)^\sigma A_{i_1j_1} & \dots & (-1)^\sigma A_{i_1j_p} \\ \vdots & \dots & \vdots \\ (-1)^\sigma A_{i_pj_1} & \dots & (-1)^\sigma A_{i_pj_p} \end{vmatrix} = ((-1)^\sigma)^{p-1} \begin{vmatrix} a_{i_{p+1},j_{p+1}} & \dots & a_{i_{p+1},j_n} \\ \vdots & \dots & \vdots \\ a_{i_n,j_{p+1}} & \dots & a_{i_n,j_n} \end{vmatrix}.$$

By dividing both parts of this equality by $((-1)^\sigma)^p$ we obtain the desired. $\square$

2.6. In addition to the adjoint matrix of A it is sometimes convenient to consider the *compound matrix* $\|M_{ij}\|_1^n$ consisting of the $(n-1)$st order minors of A. The determinant of the adjoint matrix is equal to the determinant of the compound one (see e.g. Problem 1.23).

For a matrix A of size $m \times n$ we can also consider a matrix whose elements are rth order minors $A\begin{pmatrix} i_1 & \dots & i_r \\ j_1 & \dots & j_r \end{pmatrix}$, where $r \leq \min(m,n)$. The resulting matrix $C_r(A)$ is called the *rth compound matrix of A*. For example, if $m = n = 3$ and $r = 2$, then

$$C_2(A) = \begin{pmatrix} A\begin{pmatrix}12\\12\end{pmatrix} & A\begin{pmatrix}12\\13\end{pmatrix} & A\begin{pmatrix}12\\23\end{pmatrix} \\ A\begin{pmatrix}13\\12\end{pmatrix} & A\begin{pmatrix}13\\13\end{pmatrix} & A\begin{pmatrix}13\\23\end{pmatrix} \\ A\begin{pmatrix}23\\12\end{pmatrix} & A\begin{pmatrix}23\\13\end{pmatrix} & A\begin{pmatrix}23\\23\end{pmatrix} \end{pmatrix}.$$

Making use of Binet-Cauchy's formula we can show that $C_r(AB) = C_r(A)C_r(B)$. For a square matrix A of order n we have Sylvester's identity

$$\det C_r(A) = (\det A)^p, \text{ where } p = \binom{n-1}{r-1}.$$

The simplest proof of this statement makes use of the notion of exterior power (see Theorem 28.5.3).

2.7. Let $1 \le m \le r < n$, $A = \|a_{ij}\|_1^n$. Set $A_n = |a_{ij}|_1^n$, $A_m = |a_{ij}|_1^m$. Consider the matrix $S_{m,n}^r$ whose elements are the rth order minors of A containing the left upper corner principal minor A_m. The determinant of $S_{m,n}^r$ is a minor of order $\binom{n-m}{r-m}$ of $C_r(A)$. The determinant of $S_{m,n}^r$ can be expressed in terms of A_m and A_n.

THEOREM (Generalized Sylvester's identity, [Mohr,1953]).

$$(1) \qquad |S_{m,n}^r| = A_m^p A_n^q, \text{ where } p = \binom{n-m-1}{r-m}, q = \binom{n-m-1}{r-m-1}.$$

PROOF. Let us prove identity (1) by induction on n. For $n = 2$ it is obvious.

The matrix $S_{0,n}^r$ coincides with $C_r(A)$ and since $|C_r(A)| = A_n^q$, where $q = \binom{n-1}{r-1}$ (see Theorem 28.5.3), then (1) holds for $m = 0$ (we assume that $A_0 = 1$). Both sides of (1) are continuous with respect to a_{ij} and, therefore, it suffices to prove the inductive step when $a_{11} \neq 0$.

All considered minors contain the first row and, therefore, from the rows whose numbers are $2, \dots, n$ we can subtract the first row multiplied by an arbitrary factor; this operation does not affect $\det(S_{m,n}^r)$. With the help of this operation all elements of the first column of A except a_{11} can be made equal to zero. Let $\overline{A}$ be the matrix obtained from the new one by strikinging out the first column and the first row, and let $\overline{S}_{m-1,n-1}^{r-1}$ be the matrix composed of the minors of order $r-1$ of $\overline{A}$ containing its left upper corner principal minor of order $m-1$.

Obviously, $S_{m,n}^r = a_{11}\overline{S}_{m-1,n-1}^{r-1}$ and we can apply to $\overline{S}_{m-1,n-1}^{r-1}$ the inductive hypothesis (the case $m-1=0$ was considered separately). Besides, if $\overline{A}_{m-1}$ and $\overline{A}_{n-1}$ are the left upper corner principal minors of orders $m-1$ and $n-1$ of A, respectively, then $A_m = a_{11}\overline{A}_{m-1}$ and $A_n = a_{11}\overline{A}_{n-1}$. Therefore,

$$|S_{m,n}^r| = a_{11}^t \overline{A}_{m-1}^{p_1} \overline{A}_{n-1}^{q_1} = a_{11}^{t-p_1-q_1} A_m^{p_1} A_n^{q_1},$$

where $t = \binom{n-m}{r-m}$, $p_1 = \binom{n-m-1}{r-m} = p$ and $q_1 = \binom{n-m-1}{r-m-1} = q$. Taking into account that $t = p + q$, we get the desired conclusion. $\square$

REMARK. Sometimes the term "Sylvester's identity" is applied to identity (1) not only for $m = 0$ but also for $r = m+1$, i.e., $|S_{m,n}^{m+1}| = A_m^{n-m} A_n$

2.8 THEOREM (Chebotarev). *Let p be a prime and $\varepsilon = \exp(2\pi i/p)$. Then all minors of the Vandermonde matrix $\|a_{ij}\|_0^{p-1}$, where $a_{ij} = \varepsilon^{ij}$, are nonzero.*

PROOF (Following [Reshetnyak, 1955]). Suppose that

$$\begin{vmatrix} \varepsilon^{k_1 l_1} & \dots & \varepsilon^{k_1 l_j} \\ \varepsilon^{k_2 l_1} & \dots & \varepsilon^{k_2 l_j} \\ \vdots & \dots & \vdots \\ \varepsilon^{k_j l_1} & \dots & \varepsilon^{k_j l_j} \end{vmatrix} = 0.$$

Then there exist complex numbers $c_1, \dots, c_j$ not all equal to 0 such that the linear combination of the corresponding columns with coefficients $c_1, \dots, c_j$ vanishes, i.e., the numbers $\varepsilon^{k_1}, \dots, \varepsilon^{k_j}$ are roots of the polynomial $c_1 x^{l_1} + \dots + c_j x^{l_j}$. Let

$$(1) \qquad (x - \varepsilon^{k_1}) \dots (x - \varepsilon^{k_j}) = x^j - b_1 x^{j-1} + \dots \pm b_j.$$

Then

$$(2) \qquad c_1 x^{l_1} + \dots + c_j x^{l_j} = (b_0 x^j - b_1 x^{j-1} + \dots \pm b_j)(a_s x^s + \dots + a_0),$$

where $b_0 = 1$ and $a_s \neq 0$. For convenience let us assume that $b_t = 0$ for $t > j$ and $t < 0$. The coefficient of x^{j+s-t} in the right-hand side of (2) is equal to $\pm(a_s b_t - a_{s-1}b_{t-1} + \dots \pm a_0 b_{t-s})$. The degree of the polynomial (2) is equal to $s + j$ and it is only the coefficients of the monomials of degrees $l_1, \dots, l_j$ that may be nonzero and, therefore, there are $s+1$ zero coefficients:

$$a_s b_t - a_{s-1}b_{t-1} + \dots \pm a_0 b_{t-s} = 0 \text{ for } t = t_0, t_1, \dots, t_s.$$

The numbers $a_0, \dots, a_{s-1}, a_s$ are not all zero and therefore, $|c_{kl}|_0^s = 0$ for $c_{kl} = b_t$, where $t = t_k - l$.

Formula (1) shows that b_t can be represented in the form $f_t(\varepsilon)$, where f_t is a polynomial with integer coefficients and this polynomial is the sum of $\binom{j}{t}$ powers of ε; hence, $f_t(1) = \binom{j}{t}$. Since $c_{kl} = b_t = f_t(\varepsilon)$, we have $|c_{kl}|_0^s = g(\varepsilon)$ and $g(1) = |c'_{kl}|_0^s$, where $c'_{kl} = \binom{j}{t_k - l}$. The polynomial $q(x) = x^{p-1} + \dots + x + 1$ is irreducible over $\mathbb{Z}$ (see Appendix 2) and $q(\varepsilon) = 0$. Therefore, $g(x) = q(x)\varphi(x)$, where φ is a polynomial with integer coefficients (see Appendix 1). Therefore, $g(1) = q(1)\varphi(1) = p\varphi(1)$, i.e., $g(1)$ is divisible by p.

To get a contradiction it suffices to show that the number $g(1) = |c'_{kl}|_0^s$, where $c'_{kl} = \binom{j}{t_k - l}$, $0 \le t_k \le j + s$ and $0 < j + s \le p - 1$, is not divisible by p. It is easy to verify that $\Delta = |c'_{kl}|_0^s = |a_{kl}|_0^s$, where $a_{kl} = \binom{j+l}{t_k}$ (see Problem 1.27). It is also clear that

$$\binom{j+l}{t} = \left(1 - \frac{t}{j+l+1}\right) \dots \left(1 - \frac{t}{j+s}\right) \binom{j+s}{t} = \varphi_{s-l}(t) \binom{j+s}{t}.$$

Hence,

$$\Delta = \prod_{\lambda=0}^{s} \binom{j+s}{t_\lambda} \begin{vmatrix} \varphi_s(t_0) & \varphi_{s-1}(t_0) & \dots & 1 \\ \varphi_s(t_1) & \varphi_{s-1}(t_1) & \dots & 1 \\ \vdots & \vdots & \dots & \vdots \\ \varphi_s(t_s) & \varphi_{s-1}(t_s) & \dots & 1 \end{vmatrix} = \pm \prod_{\lambda=0}^{s} \left(\binom{j+s}{t_\lambda} A_\lambda \right) \prod_{\mu > \nu} (t_\mu - t_\nu),$$

where $A_0, A_1, \dots, A_s$ are the coefficients of the highest powers of t in the polynomials $\varphi_0(t), \varphi_1(t), \dots, \varphi_s(t)$, respectively, where $\varphi_0(t) = 1$; the degree of $\varphi_i(t)$ is equal to i. Clearly, the product obtained has no irreducible fractions with numerators divisible by p, because $j + s < p$. □

Problems

2.1. Let A_n be a matrix of size $n \times n$. Prove that $|A + \lambda I| = \lambda^n + \sum_{k=1}^{n} S_k \lambda^{n-k}$, where S_k is the sum of all $\binom{n}{k}$ principal kth order minors of A.

2.2. Prove that

$$\begin{vmatrix} a_{11} & \dots & a_{1n} & x_1 \\ \vdots & \dots & \vdots & \vdots \\ a_{n1} & \dots & a_{nn} & x_n \\ y_1 & \dots & y_n & 0 \end{vmatrix} = -\sum_{i,j} x_i y_j A_{ij},$$

where A_{ij} is the cofactor of a_{ij} in $\|a_{ij}\|_1^n$.

2.3. Prove that the sum of principal k-minors of $A^T A$ is equal to the sum of squares of all k-minors of A.

2.4. Prove that

$$\begin{vmatrix} u_1 a_{11} & \dots & u_n a_{1n} \\ a_{21} & \dots & a_{2n} \\ \vdots & \dots & \vdots \\ a_{n1} & \dots & a_{nn} \end{vmatrix} + \dots + \begin{vmatrix} a_{11} & \dots & a_{1n} \\ a_{21} & \dots & a_{2n} \\ \vdots & \dots & \vdots \\ u_1 a_{n1} & \dots & u_n a_{nn} \end{vmatrix} = (u_1 + \dots + u_n)|A|.$$

Inverse and adjoint matrices

2.5. Let A and B be square matrices of order n. Compute

$$\begin{pmatrix} I & A & C \\ 0 & I & B \\ 0 & 0 & I \end{pmatrix}^{-1}.$$

2.6. Prove that the matrix inverse to an invertible upper triangular matrix is also an upper triangular one.

2.7. Give an example of a matrix of order n whose adjoint has only one nonzero element and this element is situated in the ith row and jth column for given i and j.

2.8. Let x and y be columns of length n. Prove that

$$\operatorname{adj}(I - xy^T) = xy^T + (1 - y^T x)I.$$

2.9. Let A be a skew-symmetric matrix of order n. Prove that $\operatorname{adj} A$ is a symmetric matrix for odd n and a skew-symmetric one for even n.

2.10. Let A_n be a skew-symmetric matrix of order n with elements $+1$ above the main diagonal. Calculate $\operatorname{adj} A_n$.

2.11. The matrix $\operatorname{adj}(A - \lambda I)$ can be expressed in the form $\sum_{k=0}^{n-1} \lambda^k A_k$, where n is the order of A. Prove that:

a) for any k $(1 \leq k \leq n-1)$ the matrix $A_k A - A_{k-1}$ is a scalar matrix;

b) the matrix A_{n-s} can be expressed as a polynomial of degree $s-1$ in A.

2.12. Find all matrices A with nonnegative elements such that all elements of A^{-1} are also nonnegative.

2.13. Let $\varepsilon = \exp(2\pi i/n)$; $A = \left\|a_{ij}\right\|_1^n$, where $a_{ij} = \varepsilon^{ij}$. Calculate the matrix A^{-1}.

2.14. Calculate the matrix inverse to the Vandermonde matrix V.

3. The Schur complement

3.1. Let $P = \begin{pmatrix} A & B \\ C & D \end{pmatrix}$ be a block matrix with square matrices A and D. In order to facilitate the computation of $\det P$ we can factorize the matrix P as follows:

$$\begin{pmatrix} A & B \\ C & D \end{pmatrix} = \begin{pmatrix} A & 0 \\ C & I \end{pmatrix} \begin{pmatrix} I & Y \\ 0 & X \end{pmatrix} = \begin{pmatrix} A & AY \\ C & CY + X \end{pmatrix}. \tag{1}$$

The equations $B = AY$ and $D = CY + X$ are solvable when the matrix A is invertible. In this case $Y = A^{-1}B$ and $X = D - CA^{-1}B$. The matrix $D - CA^{-1}B$ is called *the Schur complement of A in P*, and is denoted by $(P|A)$. It is clear that $\det P = \det A \det(P|A)$.

It is easy to verify that

$$\begin{pmatrix} A & AY \\ C & CY+X \end{pmatrix} = \begin{pmatrix} A & 0 \\ C & X \end{pmatrix} \begin{pmatrix} I & Y \\ 0 & I \end{pmatrix}.$$

Therefore, instead of the factorization (1) we can write

$$(2)\quad P = \begin{pmatrix} A & 0 \\ C & (P|A) \end{pmatrix} \begin{pmatrix} I & A^{-1}B \\ 0 & I \end{pmatrix} = \begin{pmatrix} I & 0 \\ CA^{-1} & I \end{pmatrix} \begin{pmatrix} A & 0 \\ 0 & (P|A) \end{pmatrix} \begin{pmatrix} I & A^{-1}B \\ 0 & I \end{pmatrix}.$$

If the matrix D is invertible we have an analogous factorization

$$P = \begin{pmatrix} I & BD^{-1} \\ 0 & I \end{pmatrix} \begin{pmatrix} A - BD^{-1}C & 0 \\ 0 & D \end{pmatrix} \begin{pmatrix} I & 0 \\ D^{-1}C & I \end{pmatrix}.$$

We have proved the following assertion.

3.1.1. THEOREM. a) *If $|A| \neq 0$, then $|P| = |A| \cdot |D - CA^{-1}B|$;*
b) *If $|D| \neq 0$, then $|P| = |A - BD^{-1}C| \cdot |D|$.*

Another application of the factorization (2) is a computation of P^{-1}. Clearly,

$$\begin{pmatrix} I & X \\ 0 & I \end{pmatrix}^{-1} = \begin{pmatrix} I & -X \\ 0 & I \end{pmatrix}.$$

This fact together with (2) gives us formula

$$P^{-1} = \begin{pmatrix} A^{-1} + A^{-1}BX^{-1}CA^{-1} & -A^{-1}BX^{-1} \\ -X^{-1}CA^{-1} & X^{-1} \end{pmatrix}, \text{ where } X = (P|A).$$

3.1.2. THEOREM. *If A and D are square matrices of order n, $|A| \neq 0$, and $AC = CA$, then $|P| = |AD - CB|$.*

PROOF. By Theorem 3.1.1

$$|P| = |A| \cdot |D - CA^{-1}B| = |AD - ACA^{-1}B| = |AD - CB|. \quad \square$$

Is the above condition $|A| \neq 0$ necessary? The answer is "no", but in certain similar situations the answer is "yes". If, for instance, $CD^T = -DC^T$, then

$$|P| = |A - BD^{-1}C| \cdot |D^T| = |AD^T + BC^T|.$$

This equality holds for any invertible matrix D. But if

$$A = \begin{pmatrix} 1 & 0 \\ 0 & 0 \end{pmatrix}, \quad B = \begin{pmatrix} 0 & 0 \\ 0 & 1 \end{pmatrix}, \quad C = \begin{pmatrix} 0 & 1 \\ 0 & 0 \end{pmatrix} \text{ and } D = \begin{pmatrix} 0 & 0 \\ 1 & 0 \end{pmatrix},$$

then

$$CD^T = -DC^T = 0 \text{ and } |AD^T + BC^T| = -1 \neq 1 = P.$$

Let us return to Theorem 3.1.2. The equality $|P| = |AD - CB|$ is a polynomial identity for the elements of the matrix P. Therefore, if there exist invertible matrices A_ε such that $\lim\limits_{\varepsilon \to 0} A_\varepsilon = A$ and $A_\varepsilon C = CA_\varepsilon$, then this equality holds for the matrix A as well. Given any matrix A, consider $A_\varepsilon = A + \varepsilon I$. It is easy to see (cf. 2.4.2) that the matrices A_ε are invertible for every sufficiently small nonzero ε, and if $AC = CA$, then $A_\varepsilon C = CA_\varepsilon$. Hence, Theorem 3.1.2 is true even if $|A| = 0$.

3.1.3. THEOREM. *Suppose u is a row, v is a column, and a is a number. Then*

$$\begin{vmatrix} A & v \\ u & a \end{vmatrix} = a\,|A| - u(\operatorname{adj} A)v.$$

PROOF. By Theorem 3.1.1

$$\begin{vmatrix} A & v \\ u & a \end{vmatrix} = |A|(a - uA^{-1}v) = a\,|A| - u(\operatorname{adj} A)v$$

if the matrix A is invertible. Both sides of this equality are polynomial functions of the elements of A. Hence, the theorem is true, by continuity, for noninvertible A as well. □

3.2. Let $A = \begin{vmatrix} A_{11} & A_{12} & A_{13} \\ A_{21} & A_{22} & A_{23} \\ A_{31} & A_{32} & A_{33} \end{vmatrix}$, $B = \begin{vmatrix} A_{11} & A_{12} \\ A_{21} & A_{22} \end{vmatrix}$ and $C = A_{11}$ be square matrices, and let B and C be invertible. The matrix $(B|C) = A_{22} - A_{21}A_{11}^{-1}A_{12}$ may be considered as a submatrix of the matrix

$$(A|C) = \begin{pmatrix} A_{22} & A_{23} \\ A_{32} & A_{33} \end{pmatrix} - \begin{pmatrix} A_{21} \\ A_{31} \end{pmatrix} A_{11}^{-1}(A_{12}\; A_{13}).$$

THEOREM (Emily Haynsworth). $(A|B) = ((A|C)|(B|C))$.

PROOF (Following [Ostrowski, 1973]). Consider two factorizations of A:

$$(1) \qquad A = \begin{pmatrix} A_{11} & 0 & 0 \\ A_{21} & I & 0 \\ A_{31} & 0 & I \end{pmatrix} \begin{pmatrix} I & * & * \\ 0 & \multicolumn{2}{c}{(A|C)} \\ 0 & & \end{pmatrix},$$

$$(2) \qquad A = \begin{pmatrix} A_{11} & A_{12} & 0 \\ A_{21} & A_{22} & 0 \\ A_{31} & A_{32} & I \end{pmatrix} \begin{pmatrix} I & 0 & * \\ 0 & I & * \\ 0 & 0 & (A|B) \end{pmatrix}.$$

For the Schur complement of A_{11} in the left factor of (2) we can write a similar factorization

$$(3)\qquad \begin{pmatrix} A_{11} & A_{12} & 0 \\ A_{21} & A_{22} & 0 \\ A_{31} & A_{32} & I \end{pmatrix} = \begin{pmatrix} A_{11} & 0 & 0 \\ A_{21} & I & 0 \\ A_{31} & 0 & I \end{pmatrix} \begin{pmatrix} I & X_1 & X_2 \\ 0 & X_3 & X_4 \\ 0 & X_5 & X_6 \end{pmatrix}.$$

Since A_{11} is invertible, we derive from (1), (2) and (3) after simplification (division by the same factors):

$$\begin{pmatrix} I & * & * \\ 0 & & \\ 0 & \multicolumn{2}{c}{(A|C)} \end{pmatrix} = \begin{pmatrix} I & X_1 & X_2 \\ 0 & X_3 & X_4 \\ 0 & X_5 & X_6 \end{pmatrix} \begin{pmatrix} I & 0 & * \\ 0 & I & * \\ 0 & 0 & (A|B) \end{pmatrix}.$$

It follows that

$$(A|C) = \begin{pmatrix} X_3 & X_4 \\ X_5 & X_6 \end{pmatrix} \begin{pmatrix} I & * \\ 0 & (A|B) \end{pmatrix}.$$

To finish the proof we only have to verify that $X_3 = (B|C)$, $X_4 = 0$ and $X_6 = I$. Equating the last columns in (3), we get $0 = A_{11}X_2$, $0 = A_{21}X_2 + X_4$ and $I = A_{31}X_2 + X_6$. The matrix A_{11} is invertible; therefore, $X_2 = 0$. It follows that $X_4 = 0$ and $X_6 = I$. Another straightforward consequence of (3) is

$$\begin{pmatrix} A_{11} & A_{12} \\ A_{21} & A_{22} \end{pmatrix} = \begin{pmatrix} A_{11} & 0 \\ A_{21} & I \end{pmatrix} \begin{pmatrix} I & X_1 \\ 0 & X_3 \end{pmatrix},$$

i.e., $X_3 = (B|C)$. □

Problems

3.1. Let u and v be rows of length n, A a square matrix of order n. Prove that

$$|A + u^T v| = |A| + v(\operatorname{adj} A)u^T.$$

3.2. Let A be a square matrix. Prove that

$$\begin{vmatrix} I & A \\ A^T & I \end{vmatrix} = 1 - \sum M_1^2 + \sum M_2^2 - \sum M_3^2 + \dots,$$

where $\sum M_k^2$ is the sum of the squares of all k-minors of A.

4. Symmetric functions, sums $x_1^k + \dots + x_n^k$, and Bernoulli numbers

In this section we will obtain determinant relations for elementary symmetric functions $\sigma_k(x_1, \dots, x_n)$, functions $s_k(x_1, \dots, x_n) = x_1^k + \dots + x_n^k$, and sums of homogeneous monomials of degree k,

$$p_k(x_1, \dots, x_n) = \sum_{i_1 + \dots + i_n = k} x_1^{i_1} \dots x_n^{i_n}.$$

4.1. Let $\sigma_k(x_1, \dots, x_n)$ be the kth elementary function, i.e., the coefficient of x^{n-k} in the standard power series expression of the polynomial $(x+x_1)\dots(x+x_n)$. We will assume that $\sigma_k(x_1, \dots, x_n) = 0$ for $k > n$. First of all, let us prove that

$$s_k - s_{k-1}\sigma_1 + s_{k-2}\sigma_2 - \dots + (-1)^k k\sigma_k = 0.$$

The product $s_{k-p}\sigma_p$ consists of terms of the form $x_i^{k-p}(x_{j_1}\dots x_{j_p})$. If $i \in \{j_1, \dots j_p\}$, then this term cancels the term $x_i^{k-p+1}(x_{j_1}\dots \widehat{x}_i \dots x_{j_p})$ of the product $s_{k-p+1}\sigma_{p-1}$, and if $i \notin \{j_1, \dots, j_p\}$, then it cancels the term $x_i^{k-p-1}(x_i x_{j_1}\dots x_{j_p})$ of the product $s_{k-p-1}\sigma_{p+1}$.

Consider the relations

$$\begin{aligned} \sigma_1 &= s_1 \\ s_1\sigma_1 - 2\sigma_2 &= s_2 \\ s_2\sigma_1 - s_1\sigma_2 + 3\sigma_3 &= s_3 \\ &\dots\dots\dots\dots \\ s_k\sigma_1 - s_{k-1}\sigma_2 + \dots + (-1)^{k+1}k\sigma_k &= s_k \end{aligned}$$

as a system of linear equations for $\sigma_1, \dots, \sigma_k$. With the help of Cramer's rule it is easy to see that

$$\sigma_k = \frac{1}{k!}\begin{vmatrix} s_1 & 1 & 0 & 0 & \dots & 0 \\ s_2 & s_1 & 1 & 0 & \dots & 0 \\ s_3 & s_2 & s_1 & 1 & \dots & 0 \\ \vdots & \vdots & \vdots & \vdots & \ddots & \vdots \\ s_{k-1} & s_{k-2} & \dots & \dots & \ddots & 1 \\ s_k & s_{k-1} & \dots & \dots & \dots & s_1 \end{vmatrix}.$$

Similarly,

$$s_k = \begin{vmatrix} \sigma_1 & 1 & 0 & 0 & \dots & 0 \\ 2\sigma_2 & \sigma_1 & 1 & 0 & \dots & 0 \\ 3\sigma_3 & \sigma_2 & \sigma_1 & 1 & \dots & 0 \\ \vdots & \vdots & \vdots & \ddots & \ddots & \vdots \\ (k-1)\sigma_{k-1} & \sigma_{k-2} & \dots & \dots & \dots & 1 \\ k\sigma_k & \sigma_{k-1} & \dots & \dots & \dots & \sigma_1 \end{vmatrix}.$$

4.2. Let us obtain first a relation between p_k and σ_k and then a relation between p_k and s_k. It is easy to verify that

$$\begin{aligned} 1 + p_1t + p_2t^2 + p_3t^3 + \dots &= (1 + x_1t + (x_1t)^2 + \dots)\dots(1 + x_nt + (x_nt)^2 + \dots) \\ &= \frac{1}{(1-x_1t)\dots(1-x_nt)} = \frac{1}{1 - \sigma_1t + \sigma_2t^2 - \dots + (-1)^n\sigma_nt^n}, \end{aligned}$$

i.e.,

$$\begin{aligned} p_1 - \sigma_1 &= 0 \\ p_2 - p_1\sigma_1 + \sigma_2 &= 0 \\ p_3 - p_2\sigma_1 + p_1\sigma_2 - \sigma_3 &= 0 \\ &\dots\dots\dots\dots \end{aligned}$$

Therefore,

$$\sigma_k = \begin{vmatrix} p_1 & 1 & 0 & \dots & 0 \\ p_2 & p_1 & 1 & \dots & 0 \\ \vdots & \vdots & \vdots & \ddots & \vdots \\ p_{k-1} & p_{k-2} & \dots & \dots & 1 \\ p_k & p_{k-1} & \dots & \dots & p_k \end{vmatrix} \text{ and } p_k = \begin{vmatrix} \sigma_1 & 1 & 0 & \dots & 0 \\ \sigma_2 & \sigma_1 & 1 & \dots & 0 \\ \vdots & \vdots & \vdots & \ddots & \vdots \\ \sigma_{k-1} & \sigma_{k-2} & \dots & \dots & 1 \\ \sigma_k & \sigma_{k-1} & \dots & \dots & \sigma_k \end{vmatrix}.$$

To get relations between p_k and s_k is a bit more difficult. Consider the function $f(t) = (1 - x_1 t)\dots(1 - x_n t)$. Then

$$-\frac{f'(t)}{f^2(t)} = \left(\frac{1}{f(t)}\right)' = \left[\left(\frac{1}{1 - x_1 t}\right)\cdots\left(\frac{1}{1 - x_n t}\right)\right]'$$
$$= \left(\frac{x_1}{1 - x_1 t} + \dots + \frac{x_n}{1 - x_n t}\right)\frac{1}{f(t)}.$$

Therefore,

$$-\frac{f'(t)}{f(t)} = \frac{x_1}{1 - x_1 t} + \dots + \frac{x_n}{1 - x_n t} = s_1 + s_2 t + s_3 t^2 + \dots.$$

On the other hand, $(f(t))^{-1} = 1 + p_1 t + p_2 t^2 + p_3 t^3 + \dots$ and, therefore,

$$-\frac{f'(t)}{f(t)} = \left(\frac{1}{f(t)}\right)' \cdot \left(\frac{1}{f(t)}\right)^{-1} = \frac{p_1 + 2p_2 t + 3p_3 t^2 + \dots}{1 + p_1 t + p_2 t^2 + p_3 t^3 + \dots},$$

i.e.,

$$(1 + p_1 t + p_2 t^2 + p_3 t^3 + \dots)(s_1 + s_2 t + s_3 t^2 + \dots) = p_1 + 2p_2 t + 3p_3 t^2 + \dots.$$

Therefore,

$$s_k = (-1)^{k-1} \begin{vmatrix} p_1 & 1 & 0 & \dots & 0 & 0 & 0 \\ 2p_2 & p_1 & 1 & \dots & 0 & 0 & 0 \\ \vdots & \vdots & \ddots & \ddots & \vdots & \vdots & \vdots \\ (k-1)p_{k-1} & p_{k-2} & \dots & \dots & p_2 & p_1 & 1 \\ kp_k & p_{k-1} & \dots & \dots & p_3 & p_2 & p_1 \end{vmatrix},$$

and

$$p_k = \frac{1}{k!} \begin{vmatrix} s_1 & -1 & 0 & \dots & 0 & 0 & 0 \\ s_2 & s_1 & -2 & \dots & 0 & 0 & 0 \\ \vdots & \vdots & \ddots & \ddots & \vdots & \vdots & \vdots \\ s_{k-1} & s_{k-2} & \dots & \dots & s_2 & s_1 & -k+1 \\ s_k & s_{k-1} & \dots & \dots & s_3 & s_2 & s_1 \end{vmatrix}.$$

4.3. In this subsection we will study properties of the sum $S_n(k) = 1^n + \dots + (k-1)^n$. Let us prove that

$$S_{n-1}(k) = \frac{1}{n!}\begin{vmatrix} k^n & \binom{n}{n-2} & \binom{n}{n-3} & \dots & \binom{n}{1} & 1 \\ k^{n-1} & \binom{n-1}{n-2} & \binom{n-1}{n-3} & \dots & \binom{n-1}{1} & 1 \\ k^{n-2} & 1 & \binom{n-2}{n-3} & \dots & \binom{n-2}{1} & 1 \\ \vdots & \vdots & \vdots & \dots & \vdots & \vdots \\ k & 0 & 0 & \dots & 0 & 1 \end{vmatrix}.$$

To this end, add up the identities

$$(x+1)^n - x^n = \sum_{i=0}^{n-1} \binom{n}{i} x^i \text{ for } x = 1, 2, \dots, k-1.$$

We get

$$k^n = \sum_{i=0}^{n-1} \binom{n}{i} S_i(k).$$

The set of these identities for $i = 1, 2, \dots, n$ can be considered as a system of linear equations for $S_i(k)$. This system yields the desired expression for $S_{n-1}(k)$.

The obtained expression for $S_{n-1}(k)$ implies that $S_{n-1}(k)$ is a polynomial in k of degree n.

4.4. Now, let us give matrix expressions for $S_n(k)$ which imply that $S_n(x)$ can be polynomially expressed in terms of $S_1(x)$ and $S_2(x)$; more precisely, the following assertion holds.

THEOREM. *Let $u = S_1(x)$ and $v = S_2(x)$; then for $k \geq 1$ there exist polynomials p_k and q_k with rational coefficients such that $S_{2k+1}(x) = u^2 p_k(u)$ and $S_{2k}(x) = vq_k(u)$.*

To get an expression for S_{2k+1} let us make use of the identity

$$(1) \quad [n(n-1)]^r = \sum_{x=1}^{n-1} (x^r(x+1)^r - x^r(x-1)^r)$$
$$= 2\left(\binom{r}{1}\sum x^{2r-1} + \binom{r}{3}\sum x^{2r-3} + \binom{r}{5}\sum x^{2r-5} + \dots\right),$$

i.e., $[n(n-1)]^{i+1} = \sum \binom{i+1}{2(i-j)+1} S_{2j+1}(n)$. For $i = 1, 2, \dots$ these equalities can be expressed in the matrix form:

$$\begin{pmatrix} [n(n-1)]^2 \\ [n(n-1)]^3 \\ [n(n-1)]^4 \\ \vdots \end{pmatrix} = 2 \begin{pmatrix} 2 & 0 & 0 & \dots \\ 1 & 3 & 0 & \dots \\ 0 & 4 & 4 & \dots \\ \vdots & \vdots & \vdots & \ddots \end{pmatrix} \begin{pmatrix} S_3(n) \\ S_5(n) \\ S_7(n) \\ \vdots \end{pmatrix}.$$

The principal minors of finite order of the obtained matrix are all nonzero and, therefore,

$$\begin{pmatrix} S_3(n) \\ S_5(n) \\ S_7(n) \\ \vdots \end{pmatrix} = \frac{1}{2}\|a_{ij}\|^{-1} \begin{pmatrix} [n(n-1)]^2 \\ [n(n-1)]^3 \\ [n(n-1)]^4 \\ \vdots \end{pmatrix}, \text{ where } a_{ij} = \binom{i+1}{2(i-j)+1}.$$

This formula implies that $S_{2k+1}(n)$ can be expressed in terms of $n(n-1) = 2u(n)$ and is divisible by $[n(n-1)]^2$.

To get an expression for S_{2k} let us make use of the identity

$$\begin{aligned} n^{r+1}(n-1)^r &= \sum_{x=1}^{n-1}(x^r(x+1)^{r+1} - (x-1)^r x^{r+1}) \\ &= \sum x^{2r}\left(\binom{r+1}{1} + \binom{r}{1}\right) + \sum x^{2r-1}\left(\binom{r+1}{2} - \binom{r}{2}\right) \\ &\quad + \sum x^{2r-2}\left(\binom{r+1}{3} + \binom{r}{3}\right) + \sum x^{2r-3}\left(\binom{r+1}{4} - \binom{r}{4}\right) + \dots \\ &= \left(\binom{r+1}{1} + \binom{r}{1}\right)\sum x^{2r} + \left(\binom{r+1}{3} + \binom{r}{3}\right)\sum x^{2r-2} + \dots \\ &\quad + \binom{r}{1}\sum x^{2r-1} + \binom{r}{3}\sum x^{2r-3} + \dots \end{aligned}$$

The sums of odd powers can be eliminated with the help of (1). As a result we get

$$\begin{aligned} n^{r+1}(n-1)^r = \frac{(n^r(n-1)^r)}{2} &+ \left(\binom{r+1}{1} + \binom{r}{1}\right)\sum x^{2r} \\ &+ \left(\binom{r+1}{3} + \binom{r}{3}\right)\sum x^{2r-3}, \end{aligned}$$

i.e.,

$$n^i(n-1)^i\left(\frac{2n-1}{2}\right) = \sum\left(\binom{i+1}{2(i-j)+1} + \binom{i}{2(i-j)+1}\right)S_{2j}(n).$$

Now, similarly to the preceding case we get

$$\begin{pmatrix} S_2(n) \\ S_4(n) \\ S_6(n) \\ \vdots \end{pmatrix} = \frac{2n-1}{2}\|b_{ij}\|^{-1} \begin{pmatrix} (n(n-1) \\ [n(n-1)]^2 \\ [n(n-1)]^3 \\ \vdots \end{pmatrix},$$

where $b_{ij} = \binom{i+1}{2(i-j)+1} + \binom{i}{2(i-j)+1}$.

Since $S_2(n) = \dfrac{2n-1}{2} \cdot \dfrac{n(n-1)}{3}$, the polynomials $S_4(n), S_6(n), \dots$ are divisible by $S_2(n) = v(n)$ and the quotient is a polynomial in $n(n-1) = 2u(n)$.

4.5. In many theorems of calculus and number theory we encounter the following *Bernoulli numbers* B_k, defined from the expansion

$$\frac{t}{e^t - 1} = \sum_{k=0}^{\infty} B_k \frac{t^k}{k!} \quad (\text{for } |t| < 2\pi).$$

It is easy to verify that $B_0 = 1$ and $B_1 = -1/2$.

With the help of the Bernoulli numbers we can represent $S_m(n) = 1^m + 2^m + \dots + (n-1)^m$ as a polynomial of n.

THEOREM. $(m+1)S_m(n) = \sum_{k=0}^{m} \binom{m+1}{k} B_k n^{m+1-k}$.

PROOF. Let us write the power series expansion of $\frac{t}{e^t - 1}(e^{nt} - 1)$ in two ways. On one hand,

$$\begin{aligned}
\frac{t}{e^t - 1}(e^{nt} - 1) &= \sum_{k=0}^{\infty} \frac{B_k t^k}{k!} \sum_{s=1}^{\infty} \frac{(nt)^s}{s!} \\
&= nt + \sum_{m=1}^{\infty} \sum_{k=0}^{m} \binom{m+1}{k} B_k n^{m+1-k} \frac{t^{m+1}}{(m+1)!}.
\end{aligned}$$

On the other hand,

$$\begin{aligned}
t\frac{e^{nt} - 1}{e^t - 1} &= t\sum_{r=0}^{n-1} e^{rt} = nt + \sum_{m=1}^{\infty} \left(\sum_{r=1}^{n-1} r^m \right) \frac{t^{m+1}}{m!} \\
&= nt + \sum_{m=1}^{\infty} (m+1) S_m(n) \frac{t^{m+1}}{(m+1)!}.
\end{aligned}$$

Let us give certain determinant expressions for B_k. Set $b_k = \frac{B_k}{k!}$. Then by definition

$$x = (e^x - 1) \left(\sum_{k=0}^{\infty} b_k x^k \right) = \left(x + \frac{x^2}{2!} + \frac{x^3}{3!} + \dots \right) (1 + b_1 x + b_2 x^2 + b_3 x^3 + \dots),$$

i.e.,

$$\begin{aligned}
b_1 &= -\frac{1}{2!} \\
\frac{b_1}{2!} + b_2 &= -\frac{1}{3!} \\
\frac{b_1}{3!} + \frac{b_2}{2!} + b_3 &= -\frac{1}{4!} \\
\dots&\dots\dots\dots
\end{aligned}$$

Solving this system of linear equations by Cramer's rule we get

$$B_k = k!b_k = (-1)^k k! \begin{vmatrix} 1/2! & 1 & 0 & \dots & 0 \\ 1/3! & 1/2! & 1 & \dots & 0 \\ 1/4! & 1/3! & 1/2! & \dots & 0 \\ \vdots & \vdots & \vdots & \ddots & \vdots \\ 1/(k+1)! & 1/k! & \dots & \dots & 1/2! \end{vmatrix}.$$

Now, let us prove that $B_{2k+1} = 0$ for $k \geq 1$. Let $\frac{x}{e^x - 1} = -\frac{x}{2} + f(x)$. Then

$$f(x) - f(-x) = \frac{x}{e^x - 1} + \frac{x}{e^{-x} - 1} + x = 0,$$

i.e., f is an even function. Let $c_k = \frac{B_{2k}}{(2k)!}$. Then

$$x = \left(x + \frac{x^2}{2!} + \frac{x^3}{3!} + \dots\right)\left(1 - \frac{x}{2} + c_1x^2 + c_2x^4 + c_3x^6 + \dots\right).$$

Equating the coefficients of x^3, x^5, x^7, ... and taking into account that $\frac{1}{2(2n)!} - \frac{1}{(2n+1)!} = \frac{2n-1}{2(2n+1)!}$ we get

$$\begin{aligned} c_1 &= \frac{1}{2 \cdot 3!} \\ \frac{c_1}{3!} + c_2 &= \frac{3}{2 \cdot 5!} \\ \frac{c_1}{5!} + \frac{c_2}{3!} + c_3 &= \frac{5}{2 \cdot 7!} \\ \dots\dots\dots\dots \end{aligned}$$

Therefore,

$$B_{2k} = (2k)!c_k = \frac{(-1)^{k+1}(2k)!}{2} \begin{vmatrix} 1/3! & 1 & 0 & \dots & 0 \\ 3/5! & 1/3! & 1 & \dots & 0 \\ 5/7! & 1/5! & 1/3! & \dots & 0 \\ \vdots & \vdots & \vdots & \ddots & \vdots \\ \frac{2k-1}{(2k+1)!} & \frac{1}{(2k-1)!} & \dots & \dots & 1/3! \end{vmatrix}. \qquad \square$$

Solutions

1.1. Since $A^T = -A$ and n is odd, it follows that $|A^T| = (-1)^n|A| = -|A|$. On the other hand, $|A^T| = |A|$.

1.2. If A is a skew-symmetric matrix of even order, then

$$\begin{pmatrix} 0 & 1 & \dots & 1 \\ -1 & & & \\ \vdots & & A & \\ -1 & & & \end{pmatrix}$$

is a skew-symmetric matrix of odd order and, therefore, its determinant vanishes. Thus,

$$|A| = \begin{vmatrix} 1 & 0 & \dots & 0 \\ -x & & & \\ \vdots & & A & \\ -x & & & \end{vmatrix} + \begin{vmatrix} 0 & 1 & \dots & 1 \\ -x & & & \\ \vdots & & A & \\ -x & & & \end{vmatrix} = \begin{vmatrix} 1 & 1 & \dots & 1 \\ -x & & & \\ \vdots & & A & \\ -x & & & \end{vmatrix}.$$

In the last matrix, subtracting the first column from all other columns we get the desired.

1.3. Add the first row to and subtract the second row from the rows 3 to $2n$. As a result, we get $|A_n| = |A_{n-1}|$.

1.4. Suppose that all terms of the expansion of an nth order determinant are positive. If the intersection of two rows and two columns of the determinant singles out a matrix $\begin{pmatrix} x & y \\ u & v \end{pmatrix}$, then the expansion of the determinant has terms of the form $xv\alpha$ and $-yu\alpha$ and, therefore, $\operatorname{sign}(xv) = -\operatorname{sign}(yu)$. Let a_i, b_i, and c_i be the first three elements of the ith row ($i = 1, 2$). Then $\operatorname{sign}(a_1b_2) = -\operatorname{sign}(a_2b_1)$, $\operatorname{sign}(b_1c_2) = -\operatorname{sign}(b_2c_1)$, and $\operatorname{sign}(c_1a_2) = -\operatorname{sign}(c_2a_1)$. By multiplying these identities we get $\operatorname{sign} p = -\operatorname{sign} p$, where $p = a_1b_1c_1a_2b_2c_2$. Contradiction.

1.5. For all $i \geq 2$ let us subtract the $(i-1)$st row multiplied by a from the ith row. As a result we get an upper triangular matrix with diagonal elements $a_{11} = 1$ and $a_{ii} = 1 - a^2$ for $i > 1$. The determinant of this matrix is equal to $(1-a^2)^{n-1}$.

1.6. Expanding the determinant Δ_{n+1} with respect to the last column we get

$$\Delta_{n+1} = x\Delta_n + h\Delta_n = (x+h)\Delta_n.$$

1.7. Let us prove that the desired determinant is equal to

$$\prod(x_i - a_ib_i)\left(1 + \sum_i \frac{a_ib_i}{x_i - a_ib_i}\right)$$

by induction on n. For $n = 2$ this statement is easy to verify. We will carry out the proof of the inductive step for $n = 3$ (in the general case the proof is similar):

$$\begin{vmatrix} x_1 & a_1b_2 & a_1b_3 \\ a_2b_1 & x_2 & a_2b_3 \\ a_3b_1 & a_3b_2 & x_3 \end{vmatrix} = \begin{vmatrix} x_1 - a_1b_1 & a_1b_2 & a_1b_3 \\ 0 & x_2 & a_2b_3 \\ 0 & a_3b_2 & x_3 \end{vmatrix} + \begin{vmatrix} a_1b_1 & a_1b_2 & a_1b_3 \\ a_2b_1 & x_2 & a_2b_3 \\ a_3b_1 & a_3b_2 & x_3 \end{vmatrix}.$$

The first determinant is computed by inductive hypothesis and to compute the second one we have to break out from the first row the factor a_1 and for all $i \geq 2$ subtract from the ith row the first row multiplied by a_i.

1.8. It is easy to verify that $\det(I - A) = 1 - c$. The matrix A is the matrix of the transformation $Ae_i = c_{i-1}e_{i-1}$ and therefore, $A^n = c_1 \dots c_n I$. Hence,

$$(I + A + \dots + A^{n-1})(I - A) = I - A^n = (1-c)I$$

and, therefore,

$$(1-c)\det(I+A+\dots+A^{n-1})=(1-c)^n.$$

For $c\neq 1$ by dividing by $1-c$ we get the required formula. The determinant of the matrix considered depends continuously on $c_1,\dots,c_n$ and, therefore, the identity holds for $c=1$ as well.

1.9. Since $(1-x_iy_j)^{-1}=(y_j^{-1}-x_i)^{-1}y_j^{-1}$, we have $|a_{ij}|_1^n=\sigma|b_{ij}|_1^n$, where $\sigma=(y_1\dots y_n)^{-1}$ and $b_{ij}=(y_j^{-1}-x_i)^{-1}$, i.e., $|b_{ij}|_1^n$ is a Cauchy determinant (see 1.3). Therefore,

$$|b_{ij}|_1^n=\sigma^{-1}\prod_{i>j}(y_i-y_i)(x_j-x_i)\prod_{i,j}(1-x_iy_j)^{-1}.$$

1.10. For a fixed m consider the matrices $A_n=\left\|a_{ij}\right\|_0^m$, $a_{ij}=\binom{n+i}{j}$. The matrix A_0 is a triangular matrix with diagonal $(1,\dots,1)$. Therefore, $|A_0|=1$. Besides, $A_{n+1}=A_nB$, where $b_{i,i+1}=1$ (for $i\le m-1$), $b_{i,i}=1$ and all other elements b_{ij} are zero.

1.11. Clearly, points $A,B,\dots,F$ with coordinates $(a^2,a),\dots,(f^2,f)$, respectively, lie on a parabola. By Pascal's theorem the intersection points of the pairs of straight lines AB and DE, BC and EF, CD and FA lie on one straight line. It is not difficult to verify that the coordinates of the intersection point of AB and DE are

$$\left(\frac{(a+b)de-(d+e)ab}{d+e-a-b},\quad\frac{de-ab}{d+e-a-b}\right).$$

It remains to note that if points $(x_1,y_1),(x_2,y_2)$ and (x_3,y_3) belong to one straight line, then

$$\begin{vmatrix} x_1 & y_1 & 1\\ x_2 & y_2 & 1\\ x_3 & y_3 & 1\end{vmatrix}=0.$$

REMARK. Recall that *Pascal's theorem* states that *the opposite sides of a hexagon inscribed in a second order curve intersect at three points that lie on one line.* Its proof can be found in the books [Berger, 1977] and [Reid, 1988].

1.12. Let $s=x_1+\dots+x_n$. Then the kth element of the last column is of the form

$$(s-x_k)^{n-1}=(-x_k)^{n-1}+\sum_{i=0}^{n-2}p_ix_k^i.$$

Therefore, adding to the last column a linear combination of the remaining columns with coefficients $-p_0,\dots,-p_{n-2}$, respectively, we obtain the determinant

$$\begin{vmatrix} 1 & x_1 & \dots & x_1^{n-2} & (-x_1)^{n-1}\\ \vdots & \vdots & \dots & \vdots & \vdots\\ 1 & x_n & \dots & x_n^{n-2} & (-x_n)^{n-1}\end{vmatrix}=(-1)^{n-1}V(x_1,\dots,x_n).$$

1.13. Let Δ be the required determinant. Multiplying the first row of the corresponding matrix by $x_1,\dots$, and the nth row by x_n we get

$$\sigma\Delta=\begin{vmatrix} x_1 & x_1^2 & \dots & x_1^{n-1} & \sigma\\ \vdots & \vdots & \dots & \vdots & \vdots\\ x_n & x_n^2 & \dots & x_n^{n-1} & \sigma\end{vmatrix},\quad\text{where }\sigma=x_1\dots x_n.$$

Therefore, $\Delta = (-1)^{n-1}V(x_1, \dots, x_n)$.

1.14. Since

$$\lambda_i^{n-k}(1+\lambda_i^2)^k = \lambda_i^n(\lambda_i^{-1}+\lambda_i)^k,$$

it follows that

$$|a_{ij}|_0^n = (\lambda_0 \dots \lambda_n)^n V(\mu_0, \dots, \mu_n), \text{ where } \mu_i = \lambda_i^{-1} + \lambda_i.$$

1.15. Augment the matrix V with an $(n+1)$st column consisting of the nth powers and then add an extra first row $(1, -x, x^2, \dots, (-x)^n)$. The resulting matrix W is also a Vandermonde matrix and, therefore,

$$\det W = (x+x_1)\dots(x+x_n)\det V = (\sigma_n + \sigma_{n-1}x + \dots + x^n)\det V.$$

On the other hand, expanding W with respect to the first row we get

$$\det W = \det V_0 + x \det V_1 + \dots + x^n \det V_{n-1}.$$

1.16. Let $x_i = in$. Then

$$a_{i1} = x_i, a_{i2} = \frac{x_i(x_i-1)}{2}, \dots, a_{ir} = \frac{x_i(x_i-1)\dots(x_i-r+1)}{r!},$$

i.e., in the kth column there stand identical polynomials of kth degree in x_i. Since the determinant does not vary if to one of its columns we add a linear combination of its other columns, the determinant can be reduced to the form $|b_{ik}|_1^r$, where $b_{ik} = \dfrac{x_i^k}{k!} = \dfrac{n^k}{k!}i^k$. Therefore,

$$|a_{ik}|_1^r = |b_{ik}|_1^r = n \cdot \frac{n^2}{2!} \dots \frac{n^r}{r!} r! V(1, 2, \dots, r) = n^{r(r+1)/2},$$

because $\prod_{1 \le j < i \le r}(i-j) = 2!3!\dots(r-1)!$

1.17. For $i = 1, \dots, n$ let us multiply the ith row of the matrix $\left\|a_{ij}\right\|_1^n$ by $m_i!$, where $m_i = k_i + n - i$. We obtain the determinant $|b_{ij}|_1^n$, where

$$b_{ij} = \frac{(k_i+n-i)!}{(k_i+j-i)!} = m_i(m_i-1)\dots(m_i+j+1-n).$$

The elements of the jth row of $\left\|b_{ij}\right\|_1^n$ are identical polynomials of degree $n-j$ in m_i and the coefficients of the highest terms of these polynomials are equal to 1. Therefore, subtracting from every column linear combinations of the preceding columns we can reduce the determinant $|b_{ij}|_1^n$ to a determinant with rows $(m_i^{n-1}, m_i^{n-2}, \dots, 1)$. This determinant is equal to $\prod_{i<j}(m_i - m_j)$. It is also clear that $|a_{ij}|_1^n = |b_{ij}|_1^n (m_1!m_2!\dots m_n!)^{-1}$.

1.18. For $n = 3$ it is easy to verify that

$$\|a_{ij}\|_0^2 = \begin{pmatrix} 1 & 1 & 1 \\ x_1 & x_2 & x_3 \\ x_1^2 & x_2^2 & x_3^2 \end{pmatrix} \begin{pmatrix} p_1 & p_1x_1 & p_1x_1^2 \\ p_2 & p_2x_2 & p_2x_2^2 \\ p_3 & p_3x_3 & p_3x_3^2 \end{pmatrix}.$$

In the general case an analogous identity holds.

1.19. The required determinant can be represented in the form of a product of two determinants:

$$\begin{vmatrix} 1 & \dots & 1 & 1 \\ x_1 & \dots & x_n & y \\ x_1^2 & \dots & x_n^2 & y^2 \\ \vdots & \dots & \vdots & \vdots \\ x_1^n & \dots & x_n^n & y^n \end{vmatrix} \cdot \begin{vmatrix} 1 & x_1 & \dots & x_1^{n-1} & 0 \\ 1 & x_2 & \dots & x_2^{n-1} & 0 \\ \vdots & \vdots & \dots & \vdots & \vdots \\ 1 & x_n & \dots & x_n^{n-1} & 0 \\ 0 & 0 & \dots & 0 & 1 \end{vmatrix}$$

and, therefore, it is equal to $\prod(y - x_i)\prod_{i>j}(x_i - x_j)^2$.

1.20. It is easy to verify that for $n = 2$

$$\|a_{ij}\|_0^2 = \begin{pmatrix} 1 & 2x_0 & x_0^2 \\ 1 & 2x_1 & x_1^2 \\ 1 & 2x_2 & x_2^2 \end{pmatrix} \begin{pmatrix} y_0^2 & y_1^2 & y_2^2 \\ y_0 & y_1 & y_2 \\ 1 & 1 & 1 \end{pmatrix}$$

and that in the general case the elements of the first matrix are the numbers $\binom{n}{k}x_i^k$.

1.21. Let us suppose that there exists a nonzero solution such that the number of pairwise distinct numbers λ_i is equal to r. By combining equal numbers λ_i into r groups we get

$$m_1\lambda_1^k + \dots + m_r\lambda_r^k = 0 \text{ for } k = 1, \dots, n.$$

Let $x_1 = m_1\lambda_1, \dots, x_r = m_r\lambda_r$ Then

$$\lambda_1^{k-1}x_1 + \dots + \lambda_r^{k-1}x_r = 0 \text{ for } k = 1, \dots, n.$$

Taking the first r of these equations we get a system of linear equations for $x_1, \dots, x_r$ and the determinant of this system is $V(\lambda_1, \dots, \lambda_r) \neq 0$. Hence, $x_1 = \dots = x_r = 0$ and, therefore, $\lambda_1 = \dots = \lambda_r = 0$. The contradiction obtained shows that there is only the zero solution.

1.22. Let us carry out the proof by induction on n. For $n = 1$ the statement is obvious.

Subtracting the first column of $\|a_{ij}\|_0^n$ from every other column we get a matrix $\|b_{ij}\|_0^n$, where $b_{ij} = \sigma_i(\widehat{x}_j) - \sigma_i(\widehat{x}_0)$ for $j \geq 1$.

Now, let us prove that

$$\sigma_k(\widehat{x}_i) - \sigma_k(\widehat{x}_j) = (x_j - x_i)\sigma_{k-1}(\widehat{x}_i, \widehat{x}_j).$$

Indeed,

$$\sigma_k(x_1, \dots, x_n) = \sigma_k(\widehat{x}_i) + x_i\sigma_{k-1}(\widehat{x}_i) = \sigma_k(\widehat{x}_i) + x_i\sigma_{k-1}(\widehat{x}_i, \widehat{x}_j) + x_ix_j\sigma_{k-2}(\widehat{x}_i, \widehat{x}_j)$$

and, therefore,

$$\sigma_k(\widehat{x}_i)+x_i\sigma_{k-1}(\widehat{x}_i,\widehat{x}_j)=\sigma_k(\widehat{x}_j)+x_j\sigma_{k-1}(\widehat{x}_i,\widehat{x}_j).$$

Hence,

$$|b_{ij}|_0^n=(x_0-x_1)\dots(x_0-x_n)|c_{ij}|_0^{n-1},\text{ where } c_{ij}=\sigma_i(\widehat{x}_0,\widehat{x}_j).$$

1.23. Let $k=[n/2]$. Let us multiply by -1 the rows $2,4,\dots,2k$ of the matrix $\|b_{ij}\|_1^n$ and then multiply by -1 the columns $2,4,\dots,2k$ of the matrix obtained. As a result we get $\|a_{ij}\|_1^n$.

1.24. It is easy to verify that both expressions are equal to the product of determinants

$$\begin{vmatrix} a_1 & a_2 & 0 & 0\\ a_3 & a_4 & 0 & 0\\ 0 & 0 & b_1 & b_2\\ 0 & 0 & b_3 & b_4\end{vmatrix}\cdot\begin{vmatrix} c_1 & 0 & c_2 & 0\\ 0 & d_1 & 0 & d_2\\ c_3 & 0 & c_4 & 0\\ 0 & d_3 & 0 & d_4\end{vmatrix}.$$

1.25. Both determinants are equal to

$$a_1a_2a_3\begin{vmatrix} a_{11} & a_{12} & a_{13}\\ a_{21} & a_{22} & a_{23}\\ a_{31} & a_{32} & a_{33}\end{vmatrix}+a_1b_2b_3\begin{vmatrix} a_{11} & b_{12} & b_{13}\\ a_{21} & b_{22} & b_{23}\\ a_{31} & b_{32} & b_{33}\end{vmatrix}+b_1a_2b_3\begin{vmatrix} b_{11} & a_{12} & b_{13}\\ b_{21} & a_{22} & b_{23}\\ b_{31} & a_{32} & b_{33}\end{vmatrix}$$

$$-a_1a_2b_3\begin{vmatrix} a_{11} & a_{12} & b_{13}\\ a_{21} & a_{22} & b_{23}\\ a_{31} & a_{32} & b_{33}\end{vmatrix}-b_1a_2a_3\begin{vmatrix} b_{11} & a_{12} & a_{13}\\ b_{21} & a_{22} & a_{23}\\ b_{31} & a_{32} & a_{33}\end{vmatrix}-b_1b_2b_3\begin{vmatrix} b_{11} & b_{12} & b_{13}\\ b_{21} & b_{22} & b_{23}\\ b_{31} & b_{32} & b_{33}\end{vmatrix}.$$

1.26. It is easy to verify the following identities for the determinants of matrices of order $n+1$:

$$\begin{vmatrix} s_1-a_{11} & \dots & s_1-a_{1n} & 0\\ \vdots & \dots & \vdots & \vdots\\ s_n-a_{n1} & \dots & s_n-a_{nn} & 0\\ -1 & \dots & -1 & 1\end{vmatrix}=\begin{vmatrix} s_1-a_{11} & \dots & s_1-a_{1n} & (n-1)s_1\\ \vdots & \dots & \vdots & \vdots\\ s_n-a_{n1} & \dots & s_n-a_{nn} & (n-1)s_1\\ -1 & \dots & -1 & 1-n\end{vmatrix}$$

$$=(n-1)\begin{vmatrix} s_1-a_{11} & \dots & s_1-a_{1n} & s_1\\ \vdots & \dots & \vdots & \vdots\\ s_n-a_{n1} & \dots & s_n-a_{nn} & s_n\\ -1 & \dots & -1 & -1\end{vmatrix}=(n-1)\begin{vmatrix} -a_{11} & \dots & -a_{1n} & s_1\\ \vdots & \dots & \vdots & \vdots\\ -a_{n1} & \dots & -a_{nn} & s_n\\ 0 & \dots & 0 & -1\end{vmatrix}.$$

1.27. Since $\binom{p}{q}+\binom{p}{q-1}=\binom{p+1}{q}$, by suitably adding columns of a matrix whose rows are of the form $\left(\binom{n}{m}\binom{n}{m-1}\dots\binom{n}{m-k}\right)$ we can get a matrix whose rows are of the form $\left(\binom{n}{m}\binom{n+1}{m}\dots\binom{n+1}{m-k+1}\right)$. And so on.

1.28. In the determinant $\Delta_n(k)$ subtract from the $(i+1)$st row the ith row for every $i=n-1,\dots,1$. As a result, we get $\Delta_n(k)=\Delta'_{n-1}(k)$, where $\Delta'_m(k)=|a'_{ij}|_0^m$, $a'_{ij}=\begin{pmatrix}k+i\\2j+1\end{pmatrix}$. Since $\binom{k+i}{2j+1}=\dfrac{k+i}{2j+1}\begin{pmatrix}k-1+i\\2j\end{pmatrix}$, it follows that

$$\Delta'_{n-1}(k)=\frac{k(k+1)\dots(k+n-1)}{1\cdot 3\dots(2n-1)}\Delta_{n-1}(k-1).$$

1.29. According to Problem 1.27 $D_n = D'_n = |a'_{ij}|_0^n$, where $a'_{ij} = \binom{n+1+i}{2j}$, i.e., in the notations of Problem 1.28 we get

$$D_n = \Delta_n(n+1) = \frac{(n+1)(n+2)\dots 2n}{1\cdot 3\dots(2n-1)}\Delta_{n-1}(n) = 2^n D_{n-1},$$

since $(n+1)(n+2)\dots 2n = \dfrac{(2n)!}{n!}$ and $1\cdot 3\dots(2n-1) = \dfrac{(2n)!}{2\cdot 4\dots 2n}$.

1.30. Let us carry out the proof for $n=2$. By Problem 1.23, $|a_{ij}|_0^2 = |a'_{ij}|_0^2$, where $a'_{ij} = (-1)^{i+j}a_{ij}$. Let us add to the last column of $\left\|a'_{ij}\right\|_0^2$ its penultimate column and to the last row of the matrix obtained add its penultimate row. As a result we get the matrix

$$\begin{pmatrix} a_0 & -a_1 & -\Delta_1 a_1 \\ -a_1 & a_2 & \Delta_1 a_2 \\ -\Delta_1 a_1 & \Delta_1 a_2 & \Delta_2 a_2 \end{pmatrix},$$

where $\Delta_1 a_k = a_k - a_{k+1}$, $\Delta_{n+1}a_k = \Delta_1(\Delta_n a_k)$. Then let us add to the 2nd row the 1st one and to the 3rd row the 2nd row of the matrix obtained; let us perform the same operation with the columns of the matrix obtained. Finally, we get the matrix

$$\begin{pmatrix} a_0 & \Delta_1 a_0 & \Delta_2 a_0 \\ \Delta_1 a_0 & \Delta_2 a_0 & \Delta_3 a_0 \\ \Delta_2 a_0 & \Delta_3 a_0 & \Delta_4 a_0 \end{pmatrix}.$$

By induction on k it is easy to verify that $b_k = \Delta_k a_0$. In the general case the proof is similar.

1.31. We can represent the matrices A and B in the form

$$A = \begin{pmatrix} P & PX \\ YP & YPX \end{pmatrix} \text{ and } B = \begin{pmatrix} WQV & WQ \\ QV & Q \end{pmatrix},$$

where $P = A_{11}$ and $Q = B_{22}$. Therefore,

$$|A+B| = \begin{vmatrix} P+WQV & PX+WQ \\ YP+QV & YPX+Q \end{vmatrix} = \begin{vmatrix} P & WQ \\ YP & Q \end{vmatrix} \cdot \begin{vmatrix} I & X \\ V & I \end{vmatrix}$$
$$= \frac{1}{|P|\cdot|Q|}\begin{vmatrix} P & WQ \\ YP & Q \end{vmatrix}\begin{vmatrix} P & PX \\ QV & Q \end{vmatrix}.$$

1.32. Expanding the determinant of the matrix

$$C = \begin{pmatrix} 0 & a_{12} & \dots & a_{1n} & b_{11} & \dots & b_{1n} \\ \vdots & \vdots & \dots & \vdots & \vdots & \dots & \vdots \\ 0 & a_{n2} & \dots & a_{nn} & b_{n1} & \dots & b_{nn} \\ a_{11} & 0 & \dots & 0 & b_{11} & \dots & b_{1n} \\ \vdots & \vdots & \dots & \vdots & \vdots & \dots & \vdots \\ a_{n1} & 0 & \dots & 0 & b_{n1} & \dots & b_{nn} \end{pmatrix}$$

with respect to the first n rows we obtain

$$|C| = \sum_{k=1}^{n}(-1)^{\varepsilon_k} \begin{vmatrix} a_{12} & \dots & a_{1n} & b_{1k} \\ \vdots & \dots & \vdots & \vdots \\ a_{n2} & \dots & a_{nn} & b_{nk} \end{vmatrix} \cdot \begin{vmatrix} a_{11} & b_{11} & \dots & \widehat{b}_{1k} & \dots & b_{1n} \\ \vdots & \vdots & \dots & \vdots & \dots & \vdots \\ a_{n1} & b_{n1} & \dots & \widehat{b}_{nk} & \dots & b_{nn} \end{vmatrix}$$
$$= \sum_{k=1}^{n}(-1)^{\varepsilon_k+\alpha_k+\beta_k} \begin{vmatrix} b_{1k} & a_{12} & \dots & a_{1n} \\ \vdots & \vdots & \dots & \vdots \\ b_{nk} & a_{n2} & \dots & a_{nn} \end{vmatrix} \cdot \begin{vmatrix} b_{11} & \dots & a_{11} & \dots & b_{1n} \\ \vdots & \dots & \vdots & \dots & \vdots \\ b_{n1} & \dots & a_{n1} & \dots & b_{nn} \end{vmatrix},$$

where $\varepsilon_k = (1+2+\dots+n)+(2+\dots+n+(k+n)) \equiv k+n+1 \pmod 2$, $\alpha_k = n-1$ and $\beta_k = k-1$, i.e., $\varepsilon_k + \alpha_k + \beta_k \equiv 1 \pmod 2$. On the other hand, subtracting from the ith row of C the $(i+n)$th row for $i = 1, \dots, n$, we get $|C| = -|A| \cdot |B|$.

2.1. The coefficient of $\lambda_{i_1} \dots \lambda_{i_m}$ in the determinant of $A + \operatorname{diag}(\lambda_1, \dots, \lambda_n)$ is equal to the minor obtained from A by striking out the rows and columns with numbers $i_1, \dots, i_m$.

2.2. Let us transpose the rows $(a_{i1} \dots a_{in} x_i)$ and $(y_1 \dots y_n 0)$. In the determinant of the matrix obtained the coefficient of $x_i y_j$ is equal to A_{ij}.

2.3. Let $B = A^T A$. Then

$$B\begin{pmatrix} i_1 \dots i_k \\ i_1 \dots i_k \end{pmatrix} = \begin{vmatrix} b_{i_1 i_1} & \dots & b_{i_1 i_k} \\ \vdots & \dots & \vdots \\ b_{i_k i_1} & \dots & b_{i_k i_k} \end{vmatrix}$$
$$= \det \left[\begin{pmatrix} a_{i_1 1} & \dots & a_{i_1 n} \\ \vdots & \dots & \vdots \\ a_{i_k 1} & \dots & a_{i_k n} \end{pmatrix} \cdot \begin{pmatrix} a_{i_1 1} & \dots & a_{i_k 1} \\ \vdots & \dots & \vdots \\ a_{i_1 n} & \dots & a_{i_k n} \end{pmatrix} \right]$$

and it remains to make use of the Binet-Cauchy formula.

2.4. The coefficient of u_1 in the sum of determinants in the left-hand side is equal to $a_{11}A_{11} + \dots a_{n1}A_{n1} = |A|$. For the coefficients of $u_2, \dots, u_n$ the proof is similar.

2.5. Answer:

$$\begin{pmatrix} I & -A & AB - C \\ 0 & I & -B \\ 0 & 0 & I \end{pmatrix}.$$

2.6. If $i < j$, then deleting the ith row and the jth column of the upper triangular matrix we get an upper triangular matrix with zeros on the diagonal at all places i to $j-1$.

2.7. Consider the unit matrix of order $n-1$. Insert a column of zeros between its $(i-1)$st and ith columns and then insert a row of zeros between the $(j-1)$st and jth rows of the matrix obtained . The minor M_{ji} of the matrix obtained is equal to 1 and all the other minors are equal to zero.

2.8. Since $x(y^T x)y^T I = xy^T I(y^T x)$, then

$$(I - xy^T)(xy^T + I(1 - y^T x)) = (1 - y^T x)I.$$

Hence,

$$(I - xy^T)^{-1} = xy^T(1 - y^T x)^{-1} + I.$$

Besides, according to Problem 8.2

$$\det(I - xy^T) = 1 - \operatorname{tr}(xy^T) = 1 - y^T x.$$

2.9. By definition $A_{ij} = (-1)^{i+j}\det B$, where B is a matrix of order $n-1$. Since $A^T = -A$, we have $A_{ji} = (-1)^{i+j}\det(-B) = (-1)^{n-1}A_{ij}$.

2.10. The answer depends on the parity of n. By Problem 1.3 we have $|A_{2k}| = 1$ and, therefore, $\operatorname{adj} A_{2k} = A_{2k}^{-1}$. For $n = 4$ it is easy to verify that

$$\begin{pmatrix} 0 & 1 & 1 & 1 \\ -1 & 0 & 1 & 1 \\ -1 & -1 & 0 & 1 \\ -1 & -1 & -1 & 0 \end{pmatrix} \cdot \begin{pmatrix} 0 & -1 & 1 & -1 \\ 1 & 0 & -1 & 1 \\ -1 & 1 & 0 & -1 \\ 1 & -1 & 1 & 0 \end{pmatrix} = I.$$

A similar identity holds for any even n.

Now, let us compute $\operatorname{adj} A_{2k+1}$. Since $|A_{2k}| = 1$, we have $\operatorname{rank} A_{2k+1} = 2k$. It is also clear that $A_{2k+1}v = 0$ if v is the column $(1, -1, 1, -1, \dots)^T$. Hence, the columns of the matrix $B = \operatorname{adj} A_{2k+1}$ are of the form λv. Besides, $b_{11} = |A_{2k}| = 1$ and, therefore, B is a symmetric matrix (cf. Problem 2.9). Therefore,

$$B = \begin{pmatrix} 1 & -1 & 1 & \dots \\ -1 & 1 & -1 & \dots \\ 1 & -1 & 1 & \dots \\ \vdots & \vdots & \vdots & \ddots \end{pmatrix}.$$

2.11. a) Since $[\operatorname{adj}(A - \lambda I)](A - \lambda I) = |A - \lambda I| \cdot I$ is a scalar matrix, we have

$$\begin{aligned}\left(\sum_{k=0}^{n-1} \lambda^k A_k\right)(A - \lambda I) &= \sum_{k=0}^{n-1} \lambda^k A_k A - \sum_{k=1}^{n} \lambda^k A_{k-1} \\ &= A_0 A - \lambda^n A_{n-1} + \sum_{k=1}^{n-1} \lambda^k (A_k A - A_{k-1})\end{aligned}$$

is also a scalar matrix.

b) $A_{n-1} = \pm I$. Besides, $A_{n-s-1} = \mu I - A_{n-s}A$.

2.12. Let $A = \|a_{ij}\|$, $A^{-1} = \|b_{ij}\|$ and $a_{ij}, b_{ij} \geq 0$. If $a_{ir}, a_{is} > 0$, then $\sum a_{ik}b_{kj} = 0$ for $i \neq j$ and, therefore, $b_{rj} = b_{sj} = 0$. In the rth row of the matrix B there is only one nonzero element, b_{ri}, and in the sth row there is only one nonzero element, b_{si}. Hence, the rth and the sth rows are proportional. Contradiction.

Therefore, every row and every column of the matrix A has precisely one nonzero element.

2.13. $A^{-1} = \|b_{ij}\|$, where $b_{ij} = n^{-1}\varepsilon^{-ij}$.

2.14. Let $\sigma_{n-k}^i = \sigma_{n-k}(x_1, \dots, \widehat{x}_i, \dots, x_n)$. Making use of the result of Problem 1.15 it is easy to verify that $(\operatorname{adj} V)^T = \|b_{ij}\|_1^n$, where

$$b_{ij} = (-1)^{i+j}\sigma_{n-j}^i V(x_1, \dots, \widehat{x}_i, \dots, x_n).$$

3.1. $|A + u^T v| = \begin{vmatrix} A & -u^T \\ v & 1 \end{vmatrix} = |A|(1 + vA^{-1}u^T)$.

3.2. $\begin{vmatrix} I & A \\ A^T & I \end{vmatrix} = |I - A^T A| = (-1)^n |A^T A - I|$. It remains to apply the results of Problem 2.1 (for $\lambda = -1$) and of Problem 2.4.

CHAPTER II

LINEAR SPACES

The notion of a linear space appeared much later than the notion of determinant. **Leibniz**'s share in the creation of this notion is considerable. He was not satisfied with the fact that the language of algebra only allowed one to describe various quantities of the existing geometry, but not the positions of points and not the directions of straight lines. Leibniz began to consider sets of points $A_1 \dots A_n$ and assumed that $\{A_1, \dots, A_n\} = \{X_1, \dots, X_n\}$ whenever the lengths of the segments A_iA_j and X_iX_j are equal for all i and j. He, certainly, used a somewhat different notation, namely, something like $A_1 \dots A_n$ ♉ $X_1 \dots X_n$; he did not use indices though.

In these terms the equation AB ♉ AY determines the sphere of radius AB and center A; the equation AY ♉ BY ♉ CY determines a straight line perpendicular to the plane ABC.

Though Leibniz did consider pairs of points, these pairs did not in any way correspond to vectors: only the lengths of segments counted, but not their directions and the pairs AB and BA were not distinguished.

These works of Leibniz were unpublished for more than 100 years after his death. They were published in 1833, and for the development of these ideas a prize was set. In 1845 Möbius informed **Grassmann** about this prize, and in a year Grassmann presented his paper and collected the prize. Grassmann's book was published but nobody was interested in it.

An important step in moulding the notion of a "vector space" was the geometric representation of complex numbers. Calculations with complex numbers urgently required the justification of their usage and a sufficiently rigorous theory of them. Already in the 17th century **John Wallis** tried to represent the complex numbers geometrically, but failed. Between 1799 and 1831 six mathematicians independently published papers containing a geometric interpretation of the complex numbers. Of these, the most influential on mathematicians' thought was the paper by **Gauss** published in 1831. Gauss himself did not consider a geometric interpretation (which appealed to the Euclidean plane) as sufficiently convincing justification of the existence of complex numbers because, at that time, he already had the idea for the development of non-Euclidean geometry.

The decisive step in the creation of the notion of an n-dimensional space was simultaneously made by two mathematicians — **Hamilton** and Grassmann. Their approaches were distinct in principle. Also distinct was the impact of their works on the development of mathematics. The works of Grassmann contained deep ideas with great influence on the development of algebra, algebraic geometry, and mathematical physics of the second half of our century. But his books were difficult to understand and the recognition of the importance of his ideas was far from immediate.

The development of linear algebra took mainly the road indicated by Hamilton.

Sir William Rowan Hamilton (1805–1865)

The Irish mathematician and astronomer Sir William Rowan Hamilton, a member of many academies, was born in 1805 in Dublin. From the age of three he was raised by his uncle, a minister. By the age 13 he had learned 13 languages and at 16 he read Laplace's *Méchanique Céleste*.

In 1823, Hamilton entered Trinity College in Dublin and when he graduated he was offered professorship in astronomy at the University of Dublin and he also became the Royal Astronomer of Ireland. Hamilton gained much publicity for his theoretical prediction of two previously unknown phenomena in optics that soon afterwards were confirmed experimentally. In 1837 he became the President of the Irish Academy of Sciences and in the same year he published his papers in which complex numbers were introduced as pairs of real numbers.

This discovery was not valued much at first. All mathematicians except, perhaps, Gauss and Bolyai were quite satisfied with the geometric interpretation of complex numbers. Only when non-Euclidean geometry was sufficiently widespread did the mathematicians become interested in the interpretation of complex numbers as pairs of real ones.

Hamilton soon realized the possibilities offered by his discovery. In 1841 he started to consider sets $\{a_1, \dots, a_n\}$, where the a_i are real numbers. This is precisely the idea on which the most common approach to the notion of a linear space is based. Hamilton was most involved in the study of triples of real numbers: he wanted to get a three-dimensional analogue of complex numbers. His excitement was transferred to his children. As Hamilton used to recollect, when he would join them for breakfast they would cry: " 'Well, Papa, can you multiply triplets?' Whereto I was always obliged to reply, with a sad shake of the head: 'No, I can only add and subtract them' ".

These frenzied studies were fruitful. On October 16, 1843, during a walk, Hamilton almost visualized the symbols i, j, k and the relations $i^2 = j^2 = k^2 = ijk = -1$. The elements of the algebra with unit generated by i, j, k are called *quaternions*. For the last 25 years of his life Hamilton worked exclusively with quaternions and their applications in geometry, mechanics and astronomy. He abandoned his brilliant study in physics and studied, for example, how to raise a quaternion to a quaternion power. He published two books and more than 100 papers on quaternions. Working with quaternions, Hamilton gave the definitions of inner and vector products of vectors in three-dimensional space.

Hermann Günther Grassmann (1809–1877)

The public side of Hermann Grassmann's life was far from being as brilliant as the life of Hamilton.

Until his death, he was a gymnasium teacher in his native town of Stettin. Several times he tried to get a university position but in vain. Hamilton, having read a book by Grassmann, called him the greatest German genius. Concerning the same book, 30 years after its publication the publisher wrote to Grassmann: "Your book *Die Ausdehnungslehre* has been out of print for some time. Since your work hardly sold at all, roughly 600 copies were used in 1864 as waste paper and the remaining few odd copies have now been sold out, with the exception of the one copy in our library".

Grassmann himself thought that his next book would enjoy even less success. Grassmann's ideas began to spread only towards the end of his life. By that time he lost his contacts with mathematicians and his interest in geometry. The last years of his life Grassmann was mainly working with Sanscrit. He made a translation of *Rig-Veda* (more than 1,000 pages) and made a dictionary for it (about 2,000 pages). For this he was elected a member of the American Oriental Society. In modern studies of Rig-Veda, Grassmann's works are often cited. In 1955, the third edition of Grassmann's dictionary to Rig-Veda was issued.

Grassmann can be described as a self-taught person. Although he did graduate from the Berlin University, he only studied philology and theology there. His father was a teacher of mathematics in Stettin, but Grassmann read his books only as a student at the University; Grassmann said later that many of his ideas were borrowed from these books and that he only developed them further.

In 1832 Grassmann actually arrived at the vector form of the laws of mechanics; this considerably simplified various calculations. He noticed the commutativity and associativity of the addition of vectors and explicitly distinguished these properties. Later on, Grassmann expressed his theory in quite a general form for arbitrary systems with certain properties. This overgenerality considerably hindered the understanding of his books; almost nobody could yet understand the importance of commutativity, associativity and the distributivity in algebra.

Grassmann defined the *geometric product* of two vectors as the parallelogram spanned by these vectors. He considered parallelograms of equal size parallel to one plane and of equal orientation equivalent. Later on, by analogy, he introduced the geometric product of r vectors in n-dimensional space. He considered this product as a geometric object whose coordinates are minors of order r of an $r \times n$ matrix consisting of coordinates of given vectors.

In Grassmann's works, the notion of a linear space with all its attributes was actually constructed. He gave a definition of a subspace and of linear dependence of vectors.

In 1840s, mathematicians were unprepared to come to grips with Grassmann's ideas. Grassmann sent his first book to Gauss. In reply he got a notice in which Gauss thanked him and wrote to the effect that he himself had studied similar things about half a century before and recently published something on this topic. Answering Grassmann's request to write a review of his book, Möbius informed Grassmann that being unable to understand the philosophical part of the book he could not read it completely. Later on, Möbius said that he knew only one mathematician who had read through the entirety of Grassmann's book. (This mathematician was Bretschneider.)

Having won the prize for developing Leibniz's ideas, Grassmann addressed the Minister of Culture with a request for a university position and his papers were sent to Kummer for a review. In the review, it was written that the papers lacked clarity. Grassmann's request was turned down.

In the 1860s and 1870s various mathematicians came, by their own ways, to ideas similar to Grassmann's ideas. His works got high appreciation by Cremona, Hankel, Clebsh and Klein, but Grassmann himself was not interested in mathematics any more.

5. The dual space. The orthogonal complement

WARNING. While reading this section the reader should keep in mind that here, as well as throughout the whole book, we consider only finite dimensional spaces.

For infinite dimensional spaces the majority of the statements of this section are false.

5.1. To a linear space V over a field K we can assign a linear space V^* whose elements are linear functions on V, i.e., maps $f : V \longrightarrow K$ such that

$$f(\lambda_1 v_1 + \lambda_2 v_2) = \lambda_1 f(v_1) + \lambda_2 f(v_2) \text{ for any } \lambda_1, \lambda_2 \in K \text{ and } v_1, v_2 \in V.$$

The space V^* is called the *dual to* V.

To a basis $e_1, \dots, e_n$ of V we can assign a basis $e_1^*, \dots, e_n^*$ of V^* setting $e_i^*(e_j) = \delta_{ij}$. Any element $f \in V^*$ can be represented in the form $f = \sum f(e_i)e_i^*$. The linear independence of the vectors e_i^* follows from the identity $(\sum \lambda_i e_i^*)(e_j) = \lambda_j$.

Thus, if a basis $e_1, \dots e_n$ of V is fixed we can construct an isomorphism $g : V \longrightarrow V^*$ setting $g(e_i) = e_i^*$. Selecting another basis in V we get another isomorphism (see 5.3), i.e., the isomorphism constructed is not a canonical one.

We can, however, construct a canonical isomorphism between V and $(V^*)^*$ assigning to every $v \in V$ an element $v' \in (V^*)^*$ such that $v'(f) = f(v)$ for any $f \in V^*$.

REMARK. The elements of V^* are sometimes called the *covectors* of V. Besides, the elements of V are sometimes called *contravariant vectors* whereas the elements of V^* are called *covariant vectors*.

5.2. To a linear operator $A : V_1 \longrightarrow V_2$ we can assign the *adjoint operator* $A^* : V_2^* \longrightarrow V_1^*$ setting $(A^* f_2)(v_1) = f_2(Av_1)$ for any $f_2 \in V_2^*$ and $v_1 \in V_1$.

It is more convenient to denote $f(v)$, where $v \in V$ and $f \in V^*$, in a more symmetric way: $\langle f, v \rangle$. The definition of A^* in this notation can be rewritten as follows

$$\langle A^* f_2, v_1 \rangle = \langle f_2, A v_1 \rangle.$$

If a basis[1] $\{e_\alpha\}$ is selected in V_1 and a basis $\{\varepsilon_\beta\}$ is selected in V_2, then to the operator A we can assign the matrix $\|a_{ij}\|$, where $Ae_j = \sum_i a_{ij}\varepsilon_i$. Similarly, to the operator A^* we can assign the matrix $\|a_{ij}^*\|$ with respect to bases $\{e_\alpha^*\}$ and $\{\varepsilon_\beta^*\}$. Let us prove that $\|a_{ij}^*\| = \|a_{ij}\|^T$. Indeed, on the one hand,

$$\langle \varepsilon_k^*, Ae_j \rangle = \sum_i a_{ij} \langle \varepsilon_k^*, \varepsilon_i \rangle = a_{kj}.$$

On the other hand

$$\langle \varepsilon_k^*, Ae_j \rangle = \langle A^* \varepsilon_k^*, e_j \rangle = \sum_p a_{pk}^* \langle \varepsilon_p^*, e_j \rangle = a_{jk}^*.$$

Hence, $a_{jk}^* = a_{kj}$.

[1] As is customary nowadays, we will, by abuse of language, briefly write $\{e_i\}$ to denote the complete set $\{e_i : i \in I\}$ of vectors of a basis and hope this will not cause a misunderstanding.

5.3. Let $\{e_\alpha\}$ and $\{\varepsilon_\beta\}$ be two bases such that $\varepsilon_j = \sum a_{ij}e_i$ and $\varepsilon_p^* = \sum b_{qp}e_q^*$. Then

$$\delta_{pj} = \varepsilon_p^*(\varepsilon_j) = \sum a_{ij}\varepsilon_p^*(e_i) = \sum a_{ij}b_{qp}\delta_{qi} = \sum a_{ij}b_{ip}, \text{ i.e., } AB^T = I.$$

The maps $f, g : V \longrightarrow V^*$ constructed from bases $\{e_\alpha\}$ and $\{\varepsilon_\beta\}$ coincide if $f(\varepsilon_j) = g(\varepsilon_j)$ for all j, i.e., $\sum a_{ij}e_i^* = \sum b_{ij}e_i^*$ and, therefore $A = B = (A^T)^{-1}$.

In other words, the bases $\{e_\alpha\}$ and $\{\varepsilon_\beta\}$ induce the same isomorphism $V \longrightarrow V^*$ if and only if the matrix A of the passage from one basis to another is an orthogonal one.

Notice that the inner product enables one to distinguish the set of orthonormal bases and, therefore, it enables one to construct a canonical isomorphism $V \longrightarrow V^*$. Under this isomorphism to a vector $v \in V$ we assign the linear function v^* such that $v^*(x) = (v, x)$.

5.4. Consider a system of linear equations

$$\text{(1)} \qquad \begin{cases} f_1(x) = b_1, \\ \dots\dots\dots\dots \\ f_m(x) = b_m. \end{cases}$$

We may assume that the covectors $f_1, \dots, f_k$ are linearly independent and $f_i = \sum_{j=1}^k \lambda_{ij}f_j$ for $i > k$. If x_0 is a solution of (1), then $f_i(x_0) = \sum_{j=1}^k \lambda_{ij}f_j(x_0)$ for $i > k$, i.e.,

$$\text{(2)} \qquad b_i = \sum_{j=1}^k \lambda_{ij}b_j \text{ for } i > k.$$

Let us prove that if conditions (2) are verified, then the system (1) is consistent. Let us complement the set of covectors $f_1, \dots, f_k$ to a basis and consider the dual basis $e_1, \dots, e_n$. For a solution we can take $x_0 = b_1e_1 + \dots + b_ke_k$. The general solution of the system (1) is of the form $x_0 + t_1e_{k+1} + \dots + t_{n-k}e_n$ where $t_1, \dots, t_{n-k}$ are arbitrary numbers.

5.4.1. THEOREM. *If the system* (1) *is consistent, then it has a solution* $x = (x_1, \dots, x_n)$, *where* $x_i = \sum_{j=1}^k c_{ij}b_j$ *and the numbers* c_{ij} *do not depend on the* b_j.

To prove it, it suffices to consider the coordinates of the vector $x_0 = b_1e_1 + \dots + b_ke_k$ with respect to the initial basis.

5.4.2. THEOREM. *If* $f_i(x) = \sum_{j=1}^n a_{ij}x_j$ *with* $a_{ij} \in \mathbb{Q}$ *and the covectors* $f_1, \dots, f_m$ *constitute a basis* (in particular it follows that $m = n$), *then the system* (1) *has a solution* $x_i = \sum_{j=1}^n c_{ij}b_j$, *where the numbers* c_{ij} *are rational and do not depend on* b_j*; this solution is unique.*

PROOF. Since $Ax = b$, where $A = \|a_{ij}\|$, we have $x = A^{-1}b$. If the elements of A are rational numbers, then the elements of A^{-1} are also rational ones. □

The results of 5.4.1 and 5.4.2 have a somewhat unexpected application.

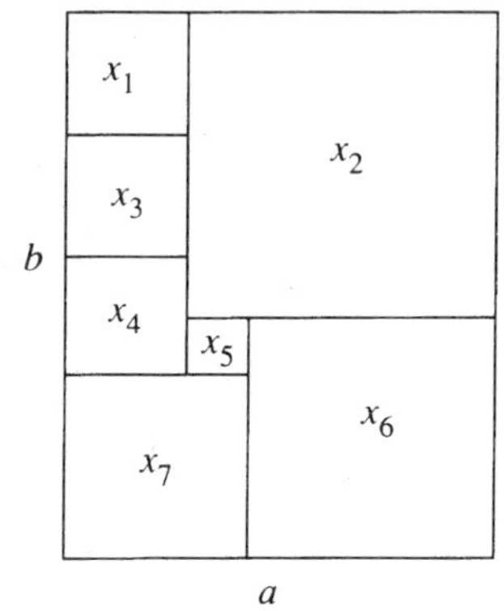

FIGURE 1

5.4.3. THEOREM. *If a rectangle with sides a and b is arbitrarily cut into squares with sides $x_1, \dots, x_n$, then $\frac{x_i}{a} \in \mathbb{Q}$ and $\frac{x_i}{b} \in \mathbb{Q}$ for all i.*

PROOF. Figure 1 illustrates the following system of equations:

$$\begin{aligned} x_1 + x_2 &= a \\ x_3 + x_2 &= a \qquad\qquad & x_1 + x_3 + x_4 + x_7 &= b \\ x_4 + x_2 &= a & x_2 + x_5 + x_7 &= b \\ x_4 + x_5 + x_6 &= a & x_2 + x_6 &= b \\ x_6 + x_7 &= a. \end{aligned} \tag{3}$$

A similar system of equations can be written for any other partition of a rectangle into squares. Notice also that if the system corresponding to a partition has another solution consisting of positive numbers, then to this solution a partition of the rectangle into squares can also be assigned, and for any partition we have the equality of areas $x_1^2 + \dots x_n^2 = ab$.

First, suppose that system (3) has a unique solution. Then

$$x_i = \lambda_i a + \mu_i b \text{ and } \lambda_i, \mu_i \in \mathbb{Q}.$$

Substituting these values into all equations of system (3) we get identities of the form $p_j a + q_j b = 0$, where $p_j, q_j \in \mathbb{Q}$. If $p_j = q_j = 0$ for all j, then system (3) is consistent for all a and b. Therefore, for any sufficiently small variation of the numbers a and b system (3) has a positive solution $x_i = \lambda_i a + \mu_i b$; therefore, there exists the corresponding partition of the rectangle. Hence, for all a and b from certain intervals we have

$$\left(\sum \lambda_i^2\right) a^2 + 2\left(\sum \lambda_i \mu_i\right) ab + \left(\sum \mu_i^2\right) b^2 = ab.$$

Thus, $\sum \lambda_i^2 = \sum \mu_i^2 = 0$ and, therefore, $\lambda_i = \mu_i = 0$ for all i. We got a contradiction; hence, in one of the identities $p_j a + q_j b = 0$ one of the numbers p_j and q_j is nonzero. Thus, $b = ra$, where $r \in \mathbb{Q}$, and $x_i = (\lambda_i + r\mu_i)a$, where $\lambda_i + r\mu_i \in \mathbb{Q}$.

Now, let us prove that the dimension of the space of solutions of system (3) cannot be greater than zero. The solutions of (3) are of the form

$$x_i = \lambda_i a + \mu_i b + \alpha_{1i} t_1 + \dots + \alpha_{ki} t_k,$$

where $t_1, \dots, t_k$ can take arbitrary values. Therefore, the identity

$$\sum (\lambda_i a + \mu_i b + \alpha_{1i} t_1 + \dots + \alpha_{ki} t_k)^2 = ab \tag{4}$$

should be true for all $t_1, \dots, t_k$ from certain intervals. The left-hand side of (4) is a quadratic function of $t_1, \dots, t_k$. This function is of the form $\sum \alpha_{pi}^2 t_p^2 + \dots$, and, therefore, it cannot be a constant for all small changes of the numbers $t_1, \dots, t_k$. □

5.5. As we have already noted, there is no canonical isomorphism between V and V^*. There is, however, a canonical one-to-one correspondence between the set of k-dimensional subspaces of V and the set of $(n-k)$-dimensional subspaces of V^*. To a subspace $W \subset V$ we can assign the set

$$W^\perp = \{f \in V^* \mid \langle f, w \rangle = 0 \quad \text{for any } w \in W\}.$$

This set is called *the annihilator* or *orthogonal complement* to the subspace W. The annihilator is a subspace of V^* and $\dim W + \dim W^\perp = \dim V$ because if $e_1, \dots, e_n$ is a basis for V such that $e_1, \dots, e_k$ is a basis for W, then $e_{k+1}^*, \dots, e_n^*$ is a basis for $W^\perp$.

The following properties of the orthogonal complement are easily verified:
a) if $W_1 \subset W_2$, then $W_2^\perp \subset W_1^\perp$;
b) $(W^\perp)^\perp = W$;
c) $(W_1 + W_2)^\perp = W_1^\perp \cap W_2^\perp$ and $(W_1 \cap W_2)^\perp = W_1^\perp + W_2^\perp$;
d) if $V = W_1 \oplus W_2$, then $V^* = W_1^\perp \oplus W_2^\perp$.

The subspace $W^\perp$ is invariantly defined and therefore, the linear span of vectors $e_{k+1}^*, \dots, e_n^*$ does not depend on the choice of a basis in V, and only depends on the subspace W itself. On the other hand, the linear span of the vectors $e_1^*, \dots, e_k^*$ does depend on the choice of the basis $e_1, \dots, e_n$; it can be any k-dimensional subspace of V^* whose intersection with $W^\perp$ is 0. Indeed, let W_1 be a k-dimensional subspace of V^* and $W_1 \cap W^\perp = 0$. Then $(W_1)^\perp$ is an $(n-k)$-dimensional subspace of V whose intersection with W is 0. Let $e_{k+1}, \dots, e_k$ be a basis of $(W_1)^\perp$. Let us complement it a basis of W to a basis $e_1, \dots, e_n$. Then $e_1^*, \dots, e_k^*$ is a basis of W_1.

THEOREM. *If $A : V \longrightarrow V$ is a linear operator and $AW \subset W$, then $A^* W^\perp \subset W^\perp$.*

PROOF. Let $x \in W$ and $f \in W^\perp$. Then $\langle A^* f, x \rangle = \langle f, Ax \rangle = 0$ since $Ax \in W$. Therefore, $A^* f \in W^\perp$. □

5.6. In the space of real matrices of size $m \times n$ we can introduce a natural inner product. This inner product can be expressed in the form

$$\operatorname{tr}(XY^T) = \sum_{i,j} x_{ij} y_{ij}.$$

THEOREM. *Let A be a matrix of size $m \times n$. If for every matrix X of size $n \times m$ we have* $\operatorname{tr}(AX) = 0$, *then* $A = 0$.

PROOF. If $A \neq 0$, then $\operatorname{tr}(AA^T) = \sum_{i,j} a_{ij}^2 > 0$. □

Problems

5.1. A matrix A of order n is such that for any traceless matrix X (i.e., $\operatorname{tr} X = 0$) of order n we have $\operatorname{tr}(AX) = 0$. Prove that $A = \lambda I$.

5.2. Let A and B be matrices of size $m \times n$ and $k \times n$, respectively, such that if $AX = 0$ for a certain column X, then $BX = 0$. Prove that $B = CA$, where C is a matrix of size $k \times m$.

5.3. All coordinates of a vector $v \in R^n$ are nonzero. Prove that the orthogonal complement of v contains vectors from all orthants except the orthants which contain v and $-v$.

5.4. Let an isomorphism $V \longrightarrow V^* (x \mapsto x^*)$ be such that the conditions $x^*(y) = 0$ and $y^*(x) = 0$ are equivalent. Prove that $x^*(y) = B(x, y)$, where B is either a symmetric or a skew-symmetric bilinear function.

6. The kernel (null space) and the image (range) of an operator. The quotient space

6.1. For a linear map $A : V \longrightarrow W$ we can consider two sets:

$\operatorname{Ker} A = \{v \in V \mid Av = 0\}$ — the *kernel* (or the *null space*) of the map;

$\operatorname{Im} A = \{w \in W \mid \text{there exists } v \in V \text{ such that } Av = w\}$ — the *image* (or *range*) of the map.

It is easy to verify that $\operatorname{Ker} A$ is a linear subspace in V and $\operatorname{Im} A$ is a linear subspace in W. Let $e_1, \dots, e_k$ be a basis of $\operatorname{Ker} A$ and $e_1, \dots, e_k, e_{k+1}, \dots, e_n$ an extension of this basis to a basis of V. Then $Ae_{k+1}, \dots, Ae_n$ is a basis of $\operatorname{Im} A$ and, therefore,

$$\dim \operatorname{Ker} A + \dim \operatorname{Im} A = \dim V.$$

Select bases in V and W and consider the matrix of A with respect to these bases. The space $\operatorname{Im} A$ is spanned by the columns of A and, therefore, $\dim \operatorname{Im} A = \operatorname{rank} A$. In particular, it is clear that the rank of the matrix of A does not depend on the choice of bases, i.e., the *rank of an operator* is well defined.

Given maps $A : U \longrightarrow V$ and $B : V \longrightarrow W$, it is possible that $\operatorname{Im} A$ and $\operatorname{Ker} B$ have a nonzero intersection. The dimension of this intersection can be computed from the following formula.

THEOREM.

$$\dim(\operatorname{Im} A \cap \operatorname{Ker} B) = \dim \operatorname{Im} A - \dim \operatorname{Im} BA = \dim \operatorname{Ker} BA - \dim \operatorname{Ker} A.$$

PROOF. Let C be the restriction of B to $\operatorname{Im} A$. Then

$$\dim \operatorname{Ker} C + \dim \operatorname{Im} C = \dim \operatorname{Im} A,$$

i.e.,

$$\dim(\operatorname{Im} A \cap \operatorname{Ker} B) + \dim \operatorname{Im} BA = \dim \operatorname{Im} A.$$

To prove the second identity it suffices to notice that

$$\dim \operatorname{Im} BA = \dim V - \dim \operatorname{Ker} BA$$

and

$$\dim \operatorname{Im} A = \dim V - \dim \operatorname{Ker} A. \quad \square$$

6.2. The kernel and the image of A and of the adjoint operator A^* are related as follows.

6.2.1. THEOREM. $\operatorname{Ker} A^* = (\operatorname{Im} A)^\perp$ *and* $\operatorname{Im} A^* = (\operatorname{Ker} A)^\perp$.

PROOF. The equality $A^*f = 0$ means that $f(Ax) = A^*f(x) = 0$ for any $x \in V$, i.e., $f \in (\operatorname{Im} A)^\perp$. Therefore, $\operatorname{Ker} A^* = (\operatorname{Im} A)^\perp$ and since $(A^*)^* = A$, we have $\operatorname{Ker} A = (\operatorname{Im} A^*)^\perp$. Hence, $(\operatorname{Ker} A)^\perp = ((\operatorname{Im} A^*)^\perp)^\perp = \operatorname{Im} A^*$. □

COROLLARY. $\operatorname{rank} A = \operatorname{rank} A^*$.

PROOF. $\operatorname{rank} A^* = \dim \operatorname{Im} A^* = \dim(\operatorname{Ker} A)^\perp = \dim V - \dim \operatorname{Ker} A = \dim \operatorname{Im} A = \operatorname{rank} A$. □

REMARK. If V is a space with an inner product, then V^* can be identified with V and we have

$$V = \operatorname{Im} A \oplus (\operatorname{Im} A)^\perp = \operatorname{Im} A \oplus \operatorname{Ker} A^*.$$

Similarly, $V = \operatorname{Im} A^* \oplus \operatorname{Ker} A$.

6.2.2. THEOREM (The Fredholm alternative). *Let $A : V \longrightarrow V$ be a linear operator. Consider the four equations*

$$\begin{array}{llll} (1) & Ax = y \quad \text{for } x, y \in V, & (3) & Ax = 0, \\ (2) & A^*f = g \quad \text{for } f, g \in V^*, & (4) & A^*f = 0. \end{array}$$

Then either equations (1) *and* (2) *are solvable for any right-hand side and in this case the solution is unique, or equations* (3) *and* (4) *have the same number of linearly independent solutions $x_1, \dots, x_k$ and $f_1, \dots, f_k$ and in this case the equation* (1) *(resp.* (2)*) is solvable if and only if $f_1(y) = \dots = f_k(y) = 0$ (resp. $g(x_1) = \dots = g(x_k) = 0$).*

PROOF. Let us show that the Fredholm alternative is essentially a reformulation of Theorem 6.2.1. Solvability of equations (1) and (2) for any right-hand sides means that $\operatorname{Im} A = V$ and $\operatorname{Im} A^* = V$, i.e., $(\operatorname{Ker} A^*)^\perp = V$ and $(\operatorname{Ker} A)^\perp = V$ and, therefore, $\operatorname{Ker} A^* = 0$ and $\operatorname{Ker} A = 0$. These identities are equivalent since $\operatorname{rank} A = \operatorname{rank} A^*$.

If $\operatorname{Ker} A \neq 0$, then $\dim \operatorname{Ker} A^* = \dim \operatorname{Ker} A$ and $y \in \operatorname{Im} A$ if and only if $y \in (\operatorname{Ker} A^*)^\perp$, i.e., $f_1(y) = \dots = f_k(y) = 0$. Similarly, $g \in \operatorname{Im} A^*$ if and only if $g(x_1) = \dots = g(x_k) = 0$. □

6.3. The image of a linear map A is connected with the solvability of the linear equation

$$(1) \qquad Ax = b.$$

This equation is solvable if and only if $b \in \operatorname{Im} A$. In case the map is given by a matrix there is a simple criterion for solvability of (1).

6.3.1. THEOREM (Kronecker-Capelli). *Let A be a matrix, and let x and b be columns such that* (1) *makes sense. Equation* (1) *is solvable if and only if* $\operatorname{rank} A = \operatorname{rank}(A, b)$, *where (A, b) is the matrix obtained from A by augmenting it with b.*

PROOF. Let $A_1, \dots, A_n$ be the columns of A. The equation (1) can be rewritten in the form $x_1A_1 + \dots + x_nA_n = b$. This equation means that the column b is a linear combination of the columns $A_1, \dots, A_n$, i.e., $\operatorname{rank} A = \operatorname{rank}(A, b)$. □

A linear equation can be of a more complicated form. Let us consider for example the matrix equation

$$C = AXB. \tag{2}$$

First of all, let us reduce this equation to a simpler form.

6.3.2. THEOREM. *Let* $a = \operatorname{rank} A$. *Then there exist invertible matrices* L *and* R *such that* $LAR = I_a$, *where* I_a *is the unit matrix of order* a *enlarged with the help of zeros to make its size the same as that of* A.

PROOF. Let us consider the map $A : V^n \longrightarrow V^m$ corresponding to the matrix A taken with respect to bases $e_1, \dots, e_n$ and $\varepsilon_1, \dots, \varepsilon_m$ in the spaces V^n and V^m, respectively. Let $y_{a+1}, \dots, y_n$ be a basis of $\operatorname{Ker} A$ and let vectors $y_1, \dots, y_a$ complement this basis to a basis of V^n. Define a map $R : V^n \longrightarrow V^n$ setting $R(e_i) = y_i$. Then $AR(e_i) = Ay_i$ for $i \leq a$ and $AR(e_i) = 0$ for $i > a$. The vectors $x_1 = Ay_1, \dots, x_a = Ay_a$ form a basis of $\operatorname{Im} A$. Let us complement them by vectors $x_{a+1}, \dots, x_m$ to a basis of V^m. Define a map $L : V^m \longrightarrow V^m$ by the formula $Lx_i = \varepsilon_i$. Then

$$LAR(e_i) = \begin{cases} \varepsilon_i & \text{for } 1 \leq i \leq a; \\ 0 & \text{for } i > a. \end{cases}$$

Therefore, the matrices of the operators L and R with respect to the bases e and ε, respectively, are the required ones. □

6.3.3. THEOREM. *Equation* (2) *is solvable if and only if one of the following equivalent conditions holds*

a) *there exist matrices* Y *and* Z *such that* $C = AY$ *and* $C = ZB$;

b) $\operatorname{rank} A = \operatorname{rank}(A, C)$ *and* $\operatorname{rank} B = \operatorname{rank} \binom{B}{C}$, *where the matrix* (A, C) *is formed from the columns of the matrices* A *and* C *and the matrix* $\binom{B}{C}$ *is formed from the rows of the matrices* B *and* C.

PROOF. The equivalence of a) and b) is proved along the same lines as Theorem 6.3.1. It is also clear that if $C = AXB$, then we can set $Y = XB$ and $Z = AX$. Now, suppose that $C = AY$ and $C = ZB$. Making use of Theorem 6.3.2, we can rewrite (2) in the form

$$D = I_a W I_b, \quad \text{where } D = L_A C R_B \text{ and } W = R_A^{-1} X L_B^{-1}.$$

Conditions $C = AY$ and $C = ZB$ take the form $D = I_a(R_A^{-1} Y R_B)$ and $D = (L_A Z L_B^{-1}) I_b$, respectively. The first identity implies that the last $n - a$ rows of D are zero and the second identity implies that the last $m - b$ columns of D are zero. Therefore, for W we can take the matrix D. □

6.4. If W is a subspace in V, then V can be stratified into subsets

$$M_v = \{x \in V \mid x - v \in W\}.$$

It is easy to verify that $M_v = M_{v'}$ if and only if $v - v' \in W$. On the set

$$V/W = \{M_v \mid v \in V\},$$

we can introduce a linear space structure setting $\lambda M_v = M_{\lambda v}$ and $M_v + M_{v'} = M_{v+v'}$. It is easy to verify that $M_{\lambda v}$ and $M_{v+v'}$ do not depend on the choice of v and v' and only depend on the sets M_v and $M_{v'}$ themselves. The space V/W is called the *quotient space of V with respect to* (or *modulo*) W; it is convenient to denote the class M_v by $v + W$.

The map $p : V \longrightarrow V/W$, where $p(v) = M_v$, is called the *canonical projection.* Clearly, $\operatorname{Ker} p = W$ and $\operatorname{Im} p = V/W$. If $e_1, \dots, e_k$ is a basis of W and $e_1, \dots, e_k$, $e_{k+1}, \dots, e_n$ is a basis of V, then $p(e_1) = \dots = p(e_k) = 0$ whereas $p(e_{k+1}), \dots$, $p(e_n)$ is a basis of V/W. Therefore, $\dim(V/W) = \dim V - \dim W$.

THEOREM. *The following canonical isomorphisms hold:*
a) $(U/W)/(V/W) \cong U/V$ *if* $W \subset V \subset U$;
b) $V/V \cap W \cong (V+W)/W$ *if* $V, W \subset U$.

PROOF. a) Let $u_1, u_2 \in U$. The classes $u_1 + W$ and $u_2 + W$ determine the same class modulo V/W if and only if $[(u_1 + W) - (u_2 + W)] \in V$, i.e., $u_1 - u_2 \in V + W = V$, and, therefore, the elements u_1 and u_2 determine the same class modulo V.

b) The elements $v_1, v_2 \in V$ determine the same class modulo $V \cap W$ if and only if $v_1 - v_2 \in W$, hence the classes $v_1 + W$ and $v_2 + W$ coincide. $\square$

Problem

6.1. Let A be a linear operator. Prove that

$$\dim \operatorname{Ker} A^{n+1} = \dim \operatorname{Ker} A + \sum_{k=1}^{n} \dim(\operatorname{Im} A^k \cap \operatorname{Ker} A)$$

and

$$\dim \operatorname{Im} A = \dim \operatorname{Im} A^{n+1} + \sum_{k=1}^{n} \dim(\operatorname{Im} A^k \cap \operatorname{Ker} A).$$

7. Bases of a vector space. Linear independence

7.1. In spaces V and W, let there be given bases $e_1, \dots, e_n$ and $\varepsilon_1, \dots, \varepsilon_m$. Then to a linear map $f : V \longrightarrow W$ we can assign a matrix $A = \|a_{ij}\|$ such that $fe_j = \sum a_{ij}\varepsilon_i$, i.e.,

$$f\left(\sum x_j e_j\right) = \sum a_{ij} x_j \varepsilon_i.$$

Let x be a column $(x_1, \dots, x_n)^T$, and let e and ε be the rows $(e_1, \dots, e_n)$ and $(\varepsilon_1, \dots, \varepsilon_m)$. Then $f(ex) = \varepsilon Ax$. In what follows a map and the corresponding matrix will be often denoted by the same letter.

How does the matrix of a map vary under a change of bases? Let $e' = eP$ and $\varepsilon' = \varepsilon Q$ be other bases. Then

$$f(e'x) = f(ePx) = \varepsilon APx = \varepsilon' Q^{-1}APx,$$

i.e.,

$$A' = Q^{-1}AP$$

is the matrix of f with respect to e' and ε'. The most important case is when $V = W$ and $P = Q$, in which case

$$A' = P^{-1}AP.$$

THEOREM. *For a linear operator A the polynomial*

$$|\lambda I - A| = \lambda^n + a_{n-1}\lambda^{n-1} + \dots + a_0$$

does not depend on the choice of a basis.

PROOF. $|\lambda I - P^{-1}AP| = |P^{-1}(\lambda I - A)P| = |P|^{-1}|P| \cdot |\lambda I - A| = |\lambda I - A|$.
The polynomial

$$p(\lambda) = |\lambda I - A| = \lambda^n + a_{n-1}\lambda^{n-1} + \dots + a_0$$

is called *the characteristic polynomial* of the operator A, its roots are called the *eigenvalues* of A. Clearly, $|A| = (-1)^n a_0$ and $\operatorname{tr} A = -a_{n-1}$ are invariants of A. □

7.2. The majority of general statements on bases are quite obvious. There are, however, several not so transparent theorems on a possibility of getting a basis by sorting vectors of two systems of linearly independent vectors. Here is one of such theorems.

THEOREM ([Green, 1973]). *Let $x_1, \dots, x_n$ and $y_1, \dots, y_n$ be two bases, $1 \le k \le n$. Then k of the vectors $y_1, \dots, y_n$ can be swapped with the vectors $x_1, \dots, x_k$ so that we get again two bases.*

PROOF. Take the vectors $y_1, \dots, y_n$ for a basis of V. For any set of n vectors z_1, $\dots$, z_n from V consider the determinant $M(z_1, \dots, z_n)$ of the matrix whose rows are composed of coordinates of the vectors $z_1, \dots, z_n$ with respect to the basis $y_1, \dots, y_n$. The vectors $z_1, \dots, z_n$ constitute a basis if and only if $M(z_1, \dots, z_n) \neq 0$. We can express the formula of the expansion of $M(x_1, \dots, x_n)$ with respect to the first k rows in the form

$$(1) \qquad M(x_1, \dots, x_n) = \sum_{A \subset Y} \pm M(x_1, \dots, x_k, A) M(Y \setminus A, x_{k+1}, \dots, x_n),$$

where the summation runs over all $(n-k)$-element subsets of $Y = \{y_1, \dots, y_n\}$. Since $M(x_1, \dots, x_n) \neq 0$, there is at least one nonzero term in (1); the corresponding subset A determines the required set of vectors of the basis $y_1, \dots, y_n$. □

7.3. THEOREM ([Aupetit, 1988]). *Let T be a linear operator in a space V such that for any $\xi \in V$ the vectors ξ, $T\xi$, $\dots$, $T^n\xi$ are linearly dependent. Then the operators I, T, $\dots$, T^n are linearly dependent.*

PROOF. We may assume that n is the maximal of the numbers such that the vectors $\xi_0, \dots, T^{n-1}\xi_0$ are linearly independent and $T^n\xi_0 \in \operatorname{Span}(\xi_0, \dots, T^{n-1}\xi_0)$ for some ξ_0. Then there exists a polynomial p_0 of degree n such that $p_0(T)\xi_0 = 0$; we may assume that the coefficient of highest degree of p_0 is equal to 1.

Fix a vector $\eta \in V$ and let us prove that $p_0(T)\eta = 0$. Let us consider

$$W = \operatorname{Span}(\xi_0, \dots, T^n\xi_0, \eta, \dots, T^n\eta).$$

It is easy to verify that $\dim W \le 2n$ and $T(W) \subset W$. For every $\lambda \in \mathbb{C}$ consider the vectors

$$f_0(\lambda) = \xi_0 + \lambda\eta,\ f_1(\lambda) = Tf_0(\lambda),\ \dots,\ f_{n-1}(\lambda) = T^{n-1}f_0(\lambda),\ g(\lambda) = T^n f_0(\lambda).$$

The vectors $f_0(0), \dots, f_{n-1}(0)$ are linearly independent and, therefore, there are linear functions $\varphi_0, \dots, \varphi_{n-1}$ on W such that $\varphi_i(f_j(0)) = \delta_{ij}$. Let

$$\Delta(\lambda) = |a_{ij}(\lambda)|_0^{n-1}, \quad \text{where } a_{ij}(\lambda) = \varphi_i(f_j(\lambda)).$$

Then $\Delta(\lambda)$ is a polynomial in λ of degree not greater than n, such that $\Delta(0) = 1$. By the hypothesis for any $\lambda \in \mathbb{C}$ there exist complex numbers $\alpha_0(\lambda), \dots, \alpha_{n-1}(\lambda)$ such that

$$(1) \qquad g(\lambda) = \alpha_0(\lambda) f_0(\lambda) + \cdots + \alpha_{n-1}(\lambda) f_{n-1}(\lambda).$$

Therefore,

$$(2) \qquad \varphi_i(g(\lambda)) = \sum_{k=0}^{n-1} \alpha_k(\lambda)\varphi_i(f_k(\lambda)) \text{ for } i = 0, \dots, n-1.$$

If $\Delta(\lambda) \neq 0$, then system (2) of linear equations for $\alpha_k(\lambda)$ can be solved with the help of Cramer's rule. Therefore, $\alpha_k(\lambda)$ is a rational function for all $\lambda \in \mathbb{C} \setminus \Delta$, where Δ is a (finite) set of roots of $\Delta(\lambda)$.

The identity (1) can be expressed in the form $p_\lambda(T) f_0(\lambda) = 0$, where

$$p_\lambda(T) = T^n - \alpha_{n-1}(\lambda) T^{n-1} - \cdots - \alpha_0(\lambda) I.$$

Let $\beta_1(\lambda), \dots, \beta_n(\lambda)$ be the roots of $p(\lambda)$. Then

$$(T - \beta_1(\lambda)I) \dots (T - \beta_n(\lambda)I) f_0(\lambda) = 0.$$

If $\lambda \notin \Delta$, then the vectors $f_0(\lambda), \dots, f_{n-1}(\lambda)$ are linearly independent, in other words, $h(T) f_0(\lambda) \neq 0$ for any nonzero polynomial h of degree $n-1$. Hence,

$$w = (T - \beta_2(\lambda)I) \dots (T - \beta_n(\lambda)I) f_0(\lambda) \neq 0$$

and $(T - \beta_1(\lambda)I)w = 0$, i.e., $\beta_1(\lambda)$ is an eigenvalue of T. The proof of the fact that $\beta_2(\lambda), \dots, \beta_n(\lambda)$ are eigenvalues of T is similar. Thus, $|\beta_i(\lambda)| \leq \|T\|_s$ (cf. 35.1).

The rational functions $\alpha_0(\lambda), \dots, \alpha_{n-1}(\lambda)$ are symmetric functions in the functions $\beta_1(\lambda), \dots, \beta_n(\lambda)$; the latter are uniformly bounded on $\mathbb{C} \setminus \Delta$ and, therefore, α_i themselves are uniformly bounded on $\mathbb{C} \setminus \Delta$. Hence, the functions $\alpha_0(\lambda), \dots, \alpha_{n-1}(\lambda)$ are bounded on $\mathbb{C}$; by the Liouville theorem[2] they are constants: $\alpha_i(\lambda) = \alpha_i$.

Let $p(T) = T^n - \alpha_{n-1} T^{n-1} - \cdots - \alpha_0 I$. Then $p(T) f_0(\lambda) = 0$ for $\lambda \in \mathbb{C} \setminus \Delta$; hence, $p(T) f_0(\lambda) = 0$ for all λ. In particular, $p(T)\xi_0 = 0$. Hence, $p = p_0$ and $p_0(T)\eta = 0$. $\square$

[2]See any textbook on complex analysis.

Problems

7.1. In V^n there are given vectors $e_1, \dots, e_m$. Prove that if $m \geq n+2$, then there exist numbers $\alpha_1, \dots, \alpha_m$ not all of them equal to zero such that $\sum \alpha_i e_i = 0$ and $\sum \alpha_i = 0$.

7.2. A *convex linear combination of vectors* $v_1, \dots, v_m$ is an arbitrary vector $x = t_1 v_1 + \dots + t_m v_m$, where $t_i \geq 0$ and $\sum t_i = 1$.

Prove that in a real space of dimension n any convex linear combination of m vectors is also a convex linear combination of no more than $n+1$ of the given vectors.

7.3. Prove that if $|a_{ii}| > \sum_{k \neq i} |a_{ik}|$ for $i = 1, \dots, n$, then $A = \left\| a_{ij} \right\|_1^n$ is an invertible matrix.

7.4. a) Given vectors $e_1, \dots, e_{n+1}$ in an n-dimensional Euclidean space, such that $(e_i, e_j) < 0$ for $i \neq j$, prove that any n of these vectors form a basis.

b) Prove that if $e_1, \dots, e_m$ are vectors in $\mathbb{R}^n$ such that $(e_i, e_j) < 0$ for $i \neq j$, then $m \leq n+1$.

8. The rank of a matrix

8.1. The columns of the matrix AB are linear combinations of the columns of A and, therefore,

$$\operatorname{rank} AB \leq \operatorname{rank} A;$$

since the rows of AB are linear combinations of rows B, we have

$$\operatorname{rank} AB \leq \operatorname{rank} B.$$

If B is invertible, then

$$\operatorname{rank} A = \operatorname{rank}(AB)B^{-1} \leq \operatorname{rank} AB$$

and, therefore, $\operatorname{rank} A = \operatorname{rank} AB$.

Let us give two more inequalities for the ranks of products of matrices.

8.1.1. THEOREM (Frobenius inequality).

$$\operatorname{rank} BC + \operatorname{rank} AB \leq \operatorname{rank} ABC + \operatorname{rank} B.$$

PROOF. If $U \subset V$ and $X : V \longrightarrow W$, then

$$\dim(\operatorname{Ker} X|_U) \leq \dim \operatorname{Ker} X = \dim V - \dim \operatorname{Im} X.$$

For $U = \operatorname{Im} BC$, $V = \operatorname{Im} B$ and $X = A$ we get

$$\dim(\operatorname{Ker} A|_{\operatorname{Im} BC}) \leq \dim \operatorname{Im} B - \dim \operatorname{Im} AB.$$

Clearly,

$$\dim(\operatorname{Ker} A|_{\operatorname{Im} BC}) = \dim \operatorname{Im} BC - \dim \operatorname{Im} ABC. \qquad \square$$

8.1.2. Theorem (The Sylvester inequality).

$$\operatorname{rank} A + \operatorname{rank} B \le \operatorname{rank} AB + n,$$

where n is the number of columns of the matrix A and also the number of rows of the matrix B.

Proof. Make use of the Frobenius inequality for matrices $A_1 = A$, $B_1 = I_n$ and $C_1 = B$. □

8.2. The *rank of a matrix* can also be defined as follows: the rank of A is equal to the least of the sizes of matrices B and C whose product is equal to A.

Let us prove that this definition is equivalent to the conventional one. If $A = BC$ and the minimal of the sizes of B and C is equal to k, then

$$\operatorname{rank} A \le \min(\operatorname{rank} B, \operatorname{rank} C) \le k.$$

It remains to demonstrate that if A is a matrix of size $m \times n$ and $\operatorname{rank} A = k$, then A can be represented as the product of matrices of sizes $m \times k$ and $k \times n$. In A, let us single out linearly independent columns $A_1, \dots, A_k$. All other columns can be linearly expressed through them and, therefore,

$$A = (x_{11}A_1 + \dots + x_{k1}A_k, \dots, x_{1n}A_1 + \dots + x_{kn}A_k)$$
$$= (A_1 \dots A_k) \begin{pmatrix} x_{11} & \dots & x_{1n} \\ \vdots & \dots & \vdots \\ x_{k1} & \dots & x_{kn} \end{pmatrix}.$$

8.3. Let $M_{n,m}$ be the space of matrices of size $n \times m$. In this space we can indicate a subspace of dimension nr, the rank of whose elements does not exceed r. For this it suffices to take matrices in the last $n - r$ rows of which only zeros stand.

Theorem ([Flanders, 1962]). *Let $r \le m \le n$, let $U \subset M_{n,m}$ be a linear subspace and let the maximal rank of elements of U be equal to r. Then* $\dim U \le nr$.

Proof. Complementing, if necessary, the matrices by zeros, we assume that all matrices are of size $n \times n$. In U, select a matrix A of rank r. The transformation $X \mapsto PXQ$, where P and Q are invertible matrices, sends A to $\begin{pmatrix} I_r & 0 \\ 0 & 0 \end{pmatrix}$ (see Theorem 6.3.2). We now perform the same transformation over all matrices of U and express them in the corresponding block form.

8.3.1. Lemma. *If $B \in U$, then $B = \begin{pmatrix} B_{11} & B_{12} \\ B_{21} & 0 \end{pmatrix}$, where $B_{21}B_{12} = 0$.*

Proof. Let $B = \begin{pmatrix} B_{11} & B_{12} \\ B_{21} & B_{22} \end{pmatrix} \in U$, where the matrix B_{21} consists of rows $u_1, \dots, u_{n-r}$ and the matrix B_{12} consists of columns $v_1, \dots, v_{n-r}$. Any minor of order $r + 1$ of the matrix $tA + B$ vanishes and, therefore,

$$\Delta(t) = \begin{vmatrix} tI_r + B_{11} & v_j \\ u_i & b_{ij} \end{vmatrix} = 0.$$

The coefficient of t^r is equal to b_{ij} and, therefore, $b_{ij} = 0$. Hence, (see Theorem 3.1.3)

$$\Delta(t) = -u_i \ \mathrm{adj}(tI_r + B_{11})v_j.$$

Since $\mathrm{adj}(tI_r + B_{11}) = t^{r-1}I_r + \dots$, the coefficient of t^{r-1} of the polynomial $\Delta(t)$ is equal to $-u_iv_j$. Hence, $u_iv_j = 0$ and, therefore $B_{21}B_{12} = 0$. $\square$

8.3.2. Lemma. *If $B, C \in U$, then $B_{21}C_{12} + C_{21}B_{12} = 0$.*

Proof. Applying Lemma 8.3.1 to the matrix $B + C \in U$ we get $(B_{21} + C_{21})(B_{12} + C_{12}) = 0$, i.e., $B_{21}C_{12} + C_{21}B_{12} = 0$. $\square$

We now turn to the proof of Theorem 8.3. Let us consider the map $f : U \longrightarrow M_{r,n}$ given by the formula $f(C) = \|C_{11}, C_{12}\|$. Then $\mathrm{Ker}\, f$ consists of matrices of the form $\begin{pmatrix} 0 & 0 \\ B_{21} & 0 \end{pmatrix}$ and by Lemma 8.3.2 $B_{21}C_{12} = 0$ for all matrices $C \in U$.

Further, consider the map $g : \mathrm{Ker}\, f \longrightarrow M_{r,n}$ given by the formula

$$g(B)\left(\|X_{11}X_{12}\|\right) = \mathrm{tr}(B_{21}X_{12}).$$

This map is a monomorphism (see 5.6) and therefore, the space $g(\mathrm{Ker}\, f) \subset M^*_{r,n}$ is of dimension $k = \dim \mathrm{Ker}\, f$. Therefore, $(g(\mathrm{Ker}\, f))^\perp$ is a subspace of dimension $nr - k$ in $M_{r,n}$. If $C \in U$, then $B_{21}C_{12} = 0$ and, therefore, $\mathrm{tr}(B_{21}C_{12}) = 0$. Hence, $f(U) \subset (g(\mathrm{Ker}\, f))^\perp$, i.e., $\dim f(U) \le nr - k$. It remains to observe that

$$\dim f(U) + k = \dim \mathrm{Im}\, f + \dim \mathrm{Ker}\, f = \dim U. \quad \square$$

In [Flanders, 1962] there is also given a description of subspaces U such that $\dim U = nr$. If $m = n$ and U contains I_r, then U either consists of matrices whose last $n - r$ columns are zeros, or of matrices whose last $n - r$ rows are zeros.

Problems

8.1. Let $a_{ij} = x_i + y_j$. Prove that $\mathrm{rank}\|a_{ij}\|_1^n \le 2$.

8.2. Let A be a square matrix such that $\mathrm{rank}\, A = 1$. Prove that $|A + I| = (\mathrm{tr}\, A) + 1$.

8.3. Prove that $\mathrm{rank}(A^*A) = \mathrm{rank}\, A$.

8.4. Let A be an invertible matrix. Prove that if $\mathrm{rank}\begin{pmatrix} A & B \\ C & D \end{pmatrix} = \mathrm{rank}\, A$, then $D = CA^{-1}B$.

8.5. Let the sizes of matrices A_1 and A_2 be equal, and let V_1 and V_2 be the spaces spanned by the rows of A_1 and A_2, respectively; let W_1 and W_2 be the spaces spanned by the columns of A_1 and A_2, respectively. Prove that the following conditions are equivalent:

1) $\mathrm{rank}(A_1 + A_2) = \mathrm{rank}\, A_1 + \mathrm{rank}\, A_2$;

2) $V_1 \cap V_2 = 0$;

3) $W_1 \cap W_2 = 0$.

8.6. Prove that if A and B are matrices of the same size and $B^TA = 0$, then $\mathrm{rank}(A + B) = \mathrm{rank}\, A + \mathrm{rank}\, B$.

8.7. Let A and B be square matrices of odd order. Prove that if $AB = 0$, then at least one of the matrices $A + A^T$ and $B + B^T$ is not invertible.

8.8 (Generalized Ptolemy theorem). Let $X_1 \dots X_n$ be a convex polygon inscribable in a circle. Consider a skew-symmetric matrix $A = \|a_{ij}\|_1^n$, where $a_{ij} = X_iX_j$ for $i > j$. Prove that $\operatorname{rank} A = 2$.

9. Subspaces. The Gram-Schmidt orthogonalization process

9.1. The dimension of the intersection of two subspaces is related with the dimension of the space spanned by them via the following relation.

THEOREM. $\dim(V + W) + \dim(V \cap W) = \dim V + \dim W$.

PROOF. Let $e_1, \dots, e_r$ be a basis of $V \cap W$; it can be complemented to a basis $e_1, \dots, e_r, v_1, \dots, v_{n-r}$ of V^n and to a basis $e_1, \dots, e_r, w_1, \dots, w_{m-r}$ of W^m. Then $e_1, \dots, e_r, v_1, \dots, v_{n-r}, w_1, \dots, w_{m-r}$ is a basis of $V + W$. Therefore,

$$\dim(V+W)+\dim(V\cap W) = (r+(n-r)+(m-r))+r = n+m = \dim V+\dim W. \quad \square$$

9.2. Let V be a subspace over $\mathbb{R}$. An *inner product* in V is a map $V \times V \longrightarrow \mathbb{R}$ which to a pair of vectors $u, v \in V$ assigns a number $(u, v) \in \mathbb{R}$ and has the following properties:

1) $(u, v) = (v, u)$;
2) $(\lambda u + \mu v, w) = \lambda(u, w) + \mu(v, w)$;
3) $(u, u) > 0$ for any $u \neq 0$; the value $|u| = \sqrt{(u, u)}$ is called the *length* of u.

A basis $e_1, \dots, e_n$ of V is called an *orthonormal* (respectively, *orthogonal*) if $(e_i, e_j) = \delta_{ij}$ (respectively, $(e_i, e_j) = 0$ for $i \neq j$).

A matrix of the passage from an orthonormal basis to another orthonormal basis is called an *orthogonal matrix*. The columns of such a matrix A constitute an orthonormal system of vectors and, therefore,

$$A^TA = I; \text{ hence, } A^T = A^{-1} \text{ and } AA^T = I.$$

If A is an orthogonal matrix, then

$$(Ax, Ay) = (x, A^TAy) = (x, y).$$

It is easy to verify that any vectors $e_1, \dots, e_n$ such that $(e_i, e_j) = \delta_{ij}$ are linearly independent. Indeed, if $\lambda_1 e_1 + \dots + \lambda_n e_n = 0$, then $\lambda_i = (\lambda_1 e_1 + \dots + \lambda_n e_n, e_i) = 0$. We can similarly prove that an orthogonal system of nonzero vectors is linearly independent.

THEOREM (The Gram-Schmidt orthogonalization). *Let $e_1, \dots, e_n$ be a basis of a vector space. Then there exists an orthogonal basis $\varepsilon_1, \dots, \varepsilon_n$ such that $\varepsilon_i \in \operatorname{Span}(e_1, \dots, e_i)$ for all $i = 1, \dots, n$.*

PROOF is carried out by induction on n. For $n = 1$ the statement is obvious. Suppose the statement holds for n vectors. Consider a basis $e_1, \dots, e_{n+1}$ of $(n+1)$-dimensional space V. By the inductive hypothesis applied to the n-dimensional subspace $W =$

$\text{Span}(e_1, \dots, e_n)$ of V there exists an orthogonal basis $\varepsilon_1, \dots, \varepsilon_n$ of W such that $\varepsilon_i \in \text{Span}(e_1, \dots, e_i)$ for $i = 1, \dots, n$. Consider a vector

$$\varepsilon_{n+1} = \lambda_1 \varepsilon_1 + \dots + \lambda_n \varepsilon_n + e_{n+1}.$$

The condition $(\varepsilon_i, \varepsilon_{n+1}) = 0$ means that $\lambda_i(\varepsilon_i, \varepsilon_i) + (e_{n+1}, \varepsilon_i) = 0$, i.e., $\lambda_i = -\dfrac{(e_{k+1}, \varepsilon_i)}{(\varepsilon_i, \varepsilon_i)}$. Taking such numbers λ_i we get an orthogonal system of vectors $\varepsilon_1, \dots, \varepsilon_{n+1}$ in V, where $\varepsilon_{n+1} \neq 0$ since $e_{n+1} \notin \text{Span}(\varepsilon_1, \dots, \varepsilon_n) = \text{Span}(e_1, \dots, e_n)$. $\square$

REMARK 1. From an orthogonal basis $\varepsilon_1, \dots, \varepsilon_n$ we can pass to an orthonormal basis $\varepsilon_1', \dots, \varepsilon_n'$, where $\varepsilon_i' = \varepsilon_i / \sqrt{(\varepsilon_i, \varepsilon_i)}$.

REMARK 2. The orthogonalization process has a rather lucid geometric interpretation: from the vector e_{n+1} we subtract its orthogonal projection to the subspace $W = \text{Span}(e_1, \dots, e_n)$ and the result is the vector ε_{n+1} orthogonal to W.

9.3. Suppose V is a space with inner product and W is a subspace in V. A vector $w \in W$ is called *the orthogonal projection of a vector $v \in V$ on the subspace W* if $v - w \perp W$.

9.3.1. THEOREM. *For any $v \in W$ there exists a unique orthogonal projection on W.*

PROOF. In W select an orthonormal basis $e_1, \dots, e_k$. Consider a vector $w = \lambda_1 e_1 + \dots + \lambda_k e_k$. The condition $w - v \perp e_i$ means that

$$0 = (\lambda_1 e_1 + \dots + \lambda_k e_k - v, e_i) = \lambda_i - (v, e_i),$$

i.e., $\lambda_i = (v, e_i)$. Taking such numbers λ_i we get the required vector; it is of the form $w = \sum_{i=1}^k (v, e_i) e_i$. $\square$

9.3.1.1. COROLLARY. *If $e_1, \dots, e_n$ is a basis of V and $v \in V$, then $v = \sum_{i=1}^n (v, e_i) e_i$.*

PROOF. The vector $v - \sum_{i=1}^n (v, e_i) e_i$ is orthogonal to the whole V. $\square$

9.3.1.2. COROLLARY. *If w and $w^\perp$ are orthogonal projections of v on W and $W^\perp$, respectively, then $v = w + w^\perp$.*

PROOF. It suffices to complement an orthonormal basis of W to an orthonormal basis of the whole space and make use of Corollary 9.3.1.1. $\square$

9.3.2. THEOREM. *If w is the orthogonal projection of v on W and $w_1 \in W$, then*

$$|v - w_1|^2 = |v - w|^2 + |w - w_1|^2.$$

PROOF. Let $a = v - w$ and $b = w - w_1 \in W$. By definition, $a \perp b$ and, therefore, $|a + b|^2 = (a + b, a + b) = |a|^2 + |b|^2$. $\square$

9.3.2.1. COROLLARY. $|v|^2 = |w|^2 + |v - w|^2$.

PROOF. In the notation of Theorem 9.3.2 set $w_1 = 0$. $\square$

9.3.2.2. COROLLARY. $|v - w_1| \geq |v - w|$ *and the equality takes place only if* $w_1 = w$.

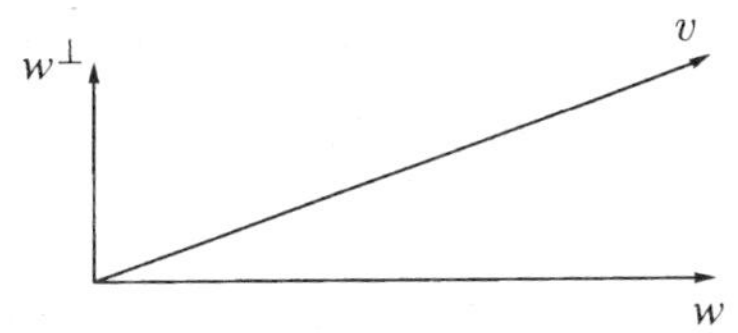

FIGURE 2

9.4. The *angle between a line l and a subspace W* is the angle between a vector v which determines l and the vector w, the orthogonal projection of v onto W (if $w = 0$, then $v \perp W$). Since $v - w \perp w$, then $(v, w) = (w, w) \geq 0$, i.e., the angle between v and w is not obtuse.

If w and $w^\perp$ are orthogonal projections of a unit vector v on W and $W^\perp$, respectively, then $\cos\angle(v, w) = |w|$ and $\cos\angle(v, w^\perp) = |w^\perp|$, see Figure 2, and therefore,

$$\cos\angle(v, W) = \sin\angle(v, W^\perp).$$

Let $e_1, \dots, e_n$ be an orthonormal basis and $v = x_1e_1 + \dots + x_ne_n$ a unit vector. Then $x_i = \cos\alpha_i$, where α_i is the angle between v and e_i. Hence, $\sum_{i=1}^n \cos^2\alpha_i = 1$ and

$$\sum_{i=1}^n \sin^2\alpha_i = \sum_{i=1}^n (1 - \cos^2\alpha_i) = n - 1.$$

THEOREM. *Let $e_1, \dots, e_k$ be an orthonormal basis of a subspace $W \subset V$, α_i the angle between v and e_i, and α the angle between v and W. Then $\cos^2\alpha = \sum_{i=1}^k \cos^2\alpha_i$.*

PROOF. Let us complement the basis $e_1, \dots, e_k$ to a basis $e_1, \dots, e_n$ of V. Then $v = x_1e_1 + \dots + x_ne_n$, where $x_i = \cos\alpha_i$ for $i = 1, \dots, k$, and the projection of v onto W is equal to $x_1e_1 + \dots + x_ke_k = w$. Hence,

$$\cos^2\alpha = |w|^2 = x_1^2 + \dots + x_k^2 = \cos^2\alpha_1 + \dots + \cos^2\alpha_k. \quad \square$$

9.5. THEOREM ([Nisnevich, Bryzgalov, 1953]). *Let $e_1, \dots, e_n$ be an orthogonal basis of V, and $d_1, \dots, d_n$ the lengths of the vectors $e_1, \dots, e_n$. An m-dimensional subspace $W \subset V$ such that the projections of these vectors on W are of equal length exists if and only if*

$$d_i^2(d_1^{-2} + \dots + d_n^{-2}) \geq m \text{ for } i = 1, \dots, n.$$

PROOF. Take an orthonormal basis in W and complement it to an orthonormal basis $\varepsilon_1, \dots, \varepsilon_n$ of V. Let $(x_{1i}, \dots, x_{ni})$ be the coordinates of e_i with respect to the basis $\varepsilon_1, \dots, \varepsilon_n$ and $y_{ki} = x_{ki}/d_i$. Then $\|y_{ki}\|$ is an orthogonal matrix and the length of the projection of e_i on W is equal to d if and only if

$$y_{1i}^2 + \dots + y_{mi}^2 = (x_{1i}^2 + \dots + x_{mi}^2)d_i^{-2} = d^2d_i^{-2}. \tag{1}$$

If the required subspace W exists, then $d \leq d_i$ and

$$m = \sum_{k=1}^m \sum_{i=1}^n y_{ki}^2 = \sum_{i=1}^n \sum_{k=1}^m y_{ki}^2 = d^2(d_1^{-2} + \dots + d_n^{-2}) \leq d_i^2(d_1^{-2} + \dots + d_n^{-2}).$$

Now, suppose that $m \leq d_i^2(d_1^{-2} + \dots + d_n^{-2})$ for $i = 1, \dots, n$ and construct an orthogonal matrix $\|y_{ki}\|_1^n$ with property (1), where $d^2 = m(d_1^{-2} + \dots + d_n^{-2})^{-1}$. We can now construct the subspace W in an obvious way.

Let us prove by induction on n that if $0 \leq \beta_i \leq 1$ for $i = 1, \dots, n$ and $\beta_1 + \dots + \beta_n = m$, then there exists an orthogonal matrix $\|y_{ki}\|_1^n$ such that $y_{1i}^2 + \dots + y_{mi}^2 = \beta_i$. For $n = 1$ the statement is obvious. Suppose the statement holds for $n-1$ and prove it for n. Consider two cases:

a) $m \leq n/2$. We can assume that $\beta_1 \geq \dots \geq \beta_n$. Then $\beta_{n-1} + \beta_n \leq 2m/n \leq 1$ and, therefore, there exists an orthogonal matrix $A = \|a_{ki}\|_1^{n-1}$ such that $a_{1i}^2 + \dots + a_{mi}^2 = \beta_i$ for $i = 1, \dots, n-2$ and $a_{1,n-1}^2 + \dots + a_{m,n-1}^2 = \beta_{n-1} + \beta_n$. Then the matrix

$$\|y_{ki}\|_1^n = \begin{pmatrix} a_{11} & \dots & a_{1,n-2} & \alpha_1 a_{1,n-1} & -\alpha_2 a_{1,n-1} \\ \vdots & \dots & \vdots & \vdots & \vdots \\ a_{n-1,1} & \dots & a_{n-1,n-2} & \alpha_1 a_{n-1,n-1} & -\alpha_2 a_{n-1,n-1} \\ 0 & \dots & 0 & \alpha_2 & \alpha_1 \end{pmatrix},$$

where $\alpha_1 = \sqrt{\dfrac{\beta_{n-1}}{\beta_{n-1} + \beta_n}}$ and $\alpha_2 = \sqrt{\dfrac{\beta_n}{\beta_{n-1} + \beta_n}}$, has orthogonal columns; besides,

$$\sum_{k=1}^{m} y_{ki}^2 = \beta_i \quad \text{for } i = 1, \dots, n-2$$

$$y_{1,n-1}^2 + \dots + y_{m,n-1}^2 = \alpha_1^2(\beta_{n-1} + \beta_n) = \beta_{n-1},$$

$$y_{1n}^2 + \dots + y_{mn}^2 = \beta_n$$

b) Let $m > n/2$. Then $n - m < n/2$, and, therefore, there exists an orthogonal matrix $\|y_{ki}\|_1^n$ such that $y_{m+1,i}^2 + \dots + y_{n,i}^2 = 1 - \beta_i$ for $i = 1, \dots, n$; hence, $y_{1i}^2 + \dots + y_{mi}^2 = \beta_i$. $\square$

9.6.1. THEOREM. *Suppose a set of k-dimensional subspaces in a space V is given so that the intersection of any two of the subspaces is of dimension $k-1$. Then either all these subspaces have a common $(k-1)$-dimensional subspace or all of them are contained in the same $(k+1)$-dimensional subspace.*

PROOF. Let $V_{ij}^{k-1} = V_i^k \cap V_j^k$ and $V_{ijl} = V_i^k \cap V_j^k \cap V_l^k$. First, let us prove that if $V_{123} \neq V_{12}^{k-1}$, then $V_3^k \subset V_1^k + V_2^k$. Indeed, if $V_{123} \neq V_{12}^{k-1}$, then V_{12}^{k-1} and V_{23}^{k-1} are distinct subspaces of V_2^k and the subspace $V_{123} = V_{12}^{k-1} \cap V_{23}^{k-1}$ is of dimension $k-2$. In V_{123}, select a basis ε and complement it by vectors e_{13} and e_{23} to bases of V_{13} and V_{23}, respectively. Then $V_3 = \operatorname{Span}(e_{13}, e_{23}, \varepsilon)$, where $e_{13} \in V_1$ and $e_{23} \in V_2$.

Suppose the subspaces V_1^k, V_2^k and V_3^k have no common $(k-1)$-dimensional subspace, i.e., the subspaces V_{12}^{k-1} and V_{23}^{k-1} do not coincide. The space V_i could not be contained in the subspace spanned by V_1, V_2 and V_3 only if $V_{12i} = V_{12}$ and $V_{23i} = V_{23}$. But then $\dim V_i \geq \dim(V_{12} + V_{23}) = k+1$ which is impossible. $\square$

If we consider the orthogonal complements to the given subspaces we get the theorem dual to Theorem 9.6.1.

9.6.2. THEOREM. *Let a set of m-dimensional subspaces in a space V be given so that any two of them are contained in a $(m+1)$-dimensional subspace. Then either all of them belong to an $(m+1)$-dimensional subspace or all of them have a common $(m-1)$-dimensional subspace.*

Problems

9.1. In an n-dimensional space V, there are given m-dimensional subspaces U and W so that $u \perp W$ for some $u \in U \setminus 0$. Prove that $w \perp U$ for some $w \in W \setminus 0$.

9.2. In an n-dimensional Euclidean space two bases $x_1, \dots, x_n$ and $y_1, \dots, y_n$ are given so that $(x_i, x_j) = (y_i, y_j)$ for all i, j. Prove that there exists an orthogonal operator U which sends x_i to y_i.

10. Complexification and realification. Unitary spaces

10.1. *The complexification of a linear space V over* $\mathbb{R}$ is the set of pairs (a, b), where $a, b \in V$, with the following structure of a linear space over $\mathbb{C}$:

$$(a, b) + (a_1, b_1) = (a + a_1, b + b_1)$$
$$(x + iy)(a, b) = (xa - yb, xb + ya).$$

Such pairs of vectors can be expressed in the form $a + ib$. The complexification of V is denoted by $V^{\mathbb{C}}$.

To an operator $A : V \longrightarrow V$ there corresponds an operator $A^{\mathbb{C}} : V^{\mathbb{C}} \longrightarrow V^{\mathbb{C}}$ given by the formula $A^{\mathbb{C}}(a + ib) = Aa + iAb$. The operator $A^{\mathbb{C}}$ is called *the complexification of A*.

10.2. A linear space V over $\mathbb{C}$ is also a linear space over $\mathbb{R}$. The space over $\mathbb{R}$ obtained is called a *realification of V*. We will denote it by $V_{\mathbb{R}}$.

A linear map $A : V \longrightarrow W$ over $\mathbb{C}$ can be considered as a linear map $A_{\mathbb{R}} : V_{\mathbb{R}} \longrightarrow W_{\mathbb{R}}$ over $\mathbb{R}$. The map $A_{\mathbb{R}}$ is called the *realification of the operator A*.

If $e_1, \dots, e_n$ is a basis of V over $\mathbb{C}$, then $e_1, \dots, e_n, ie_1, \dots, ie_n$ is a basis of $V_{\mathbb{R}}$. It is easy to verify that if $A = B + iC$ is the matrix of a linear map $A : V \longrightarrow W$ with respect to bases $e_1, \dots, e_n$ and $\varepsilon_1, \dots, \varepsilon_m$ and the matrices B and C are real, then the matrix of the linear map $A_{\mathbb{R}}$ with respect to the bases $e_1, \dots, e_n, ie_1, \dots, ie_n$ and $\varepsilon_1, \dots, \varepsilon_m, i\varepsilon_1, \dots, i\varepsilon_m$ is of the form $\begin{pmatrix} B & -C \\ C & B \end{pmatrix}$.

THEOREM. *If $A : V \longrightarrow V$ is a linear map over $\mathbb{C}$ then* $\det A_{\mathbb{R}} = |\det A|^2$.

PROOF.

$$\begin{pmatrix} I & 0 \\ -iI & I \end{pmatrix} \begin{pmatrix} B & -C \\ C & B \end{pmatrix} \begin{pmatrix} I & 0 \\ iI & I \end{pmatrix} = \begin{pmatrix} B - iC & -C \\ 0 & B + iC \end{pmatrix}.$$

Therefore, $\det A_{\mathbb{R}} = \det A \cdot \det \overline{A} = |\det A|^2$. $\square$

10.3. Let V be a linear space over $\mathbb{C}$. *An Hermitian product in V* is a map $V \times V \longrightarrow \mathbb{C}$ which to a pair of vectors $x, y \in V$ assigns a complex number (x, y) and has the following properties:

1) $(x, y) = \overline{(y, x)}$;
2) $(\alpha x + \beta y, z) = \alpha(x, z) + \beta(y, z)$;
3) (x, x) is a positive real number for any $x \neq 0$.

A space V with an Hermitian product is called *an Hermitian* (or *unitary*) space. The standard Hermitian product in $\mathbb{C}^n$ is of the form $x_1\overline{y}_1 + \cdots + x_n\overline{y}_n$.

A linear operator A^* is called the *Hermitian adjoint to A* if

$$(Ax, y) = (x, A^*y) = \overline{(A^*y, x)}.$$

(Physicists often denote the Hermitian adjoint by A^+.)

Let $\|a_{ij}\|_1^n$ and $\|b_{ij}\|_1^n$ be the matrices of A and A^* with respect to an orthonormal basis. Then

$$a_{ij} = (Ae_j, e_i) = \overline{(A^*e_j, e_i)} = \overline{b_{ji}}.$$

A linear operator A is called *unitary* if $(Ax, Ay) = (x, y)$, i.e., a unitary operator preserves the Hermitian product. If an operator A is unitary, then

$$(x, y) = (Ax, Ay) = (x, A^*Ay).$$

Therefore, $A^*A = I = AA^*$, i.e., the rows and the columns of the matrix of A constitute an orthonormal systems of vectors.

A linear operator A is called *Hermitian* (resp. *skew-Hermitian*) if $A^* = A$ (resp. $A^* = -A$). Clearly, a linear operator is Hermitian if and only if its matrix A is Hermitian with respect to an orthonormal basis, i.e., $\overline{A}^T = A$; and in this case its matrix is Hermitian with respect to any orthonormal basis.

Hermitian matrices are, as a rule, complex case analogues of real symmetric matrices. Sometimes complex symmetric or skew-symmetric matrices (that is such that satisfy the condition $A^T = A$ or $A^T = -A$, respectively) are also considered.

10.3.1. THEOREM. *Let A be a complex operator such that $(Ax, x) = 0$ for all x. Then $A = 0$.*

PROOF. Let us write the equation $(Ax, x) = 0$ twice: for $x = u + v$ and $x = u + iv$. Taking into account that $(Av, v) = (Au, u) = 0$ we get $(Av, u) + (Au, v) = 0$ and $i(Av, u) - i(Au, v) = 0$. Therefore, $(Au, v) = 0$ for all $u, v \in V$. $\square$

REMARK. For real operators the identity $(Ax, x) = 0$ means that A is a skew-symmetric operator (cf. Theorem 21.1.2).

10.3.2. THEOREM. *Let A be a complex operator such that $(Ax, x) \in \mathbb{R}$ for any x. Then A is an Hermitian operator.*

PROOF. Since $(Ax, x) = \overline{(Ax, x)} = (x, Ax)$, then

$$((A - A^*)x, x) = (Ax, x) - (A^*x, x) = (Ax, x) - (x, Ax) = 0.$$

By Theorem 10.3.1 $A - A^* = 0$. $\square$

10.3.3. THEOREM. *Any complex operator is uniquely representable in the form $A = B + iC$, where B and C are Hermitian operators.*

PROOF. If $A = B + iC$, where B and C are Hermitian operators, then $A^* = B^* - iC^* = B - iC$ and, therefore $2B = A + A^*$ and $2iC = A - A^*$. It is easy to verify that the operators $\frac{1}{2}(A + A^*)$ and $\frac{1}{2i}(A - A^*)$ are Hermitian. □

REMARK. An operator iC is skew-Hermitian if and only if the operator C is Hermitian and, therefore, any operator A is uniquely representable in the form of a sum of an Hermitian and a skew-Hermitian operator.

An operator A is called *normal* if $A^*A = AA^*$. It is easy to verify that unitary, Hermitian and skew-Hermitian operators are normal.

10.3.4. THEOREM. *An operator $A = B + iC$, where B and C are Hermitian operators, is normal if and only if $BC = CB$.*

PROOF. Since $A^* = B^* - iC^* = B - iC$, we have $A^*A = B^2 + C^2 + i(BC - CB)$ and $AA^* = B^2 + C^2 - i(BC - CB)$. Therefore, the equality $A^*A = AA^*$ is equivalent to the equality $BC - CB = 0$. □

10.4. If V is a linear space over $\mathbb{R}$, then to define on V a structure of a linear space over $\mathbb{C}$ it is necessary to determine the operation J of multiplication by i, i.e., $Jv = iv$. This linear map $J : V \longrightarrow V$ should satisfy the following property

$$J^2 v = i(iv) = -v, \text{ i.e., } J^2 = -I.$$

It is also clear that if in a space V over $\mathbb{R}$ such a linear operator J is given, then we can make V a space over $\mathbb{C}$ if we define the multiplication by a complex number $a + ib$ by the formula

$$(a + ib)v = av + bJv.$$

In particular, the dimension of V in this case must be even.

Let V be a linear space over $\mathbb{R}$. A linear (over $\mathbb{R}$) operator $J : V \longrightarrow V$ is called a *complex structure* on V if $J^2 = -I$.

The eigenvalues of the operator $J : V \longrightarrow V$ are purely imaginary and, therefore, for a more detailed study of J we will consider the complexification $V^{\mathbb{C}}$ of V. Notice that the multiplication by i in $V^{\mathbb{C}}$ has no relation whatsoever with either the complex structure J on V or its complexification $J^{\mathbb{C}}$ acting in $V^{\mathbb{C}}$.

THEOREM. *$V^{\mathbb{C}} = V_+ \oplus V_-$, where*

$$V_+ = \operatorname{Ker}(J^{\mathbb{C}} - iI) = \operatorname{Im}(J^{\mathbb{C}} + iI)$$

and

$$V_- = \operatorname{Ker}(J^{\mathbb{C}} + iI) = \operatorname{Im}(J^{\mathbb{C}} - iI).$$

PROOF. Since $(J^{\mathbb{C}} - iI)(J^{\mathbb{C}} + iI) = (J^2)^{\mathbb{C}} + I = 0$, it follows that $\operatorname{Im}(J^{\mathbb{C}} + iI) \subset \operatorname{Ker}(J^{\mathbb{C}} - iI)$. Similarly, $\operatorname{Im}(J^{\mathbb{C}} - iI) \subset \operatorname{Ker}(J^{\mathbb{C}} + iI)$. On the other hand, $-i(J^{\mathbb{C}} + iI) + i(J^{\mathbb{C}} - iI) = 2I$ and, therefore, $V^{\mathbb{C}} \subset \operatorname{Im}(J^{\mathbb{C}} + iI) + \operatorname{Im}(J^{\mathbb{C}} - iI)$. Since $\operatorname{Ker}(J^{\mathbb{C}} - iI) \cap \operatorname{Ker}(J^{\mathbb{C}} + iI) = 0$, we get the required conclusion. □

REMARK. Clearly, $V_+ = \overline{V_-}$.

Problems

10.1. Express the characteristic polynomial of the matrix $A_{\mathbb{R}}$ in terms of the characteristic polynomial of A.

10.2. Consider an $\mathbb{R}$-linear map of $\mathbb{C}$ into itself given by $Az = az + b\overline{z}$, where $a, b \in \mathbb{C}$. Prove that this map is not invertible if and only if

$$|a| = |b|.$$

10.3. Indicate in $\mathbb{C}^n$ a complex subspace of dimension $[n/2]$ on which the quadratic form $B(x, y) = x_1 y_1 + \dots + x_n y_n$ vanishes identically.

Solutions

5.1. The orthogonal complement to the space of traceless matrices is one-dimensional; it contains both matrices I and A^T.

5.2. Let $A_1, \dots, A_m$ and $B_1, \dots, B_k$ be the rows of the matrices A and B. Then

$$\operatorname{Span}(A_1, \dots, A_m)^{\perp} \subset \operatorname{Span}(B_1, \dots, B_k)^{\perp}.$$

Hence, $\operatorname{Span}(B_1, \dots, B_k) \subset \operatorname{Span}(A_1, \dots, A_m)$, i.e., $b_{ij} = \sum c_{ip} a_{pj}$.

5.3. If a vector $(w_1, \dots, w_n)$ belongs to an orthant that does not contain the vectors $\pm v$, then $v_i w_i > 0$ and $v_j w_j < 0$ for certain indices i and j. If we preserve the sign of the coordinate w_i (resp. w_j) but increase its absolute value, then the inner product (v, w) will grow (resp. decrease) and, therefore it can be made zero.

5.4. Let us express the bilinear function $x^*(y)$ in the form xBy^T. By hypothesis the conditions $xBy^T = 0$ and $yBx^T = 0$ are equivalent. Besides, $yBx^T = xB^Ty^T$. Therefore, $By^T = \lambda(y)B^Ty^T$. If vectors y and y_1 are proportional, then $\lambda(y) = \lambda(y_1)$. If the vectors y and y_1 are linearly independent, then the vectors B^Ty^T and $B^Ty_1^T$ are also linearly independent and, therefore, the equalities

$$\lambda(y + y_1)(B^Ty^T + B^Ty_1^T) = B(y^T + y_1^T) = \lambda(y)B^Ty^T + \lambda(y_1)B^Ty_1^T$$

imply $\lambda(y) = \lambda(y_1)$. Thus, $x^*(y) = B(x, y)$ and $B(x, y) = \lambda B(y, x) = \lambda^2 B(x, y)$ and, therefore, $\lambda = \pm 1$.

6.1. By Theorem 6.1

$$\dim(\operatorname{Im} A^k \cap \operatorname{Ker} A) = \dim \operatorname{Ker} A^{k+1} - \dim \operatorname{Ker} A^k \quad \text{for any } k.$$

Therefore,

$$\sum_{k=1}^{n} \dim(\operatorname{Im} A^k \cap \operatorname{Ker} A) = \dim \operatorname{Ker} A^{n+1} - \dim \operatorname{Ker} A.$$

To prove the second equality it suffices to notice that

$$\dim \operatorname{Im} A^p = \dim V - \dim \operatorname{Ker} A^p,$$

where V is the space in which A acts.

7.1. We may assume that $e_1, \dots, e_k$ $(k \le n)$ is a basis of $\operatorname{Span}(e_1, \dots, e_m)$. Then

$$e_{k+1} + \lambda_1 e_1 + \dots + \lambda_k e_k = 0 \text{ and } e_{k+2} + \mu_1 e_1 + \dots + \mu_k e_k = 0.$$

Multiply these equalities by $1 + \sum \mu_i$ and $-(1 + \sum \lambda_i)$, respectively, and add up the obtained equalities. (If $1 + \sum \mu_i = 0$ or $1 + \sum \lambda_i = 0$ we already have the required equality.)

7.2. Let us carry out the proof by induction on m. For $m \le n + 1$ the statement is obvious. Let $m \ge n + 2$. Then there exist numbers $\alpha_1, \dots, \alpha_m$ not all equal to zero such that $\sum \alpha_i v_i = 0$ and $\sum \alpha_i = 0$ (see Problem 7.1). Therefore,

$$x = \sum t_i v_i + \lambda \sum \alpha_i v_i = \sum t_i' v_i,$$

where $t_i' = t_i + \lambda \alpha_i$ and $\sum t_i' = \sum t_i = 1$. It remains to find a number λ so that all numbers $t_i + \lambda \alpha_i$ are nonnegative and at least one of them is zero. The set

$$\{\lambda \in \mathbb{R} \mid t_i + \lambda \alpha_i \ge 0 \text{ for } i = 1, \dots, m\}$$

is closed, nonempty (it contains zero) and is bounded from below (and above) since among the numbers α_i there are positive (and negative) ones; the minimal number λ from this set is the desired one.

7.3. Suppose A is not invertible. Then there exist numbers $\lambda_1, \dots, \lambda_n$ not all equal to zero such that $\sum_i \lambda_i a_{ik} = 0$ for $k = 1, \dots, n$. Let λ_s be the number among $\lambda_1, \dots, \lambda_n$ whose absolute value is the greatest (for definiteness sake let $s = 1$). Since

$$\lambda_1 a_{11} + \lambda_2 a_{12} + \dots + \lambda_n a_{1n} = 0,$$

we have

$$\begin{aligned} |\lambda_1 a_{11}| = |\lambda_2 a_{12} + \dots + \lambda_n a_{1n}| &\le |\lambda_2 a_{12}| + \dots + |\lambda_n a_{1n}| \\ &\le |\lambda_1| \left(|a_{12}| + \dots + |a_{1n}|\right) < |\lambda_1| \cdot |a_{11}|. \end{aligned}$$

Contradiction.

7.4. a) Suppose that the vectors $e_1, \dots, e_k$ are linearly dependent for $k < n + 1$. We may assume that this set of vectors is minimal, i.e., $\lambda_1 e_1 + \dots + \lambda_k e_k = 0$, where all the numbers λ_i are nonzero. Then

$$0 = (e_{n+1}, \textstyle\sum \lambda_i e_i) = \sum \lambda_i (e_{n+1}, e_i), \text{ where } (e_{n+1}, e_i) < 0.$$

Therefore, among the numbers λ_i there are both positive and negative ones. On the other hand, if

$$\lambda_1 e_1 + \dots + \lambda_p e_p = \lambda_{p+1}' e_{p+1} + \dots + \lambda_k' e_k,$$

where all numbers λ_i, λ_j' are positive, then taking the inner product of this equality with the vector at the right-hand side we get a negative number at the left-hand side and the inner product of a nonzero vector by itself, i.e., a nonnegative number, at the right-hand side.

b) Suppose that vectors $e_1, \dots, e_{n+2}$ in $\mathbb{R}^n$ are such that $(e_i, e_j) < 0$ for $i \neq j$. On the one hand, if $\alpha_1 e_1 + \dots + \alpha_{n+2} e_{n+2} = 0$, then all the numbers α_i are of the same sign (cf. solution to heading a). On the other hand, we can select the numbers $\alpha_1, \dots, \alpha_{n+2}$ so that $\sum \alpha_i = 0$ (see Problem 7.1). Contradiction.

8.1. Let

$$X = \begin{pmatrix} x_1 & 1 \\ \vdots & \vdots \\ x_n & 1 \end{pmatrix}, \quad Y = \begin{pmatrix} 1 & \dots & 1 \\ y_1 & \dots & y_n \end{pmatrix}.$$

Then $\left\| a_{ij} \right\|_1^n = XY$.

8.2. Let e_1 be a vector that generates $\operatorname{Im} A$. Let us complement it to a basis $e_1, \dots, e_n$. The matrix A with respect to this basis is of the form

$$A = \begin{pmatrix} a_1 & \dots & a_n \\ 0 & \dots & 0 \\ \vdots & \dots & \vdots \\ 0 & \dots & 0 \end{pmatrix}.$$

Therefore, $\operatorname{tr} A = a_1$ and $|A + I| = 1 + a_1$.

8.3. It suffices to show that $\operatorname{Ker} A^* \cap \operatorname{Im} A = 0$. If $A^* v = 0$ and $v = Aw$, then $(v, v) = (Aw, v) = (w, A^* v) = 0$ and, therefore, $v = 0$.

8.4. The rows of the matrix (C, D) are linear combinations of the rows of the matrix (A, B) and, therefore, $(C, D) = X(A, B) = (XA, XB)$, i.e., $D = XB = (CA^{-1})B$.

8.5. Let $r_i = \operatorname{rank} A_i$ and $r = \operatorname{rank}(A_1 + A_2)$. Then $\dim V_i = \dim W_i = r_i$ and $\dim(V_1 + V_2) = \dim(W_1 + W_2) = r$. The equality $r_1 + r_2 = r$ means that $\dim(V_1 + V_2) = \dim V_1 + \dim V_2$, i.e., $V_1 \cap V_2 = 0$. Similarly, $W_1 \cap W_2 = 0$.

8.6. The equality $B^T A = 0$ means that the columns of the matrices A and B are pair-wise orthogonal. Therefore, the space spanned by the columns of A has only zero intersection with the space spanned by the columns of B. It remains to make use of the result of Problem 8.5.

8.7. Suppose A and B are matrices of order $2m + 1$. By Sylvester's inequality,

$$\operatorname{rank} A + \operatorname{rank} B \leq \operatorname{rank} AB + 2m + 1 = 2m + 1.$$

Therefore, either $\operatorname{rank} A \leq m$ or $\operatorname{rank} B \leq m$. If $\operatorname{rank} A \leq m$, then $\operatorname{rank} A^T = \operatorname{rank} A \leq m$; hence,

$$\operatorname{rank}(A + A^T) \leq \operatorname{rank} A + \operatorname{rank} A^T \leq 2m < 2m + 1.$$

8.8. We may assume that $a_{12} \neq 0$. Let A_i be the ith row of A. Let us prove that $a_{21} A_i = a_{2i} A_1 + a_{i1} A_2$, i.e.,

$$a_{12} a_{ij} + a_{1j} a_{2i} + a_{1i} a_{j2} = 0. \tag{1}$$

The identity (1) is skew-symmetric with respect to i and j and, therefore, we can assume that $i < j$, see Figure 3.

Only the factor a_{j2} is negative in (1) and, therefore, (1) is equivalent to Ptolemy's theorem for the quadrilateral $X_1 X_2 X_i X_j$.

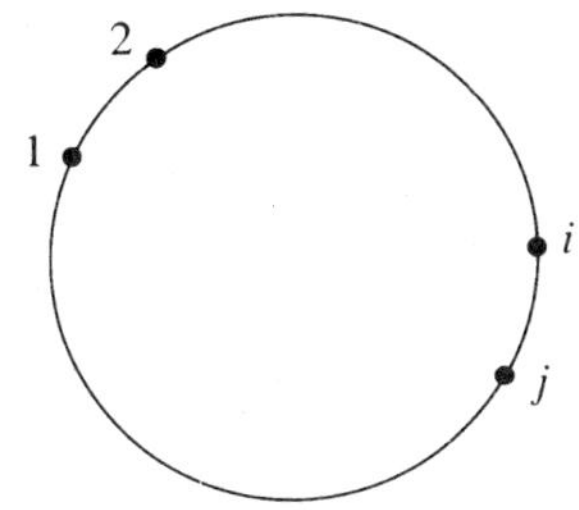

FIGURE 3

9.1. Let U_1 be the orthogonal complement of u in U. Since

$$\dim U_1^\perp + \dim W = n - (m-1) + m = n + 1,$$

then $\dim(U_1^\perp \cap W) \geq 1$. If $w \in W \cap U_1^\perp$, then $w \perp U_1$ and $w \perp u$; therefore, $w \perp U$.

9.2. Let us apply the orthogonalization process with the subsequent normalization to vectors $x_1, \dots, x_n$. As a result we get an orthonormal basis $e_1, \dots, e_n$. The vectors $x_1, \dots, x_n$ are expressed in terms of $e_1, \dots, e_n$ and the coefficients only depend on the inner products (x_i, x_j). Similarly, for the vectors $y_1, \dots, y_n$ we get an orthonormal basis $\varepsilon_1, \dots, \varepsilon_n$. The map that sends e_i to ε_i $(i = 1, \dots, n)$ is the required one.

10.1. $\det(\lambda I - A_{\mathbb{R}}) = |\det(\lambda I - A)|^2$.

10.2. Let $a = a_1 + ia_2$, $b = b_1 + ib_2$, where $a_i, b_i \in \mathbb{R}$. The matrix of the given map with respect to the basis $1, i$ is equal to $\begin{pmatrix} a_1 + b_1 & -a_2 + b_2 \\ a_2 + b_2 & a_1 - b_1 \end{pmatrix}$ and its determinant is equal to $|a|^2 - |b|^2$.

10.3. Let $p = [n/2]$. The complex subspace spanned by the vectors $e_1 + ie_2$, $e_3 + ie_4$, $\dots$, $e_{2p-1} + ie_{2p}$ possesses the required property.

CHAPTER III

CANONICAL FORMS OF MATRICES AND LINEAR OPERATORS

11. The trace and eigenvalues of an operator

11.1. The *trace* of a square matrix A is the sum of its diagonal elements; it is denoted by $\operatorname{tr} A$. It is easy to verify that

$$\operatorname{tr} AB = \sum_{i,j} a_{ij} b_{ji} = \operatorname{tr} BA.$$

Therefore,

$$\operatorname{tr} PAP^{-1} = \operatorname{tr} P^{-1}PA = \operatorname{tr} A,$$

i.e., the trace of the matrix of a linear operator does not depend on the choice of a basis.

The equality $\operatorname{tr} ABC = \operatorname{tr} ACB$ is not always true. For instance, take $A = \begin{pmatrix} 0 & 1 \\ 0 & 0 \end{pmatrix}$, $B = \begin{pmatrix} 1 & 0 \\ 0 & 0 \end{pmatrix}$ and $C = \begin{pmatrix} 0 & 0 \\ 1 & 0 \end{pmatrix}$; then $ABC = 0$ and $ACB = \begin{pmatrix} 1 & 0 \\ 0 & 0 \end{pmatrix}$.

For the trace of an operator in a Euclidean space we have the following useful formula.

THEOREM. *Let $e_1, \dots, e_n$ be an orthonormal basis. Then*

$$\operatorname{tr} A = \sum_{i=1}^{n} (Ae_i, e_i).$$

PROOF. Since $Ae_i = \sum_j a_{ij} e_j$, we have $(Ae_i, e_i) = a_{ii}$. $\square$

REMARK. The trace of an operator is invariant, but the above definition of the trace makes use of a basis and, therefore, is not invariant. One can, however, give an invariant definition of the trace of an operator (see 27.2).

11.2. A nonzero vector $v \in V$ is called an *eigenvector of the linear operator* $A : V \to V$ if $Av = \lambda v$ and this number λ is called an *eigenvalue* of A. Fix λ and consider the equation $Av = \lambda v$, i.e., $(A - \lambda I)v = 0$. This equation has a nonzero solution v if and only if $|A - \lambda I| = 0$. Therefore, the eigenvalues of A are roots of the polynomial $p(\lambda) = |\lambda I - A|$.

The polynomial $p(\lambda)$ is called the *characteristic polynomial* of A. This polynomial depends only on the operator itself and does not depend on the choice of the basis (see 7.1).

THEOREM. *If $Ae_1 = \lambda_1 e_1, \dots, Ae_k = \lambda_k e_k$ and the numbers $\lambda_1, \dots, \lambda_k$ are distinct, then $e_1, \dots, e_k$ are linearly independent.*

PROOF. Assume the contrary. Selecting a minimal linearly independent set of vectors we can assume that $e_k = \alpha_1 e_1 + \dots + \alpha_{k-1} e_{k-1}$, where $\alpha_1 \dots \alpha_{k-1} \neq 0$ and the vectors $e_1, \dots, e_{k-1}$ are linearly independent. Then $Ae_k = \alpha_1 \lambda_1 e_1 + \dots + \alpha_{k-1} \lambda_{k-1} e_{k-1}$ and $Ae_k = \lambda_k e_k = \alpha_1 \lambda_k e_1 + \dots + \alpha_{k-1} \lambda_k e_{k-1}$. Hence, $\lambda_1 = \lambda_k$. Contradiction. □

COROLLARY. *If the characteristic polynomial of an operator A over $\mathbb{C}$ has no multiple roots, then the eigenvectors of A constitute a basis.*

11.3. A linear operator A possessing a basis of eigenvectors is said to be *diagonalizable* or *semisimple*. If X is the matrix formed by the columns of the coordinates of eigenvectors $x_1, \dots, x_n$ and λ_i an eigenvalue corresponding to x_i, then $AX = X\Lambda$, where $\Lambda = \operatorname{diag}(\lambda_1, \dots, \lambda_n)$. Therefore, $X^{-1}AX = \Lambda$.

The converse is also true: if $X^{-1}AX = \operatorname{diag}(\lambda_1, \dots, \lambda_n)$, then $\lambda_1, \dots, \lambda_n$ are eigenvalues of A and the columns of X are the corresponding eigenvectors.

Over $\mathbb{C}$ only an operator with multiple eigenvalues may be nondiagonalizable and such operators constitute a set of measure zero. All normal operators (see 17.1) are diagonalizable over $\mathbb{C}$. In particular, all unitary and Hermitian operators are diagonalizable and there are orthonormal bases consisting of their eigenvectors. This can be easily proved in a straightforward way, as well with the help of the fact that for a unitary or Hermitian operator A the inclusion $AW \subset W$ implies $AW^\perp \subset W^\perp$.

The absolute value of an eigenvalue of a unitary operator A is equal to 1 since $|Ax| = |x|$. The eigenvalues of an Hermitian operator A are real since $(Ax, x) = (x, Ax) = \overline{(Ax, x)}$.

THEOREM. *For an orthogonal operator A there exists an orthonormal basis with respect to which the matrix of A is of the block-diagonal form with blocks ± 1 or* $\begin{pmatrix} \cos\varphi & -\sin\varphi \\ \sin\varphi & \cos\varphi \end{pmatrix}$.

PROOF. If ± 1 is an eigenvalue of A we can make use of the same arguments as for the complex case and therefore, let us assume that the vectors x and Ax are not parallel for all x. The function $\varphi(x) = \angle(x, Ax)$ — the angle between x and Ax — is continuous on a compact set, the unit sphere.

Let $\varphi_0 = \angle(x_0, Ax_0)$ be the minimum of $\varphi(x)$ and e the vector parallel to the bisector of the angle between x_0 and Ax_0.

Then

$$\varphi_0 \leq \angle(e, Ae) \leq \angle(e, Ax_0) + \angle(Ax_0, Ae) = \frac{\varphi_0}{2} + \frac{\varphi_0}{2}$$

and, therefore, Ae belongs to the plane $\operatorname{Span}(x_0, e)$. This plane is invariant with respect to A since $Ax_0, Ae \in \operatorname{Span}(x_0, e)$. An orthogonal transformation of a plane is either a rotation or a symmetry with respect to a straight line; the eigenvalues of a symmetry, however, are equal to ± 1 and, therefore, the matrix of the restriction of A to $\operatorname{Span}(x_0, e)$ is of the form $\begin{pmatrix} \cos\varphi & -\sin\varphi \\ \sin\varphi & \cos\varphi \end{pmatrix}$, where $\sin\varphi \neq 0$. □

11.4. The eigenvalues of the tridiagonal matrix

$$J = \begin{pmatrix} a_1 & -b_1 & 0 & \dots & 0 & 0 & 0 \\ -c_1 & a_2 & -b_2 & \dots & 0 & 0 & 0 \\ 0 & -c_2 & a_3 & \dots & 0 & 0 & 0 \\ \vdots & \vdots & \vdots & \ddots & \ddots & \ddots & \vdots \\ 0 & 0 & 0 & \dots & a_{n-2} & -b_{n-2} & 0 \\ 0 & 0 & 0 & \dots & -c_{n-2} & a_{n-1} & -b_{n-1} \\ 0 & 0 & 0 & \dots & 0 & -c_{n-1} & a_n \end{pmatrix}, \text{ where } b_i c_i > 0,$$

have interesting properties. They are real and of multiplicity one. For $J = \left\|a_{ij}\right\|_1^n$, consider the sequence of polynomials

$$D_k(\lambda) = |\lambda\delta_{ij} - a_{ij}|_1^k, \qquad D_0(\lambda) = 1.$$

Clearly, $D_n(\lambda)$ is the characteristic polynomial of J. These polynomials satisfy a recurrent relation

$$D_k(\lambda) = (\lambda - a_k)D_{k-1}(\lambda) - b_{k-1}c_{k-1}D_{k-2}(\lambda) \tag{1}$$

(cf. 1.6) and, therefore, the characteristic polynomial $D_n(\lambda)$ depends not on the numbers b_k, c_k themselves, but on their products. By replacing in J the elements b_k and c_k with $\sqrt{b_k c_k}$ we get a symmetric matrix J_1 with the same characteristic polynomial. Therefore, the eigenvalues of J are real.

A symmetric matrix has a basis of eigenvectors and therefore, it remains to demonstrate that to every eigenvalue λ of J_1 there corresponds at most one eigenvector $(x_1, \dots, x_n)$. This is also true even for J, i.e without the assumption that $b_k = c_k$. Since

$$\begin{aligned} (\lambda - a_1)x_1 - b_1x_2 &= 0 \\ -c_1x_1 + (\lambda - a_2)x_2 - b_2x_3 &= 0 \\ \dots\dots\dots\dots\dots \\ -c_{n-2}x_{n-2} + (\lambda - a_{n-1})x_{n-1} - b_{n-1}x_n &= 0 \\ -c_{n-1}x_{n-1} + (\lambda - a_n)x_n &= 0, \end{aligned}$$

it follows that the change

$$y_1 = x_1,\ y_2 = b_1x_2,\ \dots,\ y_k = b_1 \dots b_{k-1}x_k,$$

yields

$$\begin{aligned} y_2 &= (\lambda - a_1)y_1 \\ y_3 &= (\lambda - a_2)y_2 - c_1b_1y_1 \\ &\dots\dots\dots\dots\dots\dots \\ y_n &= (\lambda - a_{n-1})y_{n-1} - c_{n-2}b_{n-2}y_{n-2}. \end{aligned}$$

These relations for y_k coincide with relations (1) for D_k and, therefore, if $y_1 = c = cD_0(\lambda)$, then $y_k = cD_k(\lambda)$. Thus the eigenvector $(x_1, \dots, x_k)$ is uniquely determined up to a factor

11.5. Let us give two examples of how to calculate eigenvalues and eigenvectors. First, we observe that if λ is an eigenvalue of a matrix A and f an arbitrary polynomial, then $f(\lambda)$ is an eigenvalue of the matrix $f(A)$. This follows from the fact that $f(\lambda I) - f(A)$ is divisible by $\lambda I - A$.

a) Consider the matrix

$$P = \begin{pmatrix} 0 & 0 & 0 & \dots & 0 & 0 & 1 \\ 1 & 0 & 0 & \dots & 0 & 0 & 0 \\ 0 & 1 & 0 & \dots & 0 & 0 & 0 \\ \vdots & \vdots & \vdots & \ddots & \ddots & \vdots & \\ 0 & 0 & 0 & \dots & 0 & 1 & 0 \end{pmatrix}.$$

Since $Pe_k = e_{k+1}$, we have $P^s e_k = e_{k+s}$ and, therefore, $P^n = I$, where n is the order of the matrix. It follows that the eigenvalues are roots of the equation $x^n = 1$. Set $\varepsilon = \exp(2\pi i/n)$. Let us prove that the vector $u_s = \sum_{k=1}^n \varepsilon^{ks} e_k$ $(s = 1, \dots, n)$ is an eigenvector of P corresponding to the eigenvalue ε^{-s}. Indeed,

$$Pu_s = \sum \varepsilon^{ks} Pe_k = \sum \varepsilon^{ks} e_{k+1} = \sum \varepsilon^{-s}(\varepsilon^{s(k+1)} e_{k+1}) = \varepsilon^{-s} u_s.$$

b) Consider the matrix

$$A = \begin{pmatrix} 0 & 1 & 0 & \dots & 0 \\ 0 & 0 & 1 & \ddots & 0 \\ \vdots & \vdots & \ddots & \ddots & \vdots \\ 0 & 0 & 0 & \dots & 1 \\ p_1 & p_2 & p_3 & \dots & p_n \end{pmatrix}.$$

Let x be the column $(x_1, \dots, x_n)^T$. The equation $Ax = \lambda x$ can be rewritten in the form

$$x_2 = \lambda x_1, \ x_3 = \lambda x_2, \ \dots, \ x_n = \lambda x_{n-1}, \ p_1 x_1 + p_2 x_2 + \dots + p_n x_n = \lambda x_n.$$

Therefore, the eigenvectors of A are of the form

$$(\alpha, \lambda\alpha, \lambda^2\alpha, \dots, \lambda^{n-1}\alpha), \text{ where } p_1 + p_2\lambda + \dots + p_n\lambda^{n-1} = \lambda^n.$$

11.6. We already know that $\operatorname{tr} AB = \operatorname{tr} BA$. It turns out that a stronger statement is true: the matrices AB and BA have the same characteristic polynomials.

THEOREM. *Let A and B be $n \times n$-matrices. Then the characteristic polynomials of AB and BA coincide.*

PROOF. If A is invertible, then

$$|\lambda I - AB| = |A^{-1}(\lambda I - AB)A| = |\lambda I - BA|.$$

For a noninvertible matrix A the equality $|\lambda I - AB| = |\lambda I - BA|$ can be proved by passing to the limit. □

COROLLARY. *If A and B are $m \times n$-matrices, then the characteristic polynomials of AB^T and $B^T A$ differ by the factor λ^{n-m}.*

PROOF. Complement the matrices A and B by zeros to square matrices of equal size. □

11.7.1. THEOREM. *Let the sum of the elements of every column of a square matrix A be equal to 1, and let the column $(x_1, \dots, x_n)^T$ be an eigenvector of A such that $x_1 + \dots + x_n \neq 0$. Then the eigenvalue corresponding to this vector is equal to 1.*

PROOF. Adding up the equalities $\sum a_{1j}x_j = \lambda x_1, \dots, \sum a_{nj}x_j = \lambda x_n$ we get $\sum_{i,j} a_{ij}x_j = \lambda \sum_j x_j$. But

$$\sum_{i,j} a_{ij}x_j = \sum_j \left(x_j \sum_i a_{ij} \right) = \sum x_j$$

since $\sum_i a_{ij} = 1$. Thus, $\sum x_j = \lambda \sum x_j$, where $\sum x_j \neq 0$. Therefore, $\lambda = 1$. □

11.7.2. THEOREM. *If the sum of the absolute values of the elements of every column of a square matrix A does not exceed 1, then all its eigenvalues do not exceed 1.*

PROOF. Let $(x_1, \dots, x_n)$ be an eigenvector corresponding to an eigenvalue λ. Then

$$|\lambda x_i| = |\sum a_{ij}x_j| \leq \sum |a_{ij}||x_j|, \qquad i = 1, \dots, n.$$

Adding up these inequalities we get

$$|\lambda| \sum |x_i| \leq \sum_{i,j} |a_{ij}||x_j| = \sum_j \left(|x_j| \sum_i |a_{ij}| \right) \leq \sum_j |x_j|$$

since $\sum_i |a_{ij}| \leq 1$. Dividing both sides of this inequality by the nonzero number $\sum |x_j|$ we get $|\lambda| \leq 1$. □

REMARK. Theorem 11.7.2 remains valid also when certain of the columns of A are zero ones.

11.7.3. THEOREM. *Let $A = \|a_{ij}\|_1^n$, $S_j = \sum_{i=1}^n |a_{ij}|$. Then $\sum_{j=1}^n S_j^{-1} |a_{jj}| \leq \operatorname{rank} A$ and the summands corresponding to zero values of S_j can be replaced by zeros.*

PROOF. Multiplying the columns of A by nonzero numbers we can always make the numbers S_j for the new matrix to be either 0 or 1 and, besides, $a_{jj} \geq 0$. The rank of the matrix is not effected by these transformations. Applying Theorem 11.7.2 to the new matrix we get

$$\sum |a_{jj}| = \sum a_{jj} = \operatorname{tr} A = \sum \lambda_i \leq \sum |\lambda_i| \leq \operatorname{rank} A. \quad \square$$

Problems

11.1. a) Are there real matrices A and B such that $AB - BA = I$?

b) Prove that if $AB - BA = A$, then $|A| = 0$.

11.2. Find the eigenvalues and the eigenvectors of the matrix $A = \left\|a_{ij}\right\|_1^n$, where $a_{ij} = \lambda_i/\lambda_j$.

11.3. Prove that any square matrix A is the sum of two invertible matrices.

11.4. Prove that the eigenvalues of a matrix continuously depend on its elements. More precisely, let $A = \left\|a_{ij}\right\|_1^n$ be a given matrix. For any $\varepsilon > 0$ there exists $\delta > 0$ such that if $|a_{ij} - b_{ij}| < \delta$ and λ is an eigenvalue of A, then there exists an eigenvalue μ of $B = \left\|b_{ij}\right\|_1^n$ such that $|\lambda - \mu| < \varepsilon$.

11.5. The sum of the elements of every row of an invertible matrix A is equal to s. Prove that the sum of the elements of every row of A^{-1} is equal to $1/s$.

11.6. Prove that if the first row of the matrix $S^{-1}AS$ is of the form $(\lambda, 0, 0, \dots, 0)$, then the first column of S is an eigenvector of A corresponding to the eigenvalue λ.

11.7. Let $f(\lambda) = |\lambda I - A|$, where A is a matrix of order n. Prove that $f'(\lambda) = \sum_{i=1}^n |\lambda I - A_i|$, where A_i is the matrix obtained from A by striking out the ith row and the ith column.

11.8. Let $\lambda_1, \dots, \lambda_n$ be the eigenvalues of a matrix A. Prove that the eigenvalues of $\operatorname{adj} A$ are equal to $\prod_{i\neq 1}\lambda_i, \dots, \prod_{i\neq n}\lambda_i$.

11.9. A vector x is called *symmetric* (resp. *skew-symmetric*) if its coordinates satisfy $(x_i = x_{n-i})$ (resp. $(x_i = -x_{n-i})$). Let a matrix $A = \left\|a_{ij}\right\|_0^n$ be centrally symmetric, i.e., $a_{i,j} = a_{n-i,n-j}$. Prove that among the eigenvectors of A corresponding to any eigenvalue there is either a nonzero symmetric or a nonzero skew-symmetric vector.

11.10. The elements $a_{i,n-i+1} = x_i$ of a complex $n \times n$-matrix A can be nonzero, whereas the remaining elements are 0. What condition should the set $\{x_1, \dots, x_n\}$ satisfy for A to be diagonalizable?

11.11 ([Drazin, Haynsworth, 1962]). a) Prove that a matrix A has m linearly independent eigenvectors corresponding to real eigenvalues if and only if there exists a nonnegative definite matrix S of rank m such that $AS = SA^*$.

b) Prove that a matrix A has m linearly independent eigenvectors corresponding to eigenvalues λ such that $|\lambda| = 1$ if and only if there exists a nonnegative definite matrix S of rank m such that $ASA^* = S$.

12. The Jordan canonical (normal) form

12.1. Let A be the matrix of an operator with respect to a basis e; then $P^{-1}AP$ is the matrix of the same operator with respect to the basis eP. The matrices A and $P^{-1}AP$ are called *similar*. By selecting an appropriate basis we can reduce the matrix of an operator to a simpler form: to a Jordan normal form, cyclic form, to a matrix with equal elements on the main diagonal, to a matrix with all elements on the main diagonal, except one, being equal to zero, etc.

One might think that for a given real matrix A the set of real matrices of the form $P^{-1}AP|P$, where P is a complex matrix, is larger than the the set of real matrices of the form $P^{-1}AP|P$, where P is a real matrix. This, however, is not so.

THEOREM. *Let A and B be real matrices and $A = P^{-1}BP$, where P is a complex matrix. Then $A = Q^{-1}BQ$ for some real matrix Q.*

PROOF. We have to demonstrate that if among the solutions of the equation

$$XA = BX \tag{1}$$

there is an invertible complex matrix P, then among the solutions there is also an invertible real matrix Q. The solutions over $\mathbb{C}$ of the linear equation (1) form a linear space W over $\mathbb{C}$ with a basis $C_1, \dots, C_n$. The matrix C_j can be represented in the form $C_j = X_j + iY_j$, where X_j and Y_j are real matrices. Since A and B are real matrices, $C_jA = BC_j$ implies $X_jA = BX_j$ and $Y_jA = BY_j$. Hence, $X_j, Y_j \in W$ for all j and W is spanned over $\mathbb{C}$ by the matrices $X_1, \dots, X_n, Y_1, \dots, Y_n$ and therefore, we can select in W a basis $D_1, \dots, D_n$ consisting of real matrices.

Let $P(t_1, \dots, t_n) = |t_1D_1 + \dots + t_nD_n|$. The polynomial $P(t_1, \dots, t_n)$ is not identically equal to zero over $\mathbb{C}$ by the hypothesis and, therefore, it is not identically equal to zero over $\mathbb{R}$ either, i.e., the matrix equation (1) has a nondegenerate real solution $t_1D_1 + \dots + t_nD_n$. $\square$

12.2. A *Jordan block* of size $r \times r$ is a matrix of the form

$$J_r(\lambda) = \begin{pmatrix} \lambda & 1 & 0 & \dots & \dots & 0 \\ 0 & \lambda & 1 & \dots & \dots & 0 \\ \vdots & \vdots & \ddots & \ddots & \ddots & \vdots \\ 0 & 0 & 0 & \dots & 1 & 0 \\ 0 & 0 & 0 & \dots & \lambda & 1 \\ 0 & 0 & 0 & \dots & 0 & \lambda \end{pmatrix}.$$

A *Jordan matrix* is a block diagonal matrix with Jordan blocks $J_{r_i}(\lambda_i)$ on the diagonal.

A *Jordan basis* for an operator $A : V \to V$ is a basis of the space V in which the matrix of A is a Jordan matrix.

THEOREM (Jordan). *For any linear operator $A : V \to V$ over $\mathbb{C}$ there exists a Jordan basis and the Jordan matrix of A is uniquely determined up to a permutation of its Jordan blocks.*

PROOF (Following [Väliaho, 1986]). First, let us prove the existence of a Jordan basis. The proof will be carried out by induction on $n = \dim V$.

For $n = 1$ the statement is obvious. Let λ be an eigenvalue of A. Consider a noninvertible operator $B = A - \lambda I$. A Jordan basis for B is also a Jordan basis for $A = B + \lambda I$. The sequence $\operatorname{Im} B^0 \supset \operatorname{Im} B^1 \supset \operatorname{Im} B^2 \supset \dots$ stabilizes and, therefore, there exists a positive integer p such that $\operatorname{Im} B^{p+1} = \operatorname{Im} B^p \neq \operatorname{Im} B^{p-1}$. Then $\operatorname{Im} B^p \cap \operatorname{Ker} B = 0$ and $\operatorname{Im} B^{p-1} \cap \operatorname{Ker} B \neq 0$. Hence, $B^p(\operatorname{Im} B^p) = \operatorname{Im} B^p$.

Let $S_i = \operatorname{Im} B^{i-1} \cap \operatorname{Ker} B$. Then $\operatorname{Ker} B = S_1 \supset S_2 \supset \dots \supset S_p \neq 0$ and $S_{p+1} = 0$. Figure 4 might help to follow the course of the proof. In S_p, select a basis x_i^1 $(i = 1, \dots, n_p)$. Since $x_i^1 \in \operatorname{Im} B^{p-1}$, then $x_i^1 = B^{p-1}x_i^p$ for a vector x_i^p. Consider the vectors $x_i^k = B^{p-k}x_i^p$ $(k = 1, \dots, p)$. Let us complement the set of vectors x_i^1 to a basis of S_{p-1} by vectors y_j^1. Now, find a vector y_j^{p-1} such that $y_j^1 = B^{p-2}y_j^{p-1}$ and consider the vectors $y_j^l = B^{p-l-1}y_j^{p-1}$ $(l = 1, \dots, p-1)$. Further, let us complement the set of vectors x_i^1 and y_j^1 to a basis of S_{p-2} by vectors z_k^1, etc. The cardinality

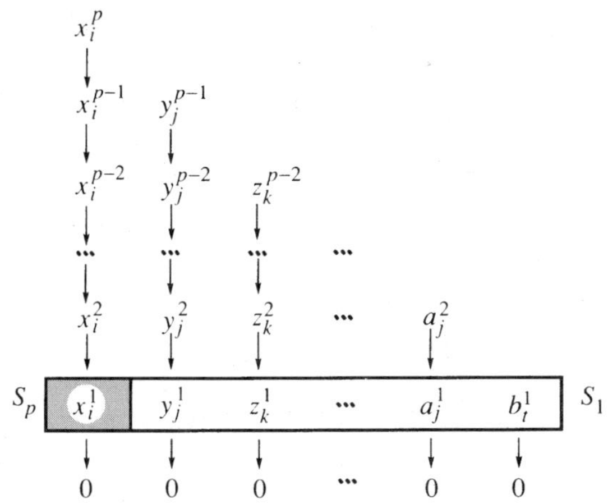

FIGURE 4

of the set of all chosen vectors $x_i^k, y_j^l, \dots, b_t^1$ is equal to $\sum_{i=1}^p \dim S_i$ since every x_i^1 contributes to p summands, every y_j^1 contributes to $p-1$ summands, etc. Since

$$\dim(\operatorname{Im} B^{i-1} \cap \operatorname{Ker} B) = \dim \operatorname{Ker} B^i - \dim \operatorname{Ker} B^{i-1}$$

(see 6.1), we have $\sum_{i=1}^p \dim S_i = \dim \operatorname{Ker} B^p$.

Let us complement the chosen vectors to a basis of $\operatorname{Im} B^p$ and prove that we obtain a basis of V. The number of these vectors indicates that it suffices to demonstrate their linear independence. Suppose that

$$(1) \qquad f + \sum \alpha_i x_i^p + \sum \beta_i x_i^{p-1} + \dots + \sum \gamma_j y_j^{p-1} + \dots + \sum \delta_t b_t^1 = 0,$$

where $f \in \operatorname{Im} B^p$. Applying the operator B^p to (1) we get $B^p(f) = 0$; hence, $f = 0$ since $B^p(\operatorname{Im} B^p) = \operatorname{Im} B^p$. Applying now the operator B^{p-1} to (1) we get $\sum \alpha_i x_i^1 = 0$ which means that all α_i are zero. Application of the operator B^{p-2} to (1) gives $\sum \beta_i x_i^1 + \sum \gamma_j y_j^1 = 0$, which means that all β_i and γ_j are zero, etc.

By the inductive hypothesis we can select a Jordan basis for B in the space $\operatorname{Im} B^p \neq V$; complementing this basis by the chosen vectors, we get a Jordan basis of V.

To prove the uniqueness of the Jordan form it suffices to verify that the number of Jordan blocks of B corresponding to eigenvalue 0 is uniquely defined. To these blocks we can associate the diagram plotted in Figure 4 and, therefore, the number of blocks of size $k \times k$ is equal to

$$\begin{aligned}
\dim S_k &- \dim S_{k+1} \\
&= (\dim \operatorname{Ker} B^k - \dim \operatorname{Ker} B^{k-1}) - (\dim \operatorname{Ker} B^{k+1} - \dim \operatorname{Ker} B^k) \\
&= 2 \dim \operatorname{Ker} B^k - \dim \operatorname{Ker} B^{k-1} - \dim \operatorname{Ker} B^{k+1} \\
&= \operatorname{rank} B^{k-1} - 2 \operatorname{rank} B^k + \operatorname{rank} B^{k+1};
\end{aligned}$$

which is invariantly defined. □

12.3. The Jordan normal form is convenient to use when we raise a matrix to some power. Indeed, if $A = P^{-1}JP$, then $A^n = P^{-1}J^nP$. To raise a Jordan block $J_r(\lambda) = \lambda I + N$ to a power we can use the Newton binomial formula

$$(\lambda I + N)^n = \sum_{k=0}^{n} \binom{n}{k} \lambda^k N^{n-k}.$$

The formula holds since $IN = NI$. The only nonzero elements of N^m are the 1's in the positions $(1, m+1)$, $(2, m+2)$, ..., $(r-m, r)$, where r is the order of N. If $m \geq r$, then $N^m = 0$.

12.4. Jordan bases always exist only over an algebraically closed field; over $\mathbb{R}$ a Jordan basis does not always exist. However, over $\mathbb{R}$ there is also a Jordan form which is a realification of the Jordan form over $\mathbb{C}$. Let us explain how it looks. First, observe that the part of a Jordan basis corresponding to real eigenvalues of A is constructed over $\mathbb{R}$ along the same lines as over $\mathbb{C}$. Therefore, only the case of nonreal eigenvalues is of interest.

Let $A^{\mathbb{C}}$ be the complexification of a real operator A (cf. 10.1).

12.4.1. THEOREM. *There is a one-to-one correspondence between the Jordan blocks of $A^{\mathbb{C}}$ corresponding to eigenvalues λ and $\overline{\lambda}$.*

PROOF. Let $B = P + iQ$, where P and Q are real operators. If x and y are real vectors, then the equations $(P + iQ)(x + iy) = 0$ and $(P - iQ)(x - iy) = 0$ are equivalent, i.e., the equations $Bz = 0$ and $\overline{B}\overline{z} = 0$ are equivalent. Since $(A - \overline{\lambda}I)^n = \overline{(A - \lambda I)^n}$, the map $z \mapsto \overline{z}$ determines a one-to-one correspondence between $\operatorname{Ker}(A - \lambda I)^n$ and $\operatorname{Ker}(A - \overline{\lambda}I)^n$. The dimensions of these spaces determine the number and the sizes of the Jordan blocks. $\square$

Let $J_n^*(\lambda)$ be the $2n \times 2n$ matrix obtained from the Jordan block $J_n(\lambda)$ by replacing each of its elements $a + ib$ by the matrix $\begin{pmatrix} a & b \\ -b & a \end{pmatrix}$.

12.4.2. THEOREM. *For an operator A over $\mathbb{R}$ there exists a basis with respect to which its matrix is of block diagonal form with blocks $J_{m_1}(t_1), \dots, J_{m_k}(t_k)$ for real eigenvalues t_i and blocks $J_{n_1}^*(\lambda_1), \dots, J_{n_s}^*(\lambda_s)$ for nonreal eigenvalues λ_i and $\overline{\lambda}_i$.*

PROOF. If λ is an eigenvalue of A, then by Theorem 12.4.1 $\overline{\lambda}$ is also an eigenvalue of A and to every Jordan block $J_n(\lambda)$ of A there corresponds the Jordan block $J_n(\overline{\lambda})$. Besides, if $e_1, \dots, e_n$ is the Jordan basis for $J_n(\lambda)$, then $\overline{e}_1, \dots, \overline{e}_n$ is the Jordan basis for $J_n(\overline{\lambda})$. Therefore, the real vectors x_1, $y_1, \dots$, x_n, y_n, where $e_k = x_k + iy_k$, are linearly independent. In the basis $x_1, y_1, \dots, x_n, y_n$ the matrix of the restriction of A to $\operatorname{Span}(x_1, y_1, \dots, x_n, y_n)$ is of the form $J_n^*(\lambda)$. $\square$

12.5. The Jordan decomposition shows that any linear operator A over $\mathbb{C}$ can be represented in the form $A = A_s + A_n$, where A_s is a semisimple (diagonalizable) operator and A_n is a nilpotent operator, such that $A_sA_n = A_nA_s$.

12.5.1. THEOREM. *The operators A_s and A_n are uniquely defined; moreover, $A_s = S(A)$ and $A_n = N(A)$, where S and N are certain polynomials.*

PROOF. First, consider one Jordan block $A = \lambda I + N_k$ of size $k \times k$. Let $S(t) = \sum_{i=1}^m s_i t^i$. Then

$$S(A) = \sum_{i=1}^{m} s_i \sum_{j=0}^{i} \binom{i}{j} \lambda^j N_k^{i-j}.$$

The coefficient at N_k^p is equal to

$$\sum_i s_i \binom{i}{i-p} \lambda^{i-p} = \frac{1}{p!} S^{(p)}(\lambda),$$

where $S^{(p)}$ is the pth derivative of S. Therefore, we have to select a polynomial S so that $S(\lambda) = \lambda$ and $S^{(1)}(\lambda) = \dots = S^{(k-1)}(\lambda) = 0$, where k is the order of the Jordan block. If $\lambda_1, \dots, \lambda_n$ are distinct eigenvalues of A and $k_1, \dots, k_n$ are the sizes of the maximal Jordan blocks corresponding to them, then S should take value λ_i at λ_i and have at λ_i zero derivatives from order 1 to order $k_i - 1$ inclusive. Such a polynomial can always be constructed (see Appendix 3). It is also clear that if $A_s = S(A)$, then $A_n = A - S(A)$, i.e., $N(A) = A - S(A)$.

Now, let us prove the uniqueness of the decomposition. Let $A_s + A_n = A = A'_s + A'_n$, where $A_s A_n = A_n A_s$ and $A'_s A'_n = A'_n A'_s$. If $AX = XA$, then $S(A)X = XS(A)$ and $N(A)X = XN(A)$. Therefore, $A_s A'_s = A'_s A_s$ and $A_n A'_n = A'_n A_n$. The operator $B = A'_s - A_s = A_n - A'_n$ is a difference of commuting diagonalizable operators and, therefore, is diagonalizable itself, cf. Problem 39.6 b). On the other hand, the operator B is the difference of commuting nilpotent operators and therefore, is nilpotent itself, cf. Problem 39.6 a). A diagonalizable nilpotent operator is equal to zero. $\square$

The *additive Jordan decomposition* $A = A_s + A_n$ enables us to get for an invertible operator A a *multiplicative Jordan decomposition* $A = A_s A_u$, where A_u is a *unipotent* operator, i.e., the sum of the identity operator and a nilpotent one.

12.5.2. THEOREM. *Let A be an invertible operator over $\mathbb{C}$. Then A can be represented in the form $A = A_s A_u = A_u A_s$, where A_s is a semisimple operator and A_u is a unipotent operator. Such a representation is unique.*

PROOF. If A is invertible, then so is A_s. Then $A = A_s + A_n = A_s A_u$ where $A_u = A_s^{-1}(A_s + A_n) = I + A_s^{-1} A_n$. Since A_s^{-1} and A_n commute, $A_s^{-1} A_n$ is a nilpotent operator which commutes with A_s.

Now, let us prove the uniqueness. If $A = A_s A_u = A_u A_s$ and $A_u = I + N$, where N is a nilpotent operator, then $A = A_s(I + N) = A_s + A_s N$, where $A_s N$ is a nilpotent operator commuting with A. Such an operator $A_s N = A_n$ is unique. $\square$

12.6. Two linear operators $A_1, A_2 \colon V \to V$ are said to be equivalent if one is obtained from the other by a change of basis in the space V, i.e., $PA_1 = A_2 P$, where P is a nondegenerate operator. The Jordan theorem allows us to choose one operator in each equivalence class and thereby solve the classification problem for linear operators.

The more complicated *classification problem for pairs of linear maps* was solved by Kronecker. In it a pair of maps $A_1, B_1 \colon V \to W$ is considered equivalent to another pair $A_2, B_2 \colon V \to W$ if one pair is obtained from the other by changing the bases of V and W, i.e., if $PA_1 = A_2 Q$ and $PB_1 = B_2 Q$, where P and Q are nondegenerate linear operators. In order to solve the classification problem for pairs of linear maps, they must be transformed to canonical form.

For the case in which at least one of the maps A and B is nondegenerate, the Kronecker classification follows immediately from the Jordan theorem. Indeed, by using an invertible map we can identify the spaces V and W (i.e., for the basis of W take the image of the basis of V). Then the matrix of the invertible map will be the unit matrix, and the matrix of the other map can be transformed to the Jordan form by an appropriate choice of basis for the space V.

Despite such a promising start, the classification of pairs of linear maps turns out to be rather complicated for the case in which both maps are degenerate.

Let us call a pair of linear maps $A, B\colon V \to W$ *decomposable* if there exist decompositions $V = V_1 \oplus V_2$ and $W = W_1 \oplus W_2$ such that $AV_i = W_i$, $BV_i = W_i$ and at least one of the decompositions is nontrivial. In the case of a decomposable pair of maps, we actually obtain two pairs of maps $A_1, B_1\colon V_1 \to W_1$ each of which can be transformed to canonical form independently of the other. In what follows we shall only consider undecomposable pairs of linear maps.

Let I_k be the unit matrix of order k, let $P_k = (I_k, 0)$ be the unit matrix with a column of zeros adjoined to it from the right, and $Q_k = (0, I_k)$. It can be shown that the pair of linear maps $P_k, Q_k\colon V^{k+1} \to W^k$ is undecomposable.

THEOREM. *Any undecomposable pair of linear maps $A, B\colon V \to W$ over $\mathbb{C}$ is equivalent to one of the following pairs*:

1) $(I_k, J_k(\lambda))$, $k \geq 1$;
2) $(J_k(\lambda)), I_k)$, $k \geq 1$;
3) (P_k, Q_k), $k \geq 0$;
4) (P_k^T, Q_k^T), $k \geq 0$.

PROOF (Following [Burgmeister, 1986]). Passing if necessary to transposed matrices (i.e., to maps of dual spaces), we can assume that $\dim V \leq \dim W$. Suppose $\dim = n$ and $\dim W = n + l$, where $l \geq 0$.

In the case $\operatorname{Ker} A \cap \operatorname{Ker} B = U \neq 0$, the undecomposable pair of maps is of the form $A, B\colon U \to 0$, where $\dim U = 1$. Further, in the case $\operatorname{Im} A + \operatorname{Im} B = U \neq W$, the undecomposable pair is of the form $A, B\colon 0 \to U^\perp$, where $U^\perp$ is the complement to U in W. Hence in what follows we shall assume that $\operatorname{Ker} A \cap \operatorname{Ker} B = 0$ and $\operatorname{Im} A + \operatorname{Im} B = W$.

Consider the subspaces $K = \operatorname{Im} A \cap \operatorname{Im} B$ and $L = A^{-1}(K) \cap B^{-1}(K)$. If $\dim \operatorname{Ker} A = a$ and $\dim \operatorname{Ker} B = b$, then $\dim \operatorname{Im} A = n - a$ and $\dim \operatorname{Im} B = n - b$. Hence $\dim K + (n + l) = (n - a) + (n - b)$, i.e., $\dim K = n - l - a - b$. It is also obvious that

$$\dim A^{-1}(K) = \dim \operatorname{Ker} A + \dim K = a + (n - l - a - b) = n - l - b$$

and $\dim B^{-1}(K) = n - l - a$. Therefore,

$$\dim L \geq \dim A^{-1}(K) + \dim B^{-1}(K) - \dim V = n - a - b - 2l.$$

Set $\dim L = n - a - b - 2l + r$, where $r \geq 0$.

Consider the subspace $(\operatorname{Ker} A)_L = L \cap \operatorname{Ker} A$ and take its complement $(\operatorname{Ker} A)_N$ in the space $\operatorname{Ker} A$. Define the subspaces $(\operatorname{Ker} B)_L$ and $(\operatorname{Ker} B)_N$ in a similar way.

In the space $A^{-1}(K)$ take the complement $A^{-1}(K)_N$ to $L \oplus (\operatorname{Ker} A)_N$. Define the subspace $B^{-1}(K)_N$ in a similar way. It is clear that

$$\begin{aligned}\dim A^{-1}(K)_N + \dim(\operatorname{Ker} A)_N &= \dim A^{-1}(K) - \dim L \\ &= (n-l-b) - (n-a-b-2l+r) = l+a-r.\end{aligned}$$

Similarly,

$$\dim B^{-1}(K)_N + \dim(\operatorname{Ker} B)_N = l + b - r.$$

Let us show that the sum of subspaces

$$L \oplus (\operatorname{Ker} A)_N \oplus A^{-1}(K)_N \quad \text{and} \quad (\operatorname{Ker} B)_N \oplus B^{-1}(K)_N$$

is direct. Assume that the vector x lies in the first of these subspaces. Then $Ax \in A(L) + K \subset K$. If the vector x lies in the second one, we have $Bx \in K$. Thus $x \in A^{-1}(K) \cap B^{-1}(K) = L$. It remains to note that $(\operatorname{Ker} B)_N \oplus B^{-1}(K)_N \cap L = 0$.

In the space V choose a subspace X so that

$$V = L \oplus (\operatorname{Ker} A)_N \oplus A^{-1}(K)_N \oplus (\operatorname{Ker} B)_N \oplus B^{-1}(K)_N \oplus X.$$

Then we have

$$\dim X = n - (n-a-b-2l+r) - (l+a-r) - (l+b-r) = r.$$

Consider the subspace $Y = AX + BX \subset W$; clearly, $\dim Y \leq 2r$. Since

$$AL \subset K, \quad A(\operatorname{Ker} A)_N = 0, \quad A(A^{-1}(K)_N) \subset K,$$

and so on, it follows that the space $W = \operatorname{Im} A + \operatorname{Im} B$ may be written in the form

$$W = K + B(\operatorname{Ker} A)_N + B(A^{-1}(K)_N) + A(\operatorname{Ker} B)_N + A(B^{-1}(K)_N) + Y.$$

The sum of dimensions of the summands is then no greater than

$$(n-l-a-b) + (l+a-r) + (l+b-r) + 2r = n + l = \dim W.$$

This means that the sum of subspaces under consideration is direct and, furthermore, the restrictions of the map A to $(\operatorname{Ker} B)_N$ and to $B^{-1}(K)_N$ are invertible, the restrictions of the map B to $(\operatorname{Ker} A)_N$ and to $A^{-1}(K)_N$ are invertible and $\dim Y = 2r$.

The maps A and B take the subspace $L \oplus A^{-1}(K)_N \oplus B^{-1}(K)_N$ to the subspace $K \oplus B(A^{-1}(K)_N) \oplus A(B^{-1}(K)_N)$, the subspace $(\operatorname{Ker} A)_N$ to $B(\operatorname{Ker} A)_N$, the subspace $(\operatorname{Ker} B)_N$ to $A(\operatorname{Ker} B)_N$, and X to Y. An undecomposable pair is obtained only if the spaces V and W reduce to one of these four direct summands.

It is easy to verify that the indecomposable maps

$$(\operatorname{Ker} A)_N \to B(\operatorname{Ker} A)_N, \quad (\operatorname{Ker} B)_N \to A(\operatorname{Ker} B)_N \quad \text{and} \quad X \to Y$$

are of the form $(0, I_1)$, $(I_1, 0)$, and P_1^T, Q_1^T, respectively.

It remains to investigate the case of the undecomposable pair of maps

$$A, B\colon V = L \oplus A^{-1}(K)_N \oplus B^{-1}(K)_N \to K \oplus B(A^{-1}(K)_N) \oplus A(B^{-1}(K)_N = W.$$

LEMMA. *Let $A_L, B_L: L \to K$ be the restrictions of the maps A and B. Then the pair of maps A_L, B_L is undecomposable.*

PROOF. Assume the converse, and let $L = L' \oplus L''$, $K = K' \oplus K''$ be the corresponding decompositions of L and K.

For the map $A: U \to K' \oplus K''$, we can construct a decomposition $U = U' \oplus U''$ so that $AU' \subset K'$ and $AU'' \subset K''$. Indeed, we have $A^{-1}(K') = \operatorname{Ker} A \oplus U_1'$ and $A^{-1}(K'') = \operatorname{Ker} A \oplus U_1''$, where $U = \operatorname{Ker} A \oplus U_1' \oplus U_1''$. Put $U' = \operatorname{Ker} A \oplus U_1'$ and $U'' = U_1''$. Then $U = U' \oplus U''$ is the required decomposition.

In this way, let us construct the decomposition

$$A^{-1}(K)_N = A^{-1}(K)_N' \oplus A^{-1}(K)_N''$$

for the map A and the decomposition

$$B^{-1}(K)_N = B^{-1}(K)_N' \oplus B^{-1}(K)_N''$$

for the map B. Then the image of the space $L' \oplus A^{-1}(K)_N' \oplus B^{-1}(K)_N'$ under the map A is contained in $K' \oplus A(B^{-1}(K)_N')$ and under the map B, in $K' \oplus B(A^{-1}(K)_N')$.

Consider the decompositions of the spaces $V = V' \oplus V''$ and $W = W' \oplus W''$, where $V' = L' \oplus A^{-1}(K)_N' \oplus B^{-1}(K)_N'$ and $W' = K' \oplus B(A^{-1}(K)_N') \oplus A(B^{-1}(K)_N')$ and the spaces V'' and W'' are defined in a similar way. These decompositions show that the pair of maps $A,\ B$ is decomposable. □

Recall that in the situation under consideration we have $r = \dim X = 0$, hence $\dim L = n - m - 2l$ and $\dim K = n - m - l$ where $m = a + b$. Let us consider three cases separately: (1) $l = 0$; (2) $l = 1$; (3) $l \geq 2$.

(1) $l = 0$. We claim that in this case one of the maps A and B is one-to-one, i.e., the pair $A,\ B$ is equivalent to one of the pairs $I,\ J$ and $J,\ I$, where I is the unit matrix and J a Jordan matrix. It is clear that for an undecomposable pair of maps the Jordan matrix J consists of one Jordan block $J_k(\lambda)$.

The proof will be by induction on n. Recall that for $l = 0$ we have

$$\dim V = \dim W = n, \quad \dim L = \dim K = n - m.$$

We start with the base of the induction $n = 1$. The relation $W = \operatorname{Im} A + \operatorname{Im} B \neq 0$ implies that at least one of the maps A and B is nonzero. But a nonzero map of a one-dimensional space is always one-to-one.

For the induction step, we shall need another induction, on m this time. For $m = 0$, i.e., for $\dim \operatorname{Ker} A + \dim \operatorname{Ker} B = 0$, both maps A and B are one-to-one. In the case $m > 0$, consider the restrictions A_L and B_L of the maps A and B to the space L. According to the lemma, the pair of maps A_L, B_L is undecomposable. But then by the induction assumption at least one of the maps A_L and B_L takes L onto K bijectively. To be definite, suppose the map $A_L: L \to K$ is bijective. Then

$$L = A^{-1}(K) = L \oplus (\operatorname{Ker} A)_N \oplus A^{-1}(K)_N,$$

so that $A^{-1}(K)_N = 0$. Therefore, the map

$$A: V = L \oplus B^{-1}(K)_N \to K \oplus A(B^{-1}(K)_N) = W$$

is one-to-one.

(2) $l = 1$. In this case the pair A, B is equivalent to $\begin{pmatrix} I_n \\ 0 \end{pmatrix}$, $\begin{pmatrix} 0 \\ I_n \end{pmatrix}$, i.e., to the pair P_n^T, Q_n^T.

The proof will be by induction on n. First let us recall that for $l = 1$ we have

$$\dim V = n, \quad \dim W = n+1, \quad \dim L = n-m-2, \quad \dim K = n-m-1.$$

When $n = 1$, for the basis in the space V choose an arbitrary nonzero vector e_1, and for the basis in the space W choose the vectors $\varepsilon_1 = Ae_1$ and $\varepsilon_2 = Be_1$.

When $n = 2$, we have $\dim L = n - m - 2 = -m$, hence $m = 0$, i.e., $\dim L = 0$ and $\dim K = 1$. In the space K choose a nonzero vector ε_2 and consider two vectors e_1 and e_2 such that $\varepsilon_2 = Be_1$ and $\varepsilon_2 = Ae_2$. Further, put $\varepsilon_1 = Ae_1$ and $\varepsilon_3 = Be_2$:

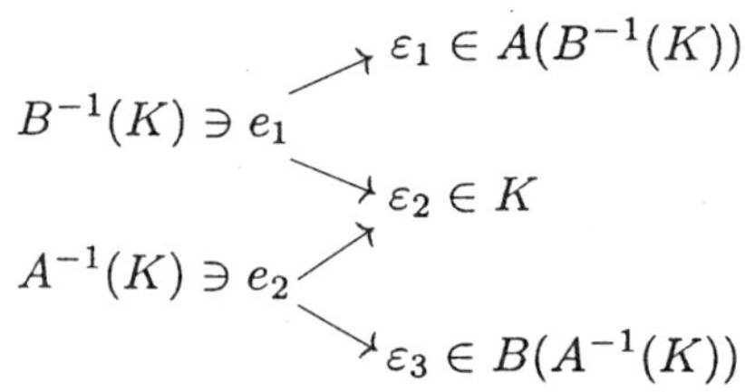

(the arrows pointing upwards correspond to the map A, those pointing downwards, to B).

When $n > 2$, consider the restrictions $A_L, B_L \colon L \to K$ of the maps A and B. According to the lemma, the pair of maps A_L, B_L is undecomposable. Here we have $\dim L = n - m - 2 = t$ and $\dim K = t + 1$, where $t < n$. By the induction assumption, the pair A_L, B_L is equivalent to the pair $\begin{pmatrix} I_t \\ 0 \end{pmatrix}$, $\begin{pmatrix} 0 \\ I_t \end{pmatrix}$. In particular, $\operatorname{Ker} A_L = \operatorname{Ker} B_L = 0$ and therefore $t = n - 2$. By the induction assumption we can construct bases $e_1, \dots, e_{n-2}$ and $\varepsilon_1, \dots, \varepsilon_{n-1}$ in the spaces L and K, and then use them to construct the required bases $e_0, e_1, \dots, e_{n-2}, e_{n-1}$ and $\varepsilon_0, \varepsilon_1, \dots, \varepsilon_{n-1}, \varepsilon_n$:

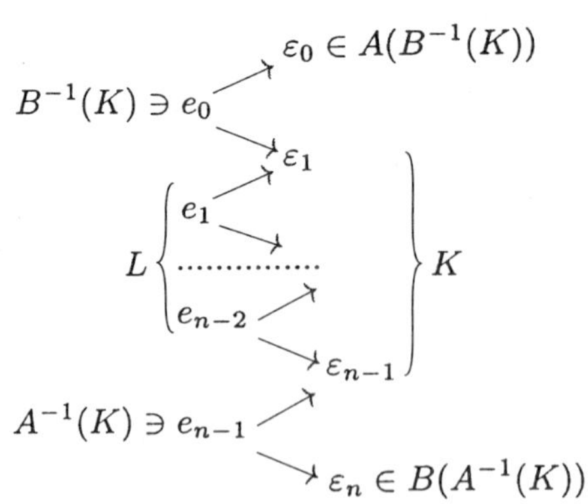

(3) $l \geq 2$. In this case the pair of maps A, B is decomposable. Let us prove this by induction on n.

For $n = 1$ we can indicate the direct summands $V \to \operatorname{Im} A + \operatorname{Im} B$ and $0 \to (\operatorname{Im} A + \operatorname{Im} B)^{\perp} \neq 0$.

For $n > 1$ let us consider the restrictions $A_L, B_L : L \to K$ of the maps A and B. Since $\dim L = n - m - 2l = t < n$ and $\dim K = t + l$, it follows from the induction assumption that the pair A_L, B_L is decomposable. But then by the lemma so is the pair A, B. □

Problems

12.1. Prove that A and A^T are similar matrices.

12.2. Let $\sigma(i)$, where $i = 1, \dots, n$, be an arbitrary permutation and $P = \left\| p_{ij} \right\|_1^n$, where $p_{ij} = \delta_{i\sigma(j)}$. Prove that the matrix $P^{-1}AP$ is obtained from A by the permutation σ of the rows and the same permutation of the columns of A.

The matrix P is called the *permutation matrix* corresponding to σ.

12.3. Let the number of distinct eigenvalues of a matrix A be equal to m, where $m > 1$. Let $b_{ij} = \operatorname{tr}(A^{i+j})$. Prove that $|b_{ij}|_0^{m-1} \neq 0$ and $|b_{ij}|_0^m = 0$.

12.4. Prove that $\operatorname{rank} A = \operatorname{rank} A^2$ if and only if $\lim\limits_{\lambda \to 0}(A + \lambda I)^{-1} A$ exists.

13. The minimal polynomial and the characteristic polynomial

13.1. Let $p(t) = \sum_{k=0}^{n} a_k t^k$ be a polynomial of degree n. For any square matrix A we can consider the matrix $p(A) = \sum_{k=0}^{n} a_k A^k$. The polynomial $p(t)$ is called an *annihilating polynomial* of A if $p(A) = 0$. (The zero on the right-hand side is the zero matrix.)

If A is an order n matrix, then the matrices $I, A, \dots, A^{n^2}$ are linearly dependent since the dimension of the space of matrices of order n is equal to n^2. Therefore, for any matrix of order n there exists an annihilating polynomial whose degree does not exceed n^2. The annihilating polynomial of A of the minimal degree and with leading coefficient 1 is called the *minimal polynomial* of A.

Let us prove that the minimal polynomial is well defined. Indeed, if $p_1(A) = A^m + \dots = 0$ and $p_2(A) = A^m + \dots = 0$, then the polynomial $p_1 - p_2$ annihilates A and its degree is smaller than m. Hence, $p_1 - p_2 = 0$.

It is easy to verify that if $B = X^{-1}AX$, then $B^n = X^{-1}A^nX$ and, therefore, $p(B) = X^{-1}p(A)X$; thus, the *minimal polynomial of an operator*, not only of a matrix, is well defined.

13.1.1. THEOREM. *Any annihilating polynomial of a matrix A is divisible by its minimal polynomial.*

PROOF. Let p be the minimal polynomial of A and q an annihilating polynomial. Dividing q by p with a remainder we get $q = pf + r$, where $\deg r < \deg p$, and $r(A) = q(A) - p(A)f(A) = 0$, and so r is an annihilating polynomial. Hence, $r = 0$. □

13.1.2. An *annihilating polynomial of a vector $v \in V$ (with respect to an operator $A : V \to V$)* is a polynomial p such that $p(A)v = 0$. The annihilating polynomial of v of minimal degree and with coefficient of the highest term equal to 1 is called the *minimal polynomial* of v. Similarly to the proof of Theorem 13.1.1, we can demonstrate that the minimal polynomial of A is divisible by the minimal polynomial of a vector.

THEOREM. *For any operator $A : V \to V$ there exists a vector whose minimal polynomial* (with respect to A) *coincides with the minimal polynomial of the operator A.*

PROOF. Any ideal I in the ring of polynomials in one indeterminate is generated by a polynomial f of minimal degree. Indeed, if $g \in I$ and $f \in I$ is a polynomial of minimal degree, then $g = fh + r$, hence, $r \in I$ since $fh \in I$.

For any vector $v \in V$ consider the ideal $I_v = \{p | p(A)v = 0\}$; this ideal is generated by a polynomial p_v with leading coefficient 1. If p_A is the minimal polynomial of A, then $p_A \in I_v$ and, therefore, p_A is divisible by p_v. Hence, when v runs over the whole of V we get only a finite number of polynomials p_v. Let these be $p_1, \dots, p_k$. The space V is contained in the union of its subspaces $V_i = \{x \in V \mid p_i(A)x = 0\}$ $(i = 1, \dots, k)$ and, therefore, $V = V_i$ for a certain i. Then $p_i(A)V = 0$; in other words p_i is divisible by p_A and, therefore, $p_i = p_A$. □

13.2. Simple considerations show that the degree of the minimal polynomial of a matrix A of order n does not exceed n^2. It turns out that the degree of the minimal polynomial does not actually exceed n, since the characteristic polynomial of A is an annihilating polynomial.

THEOREM (Cayley-Hamilton). *Let $p(t) = |tI - A|$. Then $p(A) = 0$.*

PROOF. For the Jordan form of the operator the proof is obvious because $(t - \lambda)^n$ is an annihilating polynomial of $J_n(\lambda)$. Let us, however, give a proof which does not make use of the Jordan theorem.

We may assume that A is a matrix (in a basis) of an operator over $\mathbb{C}$. Let us carry out the proof by induction on the order n of A. For $n = 1$ the statement is obvious.

Let λ be an eigenvalue of A and e_1 the corresponding eigenvector. Let us complement e_1 to a basis $e_1, \dots, e_n$. In the basis $e_1, \dots, e_n$ the matrix A is of the form $\begin{pmatrix} \lambda & * \\ 0 & A_1 \end{pmatrix}$, where A_1 is the matrix of the operator in the quotient space $V/\operatorname{Span}(e_1)$. Therefore, $p(t) = (t - \lambda)|tI - A_1| = (t - \lambda)p_1(t)$. By inductive hypothesis $p_1(A_1) = 0$ in $V/\operatorname{Span}(e_1)$, i.e., $p_1(A_1)V \subset \operatorname{Span}(e_1)$. It remains to observe that $(\lambda I - A)e_1 = 0$. □

REMARK. Making use of the Jordan normal form it is easy to verify that the minimal polynomial of A is equal to $\prod_i (t - \lambda_i)^{n_i}$, where the product is taken over all distinct eigenvalues λ_i of A and n_i is the order of the maximal Jordan block corresponding to λ_i. In particular, the matrix A is diagonalizable if and only if the minimal polynomial has no multiple roots and all its roots belong to the ground field.

13.3. By the Cayley-Hamilton theorem the characteristic polynomial of a matrix of order n coincides with its minimal polynomial if and only if the degree of the minimal polynomial is equal to n. The minimal polynomial of a matrix A is the minimal polynomial for a certain vector v (cf. Theorem 13.1.2). Therefore, the characteristic polynomial coincides with the minimal polynomial if and only if for a certain vector v the vectors $v, Av, \dots, A^{n-1}v$ are linearly independent.

THEOREM ([Farahat, Lederman, 1958]). *The characteristic polynomial of a matrix A of order n coincides with its minimal polynomial if and only if for any vector $(x_1, \dots, x_n)$ there exist columns P and Q of length n such that $x_k = Q^T A^k P$.*

PROOF. First, suppose that the degree of the minimal polynomial of A is equal to n. Then there exists a column P such that the columns $P, AP, \dots, A^{n-1}P$ are linearly independent, i.e., the matrix K formed by these columns is invertible. Any vector $X = (x_1, \dots, x_n)$ can be represented in the form $X = (XK^{-1})K = (Q^T P, \dots, Q^T A^{n-1} P)$, where $Q^T = XK^{-1}$.

Now, suppose that for any vector $(x_1, \dots, x_n)$ there exist columns P and Q such that $x_k = Q^T A^k P$. Then there exist columns $P_1, \dots, P_n, Q_1, \dots, Q_n$ such that the matrix

$$B = \begin{pmatrix} Q_1^T P_1 & \dots & Q_1^T A^{n-1} P_1 \\ \vdots & \dots & \vdots \\ Q_n^T P_n & \dots & Q_n^T A^{n-1} P_n \end{pmatrix}$$

is invertible. The matrices $I, A, \dots, A^{n-1}$ are linearly independent because otherwise the columns of B would be linearly dependent. □

13.4. The Cayley-Hamilton theorem has several generalizations. We will confine ourselves to one of them.

13.4.1. THEOREM ([Greenberg, 1984]). *Let $p_A(t)$ be the characteristic polynomial of a matrix A, and let a matrix X commute with A. Then $p_A(X) = M(A - X)$, where M is a matrix that commutes with A and X.*

PROOF. Since $B \cdot \operatorname{adj} B = |B| \cdot I$ (see 2.4),

$$p_A(\lambda) \cdot I = [\operatorname{adj}(\lambda I - A)](\lambda I - A) = \left(\sum_{k=0}^{n-1} A_k \lambda^k\right)(\lambda I - A) = \sum_{k=0}^{n} \lambda^k A'_k.$$

All matrices A'_k are diagonal, since so is $p_A(\lambda)I$. Hence, $p_A(X) = \sum_{k=0}^{n} X^k A'_k$. If X commutes with A and A_k, then $p_A(X) = (\sum_{k=0}^{n-1} A_k X^k)(X - A)$. But the matrices A_k can be expressed as polynomials of A (see Problem 2.11) and, therefore, if X commutes with A, then X commutes with A_k. □

Problems

13.1. Let A be a matrix of order n and

$$f_1(A) = A - (\operatorname{tr} A)I, \quad f_{k+1}(A) = f_k(A)A - \frac{1}{k+1} \operatorname{tr}(f_k(A)A)I.$$

Prove that $f_n(A) = 0$.

13.2. Let A and B be matrices of order n. Prove that if $\operatorname{tr} A^m = \operatorname{tr} B^m$ for $m = 1, \dots, n$, then the eigenvalues of A and B coincide.

13.3. Let a matrix A be invertible and let its minimal polynomial $p(\lambda)$ coincide with its characteristic polynomial. Prove that the minimal polynomial of A^{-1} is equal to $p(0)^{-1}\lambda^n p(\lambda^{-1})$.

13.4. Let the minimal polynomial of a matrix A be equal to $\prod(x - \lambda_i)^{n_i}$. Prove that the minimal polynomial of $\begin{pmatrix} A & I \\ 0 & A \end{pmatrix}$ is equal to $\prod(x - \lambda_i)^{n_i+1}$.

14. The Frobenius canonical form

14.1. The Jordan form is just one of several canonical forms of matrices of linear operators. An example of another canonical form is the *cyclic form* also known as the *Frobenius canonical form*.

A *Frobenius* or *cyclic block* is a matrix of the form

$$\begin{pmatrix} 0 & 0 & 0 & \dots & 0 & -a_0 \\ 1 & 0 & 0 & \dots & 0 & -a_1 \\ 0 & 1 & 0 & \dots & 0 & -a_2 \\ \vdots & \vdots & \vdots & \ddots & \ddots & \vdots \\ 0 & 0 & 0 & \dots & 1 & -a_{n-1} \end{pmatrix}.$$

If $A : V^n \to V^n$ and $Ae_1 = e_2, \dots, Ae_{n-1} = e_n$, then the matrix of the operator A with respect to $e_1, \dots, e_n$ is a cyclic block.

THEOREM. *For any linear operator $A : V \to V$ (over $\mathbb{C}$ or $\mathbb{R}$) there exists a basis in which the matrix of A is of block diagonal form with cyclic diagonal blocks.*

PROOF (Following [Jacob, 1973]). We apply induction on $\dim V$. If the degree of the minimal polynomial of A is equal to k, then there exists a vector $y \in V$ whose minimal polynomial also has degree k (see Theorem 13.1.2). Let $y_i = A^{i-1}y$. Let us complement the basis $y_1, \dots, y_k$ of $W = \operatorname{Span}(y_1, \dots, y_k)$ to a basis of V and consider $W_1^* = \operatorname{Span}(y_k^*, A^*y_k^*, \dots, A^{*k-1}y_k^*)$. Let us prove that $V = W \oplus W_1^{*\perp}$ is an A-invariant decomposition of V.

The degree of the minimal polynomial of A^* is also equal to k and, therefore, W_1^* is invariant with respect to A^*; hence, $(W_1^*)^\perp$ is invariant with respect to A. It remains to prove that $W_1^* \cap W^\perp = 0$ and $\dim W_1^* = k$. Suppose that $a_0y_k^* + \dots + a_sA^{*s}y_k^* \in W^\perp$ for $0 \le s \le k-1$ and $a_s \ne 0$. Then $A^{*k-s-1}(a_0y_k^* + \dots + a_sA^{*s}y_k^*) \in W^\perp$; hence,

$$\begin{aligned} 0 &= \langle a_0A^{*k-s-1}y_k^* + \dots + a_sA^{k-1}y_k^*, y\rangle \\ &= a_0\langle y_k^*, A^{k-s-1}y\rangle + \dots + a_s\langle y_k^*, A^{k-1}y\rangle \\ &= a_0\langle y_k^*, y_{k-s}\rangle + \dots + a_s\langle y_k^*, y_k\rangle = a_s. \end{aligned}$$

Contradiction.

The matrix of the restriction of A to W in the basis $y_1, \dots, y_k$ is a cyclic block. The restriction of A to $W_1^{*\perp}$ can be represented in the required form by the inductive hypothesis. $\square$

REMARK. In the process of the proof we have found a basis in which the matrix of A is of block diagonal form with cyclic blocks on the diagonal whose characteristic polynomials are $p_1, p_2, \dots, p_k$, where p_1 is the minimal polynomial for A, p_2 is the minimal polynomial of the restriction of A to a subspace, and, therefore, p_2 is a divisor of p_1. Similarly, p_{i+1} is a divisor of p_i.

14.2. Let us prove that the characteristic polynomial of the cyclic block

$$A = \begin{pmatrix} 0 & 0 & \dots & 0 & 0 & 0 & -a_0 \\ 1 & 0 & \dots & 0 & 0 & 0 & -a_1 \\ 0 & 1 & \dots & 0 & 0 & 0 & -a_2 \\ \vdots & \vdots & \ddots & \vdots & \vdots & & \\ 0 & 0 & \dots & 1 & 0 & 0 & -a_{n-3} \\ 0 & 0 & \dots & 0 & 1 & 0 & -a_{n-2} \\ 0 & 0 & \dots & 0 & 0 & 1 & -a_{n-1} \end{pmatrix}$$

is equal to $\lambda^n + \sum_{k=0}^{n-1} a_k \lambda_k$. Indeed, since $Ae_1 = e_2, \dots, Ae_{n-1} = e_n$, and $Ae_n = -\sum_{k=0}^{n-1} a_k e_{k+1}$, it follows that

$$\left(A^n + \sum_{k=0}^{n-1} a_k A^k\right) e_1 = 0.$$

Taking into account that $e_i = A^{i-1}e_1$ we see that $\lambda^n + \sum_{k=0}^{n-1} a_k \lambda^k$ is an annihilating polynomial of A. It remains to notice that the vectors $e_1, Ae_1, \dots, A^{n-1}e_1$ are linearly independent and, therefore, the degree of the minimal polynomial of A is no less than n.

As a byproduct we have proved that *the characteristic polynomial of a cyclic block coincides with its minimal polynomial.*

Problems

14.1. The matrix of an operator A is block diagonal and consists of two cyclic blocks with relatively prime characteristic polynomials, p and q. Prove that it is possible to select a basis so that the matrix becomes one cyclic block.

14.2. Let A be a Jordan block, i.e., there exists a basis $e_1, \dots, e_n$ such that $Ae_1 = \lambda e_1$ and $Ae_k = e_{k-1} + \lambda e_k$ for $k = 2, \dots, n$. Prove that there exists a vector v such that the vectors $v, Av, \dots, A^{n-1}v$ constitute a basis (then the matrix of the operator A with respect to the basis $v, Av, \dots, A^{n-1}v$ is a cyclic block).

14.3. For a cyclic block A indicate a symmetric matrix S such that $A = SA^TS^{-1}$.

15. How to reduce the diagonal to a convenient form

15.1. The transformation $A \mapsto XAX^{-1}$ preserves the trace and, therefore, the diagonal elements of the matrix XAX^{-1} cannot be made completely arbitrary. We can, however, reduce the diagonal of A to a more convenient form; for example, a matrix $A \neq \lambda I$ is similar to a matrix whose diagonal elements are $(0, \dots, 0, \operatorname{tr} A)$; any matrix is similar to a matrix all diagonal elements of which are equal.

THEOREM ([Gibson, 1975]). *Let $A \neq \lambda I$. Then A is similar to a matrix with the diagonal* $(0, \dots, 0, \operatorname{tr} A)$.

PROOF. The diagonal of a cyclic block is of the needed form. Therefore, the statement is true for any matrix whose characteristic and minimal polynomials coincide (cf. 14.1).

For a matrix of order 2 the characteristic polynomial does not coincide with the minimal one only for matrices of the form λI. Let now A be a matrix of order 3 such that $A \neq \lambda I$ and the characteristic polynomial of A does not coincide with its minimal polynomial. Then the minimal polynomial of A is of the form $(x-\lambda)(x-\mu)$ whereas the characteristic polynomial is $(x-\lambda)^2(x-\mu)$ and the case $\lambda = \mu$ is not excluded. Therefore, the matrix A is similar to the matrix $C = \begin{pmatrix} 0 & a & 0 \\ 1 & b & 0 \\ 0 & 0 & \lambda \end{pmatrix}$ and the characteristic polynomial of $\begin{pmatrix} 0 & a \\ 1 & b \end{pmatrix}$ is divisible by $x-\lambda$, i.e., $\lambda^2 - b\lambda - a = 0$.

If $b = \lambda = 0$, then the theorem holds.

If $b = \lambda \neq 0$, then $b^2 - b^2 - a = 0$, i.e., $a = 0$. In this case

$$\begin{pmatrix} 0 & 0 & 0 \\ 1 & b & 0 \\ 0 & 0 & b \end{pmatrix} 1 \begin{pmatrix} b & b & b \\ -1 & 0 & 0 \\ b & 0 & b \end{pmatrix} = \begin{pmatrix} 0 & 0 & 0 \\ 0 & b & b \\ b^2 & 0 & b^2 \end{pmatrix} = \begin{pmatrix} b & b & b \\ -1 & 0 & 0 \\ b & 0 & b \end{pmatrix} \begin{pmatrix} 0 & -b & -b \\ -b & 0 & -b \\ b & b & 2b \end{pmatrix},$$

and $\det \begin{pmatrix} b & b & b \\ -1 & 0 & 0 \\ b & 0 & b \end{pmatrix} \neq 0$; therefore, A is similar to $\begin{pmatrix} 0 & -b & -b \\ -b & 0 & -b \\ b & b & 2b \end{pmatrix}$.

Let, finally, $b \neq \lambda$. Then for the matrix $D = \operatorname{diag}(b, \lambda)$ the theorem is true and, therefore, there exists a matrix P such that $PDP^{-1} = \begin{pmatrix} 0 & * \\ * & * \end{pmatrix}$. The matrix

$$\begin{pmatrix} 1 & 0 \\ 0 & P \end{pmatrix} C \begin{pmatrix} 1 & 0 \\ 0 & P^{-1} \end{pmatrix} = \begin{pmatrix} 1 & 0 \\ 0 & P \end{pmatrix} \begin{pmatrix} 0 & * \\ * & D \end{pmatrix} \begin{pmatrix} 1 & 0 \\ 0 & P^{-1} \end{pmatrix} = \begin{pmatrix} 0 & * \\ * & PDP^{-1} \end{pmatrix}$$

is of the required form.

Now, suppose our theorem holds for matrices of order m, where $m \geq 3$. A matrix A of order $m+1$ is of the form $\begin{pmatrix} A_1 & * \\ * & * \end{pmatrix}$, where A_1 is a matrix of order m. Since $A \neq \lambda I$, we can assume that $A_1 \neq \lambda I$ (otherwise we perform a permutation of rows and columns, cf. Problem 12.2). By the inductive hypothesis there exists a matrix P such that the diagonal of the matrix PA_1P^{-1} is of the form $(0, 0, \dots, 0, \alpha)$ and, therefore, the diagonal of the matrix

$$X = \begin{pmatrix} P & 0 \\ 0 & 1 \end{pmatrix} \begin{pmatrix} A_1 & * \\ * & * \end{pmatrix} \begin{pmatrix} P^{-1} & 0 \\ 0 & 1 \end{pmatrix} = \begin{pmatrix} PA_1P^{-1} & * \\ * & * \end{pmatrix}$$

is of the form $(0, \dots, 0, \alpha, \beta)$. If $\alpha = 0$ we are done.

Let $\alpha \neq 0$. Then $X = \begin{pmatrix} 0 & * \\ * & C_1 \end{pmatrix}$, where the diagonal of the matrix C_1 of order m is of the form $(0, 0, \dots, \alpha, \beta)$ and, therefore, $C_1 \neq \lambda I$. Hence, there exists a matrix Q such that the diagonal of QCQ^{-1} is of the form $(0, \dots, 0, x)$. Therefore, the diagonal of $\begin{pmatrix} 1 & 0 \\ 0 & Q \end{pmatrix} \begin{pmatrix} 0 & * \\ * & C_1 \end{pmatrix} \begin{pmatrix} 1 & 0 \\ 0 & Q^{-1} \end{pmatrix}$ is of the required form. □

REMARK. The proof holds for a field of any characteristic.

15.2. THEOREM. *Let A be an arbitrary complex matrix. Then there exists a unitary matrix U such that the diagonal elements of UAU^{-1} are equal.*

PROOF. On the set of unitary matrices, consider a function f whose value at U is equal to the maximal absolute value of the difference of the diagonal elements of UAU^{-1}. This function is continuous and is defined on a compact set and, therefore, it attains its minimum on this compact set. Therefore, to prove the theorem it suffices to show that with the help of the transformation $A \mapsto UAU^{-1}$ one can always diminish the maximal absolute value of the difference of the diagonal elements unless it is already equal to zero.

Let us begin with matrices of size 2×2. Let $u = e^{i\varphi}\cos\alpha$, $v = e^{i\psi}\sin\alpha$. Then in the $(1, 1)$ entry of the matrix

$$\begin{pmatrix} u & v \\ -\overline{v} & \overline{u} \end{pmatrix} \begin{pmatrix} a_1 & b \\ c & a_2 \end{pmatrix} \begin{pmatrix} \overline{u} & -v \\ \overline{v} & u \end{pmatrix}$$

is

$$a_1 \cos^2\alpha + a_2 \sin^2\alpha + (be^{i\beta} + ce^{-i\beta})\cos\alpha\sin\alpha, \text{ where } \beta = \varphi - \psi.$$

When β varies from 0 to 2π the points $be^{i\beta} + ce^{-i\beta}$ form an ellipse (or an interval) centred at $0 \in \mathbb{C}$. Indeed, the points $e^{i\beta}$ belong to the unit circle and the map $z \mapsto bz + c\overline{z}$ determines a (possibly singular) $\mathbb{R}$-linear transformation of $\mathbb{C}$. Therefore, the number

$$p = (be^{i\beta} + ce^{-i\beta})/(a_1 - a_2)$$

is real for a certain β. Hence, $t = \cos^2\alpha + p\sin\alpha\cos\alpha$ is also real and

$$a_1 \cos^2\alpha + a_2 \sin^2\alpha + (be^{i\beta} + ce^{-i\beta})\cos\alpha\sin\alpha = ta_1 + (1 - t)a_2.$$

As α varies from 0 to $\frac{\pi}{2}$, the variable t varies from 1 to 0. In particular, t takes the value $\frac{1}{2}$. In this case the both diagonal elements of the transformed matrix are equal to $\frac{1}{2}(a_{11} + a_{22})$.

Let us treat matrices of size $n \times n$, where $n \geq 3$, as follows. Select a pair of diagonal elements the absolute value of whose difference is maximal (there could be several such pairs). With the help of a permutation matrix this pair can be placed in the positions (1, 1) and (2, 2) thanks to Problem 12.2. For the matrix $A' = \left\|a_{ij}\right\|_1^2$ there exists a unitary matrix U such that the diagonal elements of $UA'U^{-1}$ are equal to $\frac{1}{2}(a_{11} + a_{22})$. It is also clear that the transformation $A \mapsto U_1AU_1^{-1}$, where U_1 is the unitary matrix $\begin{pmatrix} U & 0 \\ 0 & I \end{pmatrix}$, preserves the diagonal elements $a_{33}, \dots, a_{nn}$. Thus, we have managed to replace two fartherst apart diagonal elements a_{11} and a_{22} by their arithmetic mean. We do not increase in this way the maximal distance between points nor do we create new pairs with the distance equal to $|a_{11} - a_{22}|$ since

$$\left| x - \frac{a_{11} + a_{22}}{2} \right| \leq \frac{|x - a_{11}|}{2} + \frac{|x - a_{22}|}{2}.$$

After a finite number of such steps we get rid of all pairs of diagonal elements with the distance $|a_{11} - a_{22}|$. $\square$

REMARK. If A is a real matrix, then we can assume that $u = \cos\alpha$ and $v = \sin\alpha$. The number p is real in such a case. Therefore, if A is real, then U can be considered to be an orthogonal matrix.

15.3. THEOREM ([Marcus, Purves, 1959]). *Any nonzero square matrix A is similar to a matrix all diagonal elements of which are nonzero.*

PROOF. Any matrix A of order n is similar to a matrix with all diagonal elements being equal to $\frac{1}{n}\operatorname{tr} A$ (see 15.2), and, therefore, it suffices to consider the case when $\operatorname{tr} A = 0$. We can assume that A is a Jordan block.

First, let us consider a matrix $A = \left\|a_{ij}\right\|_1^n$ such that $a_{ij} = \delta_{1i}\delta_{2j}$. If $U = \left\|u_{ij}\right\|$ is a unitary matrix, then $UAU^{-1} = UAU^* = B$, where $b_{ii} = u_{i1}\overline{u}_{i2}$. We can select U so that all elements u_{i1}, u_{i2} are nonzero.

The rest of the proof will be carried out by induction on n; for $n = 2$ the statement is proved.

Recall that we assume A to be in the Jordan form. First, suppose that A is a diagonal matrix and $a_{11} \neq 0$. Then $A = \begin{pmatrix} a_{11} & 0 \\ 0 & \Lambda \end{pmatrix}$, where Λ is a nonzero diagonal matrix. Let U be a matrix such that all elements of $U\Lambda U^{-1}$ are nonzero. Then the diagonal elements of the matrix

$$\begin{pmatrix} 1 & 0 \\ 0 & U \end{pmatrix} \begin{pmatrix} a_{11} & 0 \\ 0 & \Lambda \end{pmatrix} \begin{pmatrix} 1 & 0 \\ 0 & U^{-1} \end{pmatrix} = \begin{pmatrix} a_{11} & 0 \\ 0 & U\Lambda U^{-1} \end{pmatrix}$$

are nonzero.

Now, suppose that a matrix A is not diagonal. We can assume that $a_{12} = 1$ and the matrix C obtained from A by crossing out the first row and the first column is a nonzero matrix. Let U be a matrix such that all diagonal elements of UCU^{-1} are nonzero. Consider the matrix

$$D = \begin{pmatrix} 1 & 0 \\ 0 & U \end{pmatrix} A \begin{pmatrix} 1 & 0 \\ 0 & U^{-1} \end{pmatrix} = \begin{pmatrix} a_{11} & * \\ 0 & UCU^{-1} \end{pmatrix}.$$

The only zero diagonal element of D could be a_{11}. If $a_{11} = 0$, then for $\begin{pmatrix} 0 & * \\ 0 & d_{22} \end{pmatrix}$ select a matrix V such that the diagonal elements of $V\begin{pmatrix} 0 & * \\ 0 & d_{22} \end{pmatrix}V^{-1}$ are nonzero. Then the diagonal elements of $\begin{pmatrix} V & 0 \\ 0 & I \end{pmatrix} D \begin{pmatrix} V^{-1} & 0 \\ 0 & I \end{pmatrix}$ are also nonzero. □

Problem

15.1. Prove that for any nonzero square matrix A there exists a matrix X such that the matrices X and $A + X$ have no common eigenvalues.

16. The polar decomposition

16.1. Any complex number z can be represented in the form $z = |z|e^{i\varphi}$. An analogue of such a representation is the *polar decomposition of a matrix*, $A = SU$, where S is an Hermitian and U is a unitary matrix.

THEOREM. *Any square matrix A over $\mathbb{R}$ (or $\mathbb{C}$) can be represented in the form $A = SU$, where S is a symmetric (Hermitian) nonnegative definite matrix and U is an orthogonal (unitary) matrix. If A is invertible such a representation is unique.*

PROOF. If $A = SU$, where S is an Hermitian nonnegative definite matrix and U is a unitary matrix, then $AA^* = SUU^*S = S^2$. To find S, let us do the following. The Hermitian matrix AA^* has an orthonormal eigenbasis and $AA^*e_i = \lambda_i^2 e_i$, where $\lambda_i \geq 0$. Set $Se_i = \lambda_i e_i$. The Hermitian nonnegative definite matrix S is uniquely determined by A. Indeed, let $e_1', \dots, e_n'$ be an orthonormal eigenbasis for S and $Se_i' = \lambda_i' e_i'$, where $\lambda_i' \geq 0$. Then $(\lambda_i')^2 e_i' = S^2 e_i' = AA^* e_i'$ and this equation uniquely determines λ_i'.

Let $v_1, \dots, v_n$ be an orthonormal basis of eigenvectors of the Hermitian operator A^*A and $A^*Av_i) = \mu_i^2 v_i$, where $\mu_i \geq 0$. Since $(Av_i, Av_j) = (v_i, A^*Av_j) = \mu_i^2(v_i, v_j)$, we see that the vectors $Av_1, \dots, Av_n$ are pairwise orthogonal and $|Av_i| = \mu_i$. Therefore, there exists an orthonormal basis $w_1, \dots, w_n$ such that $Av_i = \mu_i w_i$. Set $Uv_i = w_i$ and $Sw_i = \mu_i w_i$. Then $SUv_i = Sw_i = \mu_i w_i = Av_i$, i.e., $A = SU$.

In the decomposition $A = SU$ the matrix S is uniquely defined. If S is invertible, then $U = S^{-1}A$ is also uniquely defined. □

REMARK. We can similarly construct a decomposition $A = U_1 S_1$, where S_1 is a symmetric (Hermitian) nonnegative definite matrix and U_1 is an orthogonal (unitary) matrix. Here $S_1 = S$ if and only if $AA^* = A^*A$, i.e., the matrix A is normal.

16.2.1. THEOREM. *Any matrix A can be represented in the form $A = UDW$, where U and W are unitary matrices and D is a diagonal matrix.*

PROOF. Let $A = SV$, where S is Hermitian and V unitary. For S there exists a unitary matrix U such that $S = UDU^*$, where D is a diagonal matrix. The matrix $W = U^*V$ is unitary and $A = SV = UDW$. □

16.2.2. THEOREM. *If $A = S_1U_1 = U_2S_2$ are the polar decompositions of an invertible matrix A, then $U_1 = U_2$.*

PROOF. Let $A = UDW$, where $D = \operatorname{diag}(d_1, \dots, d_n)$ is a diagonal matrix, and U and W are unitary matrices. Consider the matrix $D_+ = \operatorname{diag}(|d_1|, \dots, |d_n|)$; then $DD_+ = D_+D$ and, therefore,

$$A = (UD_+U^*)(UD_+^{-1}DW) = (UD_+^{-1}DW)(W^*D_+W).$$

The matrices UD_+U^* and W^*D_+W are positive definite and $D_+^{-1}D$ is unitary. The uniqueness of the polar decomposition of an invertible matrix implies that $S_1 = UD_+U^*$, $S_2 = W^*D_+W$ and $U_1 = UD_+^{-1}DW = U_2$. □

Problems

16.1. Prove that any linear transformation of $\mathbb{R}^n$ is the composition of an orthogonal transformation and a dilation along perpendicular directions (with distinct coefficients).

16.2. Let $A : \mathbb{R}^n \to \mathbb{R}^n$ be a contraction operator, i.e., $|Ax| \leq |x|$. The space $\mathbb{R}^n$ can be considered as a subspace of $\mathbb{R}^{2n}$. Prove that A is the restriction to $\mathbb{R}^n$ of the composition of an orthogonal transformation of $\mathbb{R}^{2n}$ and the projection on $\mathbb{R}^n$.

17. Factorizations of matrices

17.1. The Schur decomposition.

THEOREM (Schur). *Any square matrix A over $\mathbb{C}$ can be represented in the form $A = UTU^*$, where U is a unitary and T a triangular matrix; moreover, A is normal if and only if T is a diagonal matrix.*

PROOF. Let us prove by induction on the order of A. Let x be an eigenvector of A, i.e., $Ax = \lambda x$. We may assume that $|x| = 1$. Let W be a unitary matrix whose first column is made of the coordinates of x (to construct such a matrix it suffices to complement x to an orthonormal basis). Then

$$W^*AW = \begin{pmatrix} \lambda & * & * & * \\ 0 & & & \\ \vdots & & A_1 & \\ 0 & & & \end{pmatrix}.$$

By the inductive hypothesis there exists a unitary matrix V such that V^*A_1V is a triangular matrix. Then $U = \begin{pmatrix} 1 & 0 \\ 0 & V \end{pmatrix}$ is the desired matrix.

It is easy to verify that the equations $T^*T = TT^*$ and $A^*A = AA^*$ are equivalent. It remains to prove that a triangular normal matrix is a diagonal matrix. Let

$$T = \begin{pmatrix} t_{11} & t_{12} & \dots & t_{1n} \\ 0 & t_{21} & \dots & t_{1n} \\ \vdots & \ddots & \ddots & \vdots \\ 0 & 0 & \dots & t_{nn} \end{pmatrix}.$$

Then $(TT^*)_{11} = |t_{11}|^2 + |t_{12}|^2 + \dots + |t_{1n}|^2$ and $(T^*T)_{11} = |t_{11}|^2$. Therefore, the identity $TT^* = T^*T$ implies that $t_{12} = \dots = t_{1n} = 0$.

Now, delete the first row and the first column in T and repeat the arguments. □

17.2. The Lanczos decomposition.

THEOREM ([Lanczos, 1958]). *Any real $m \times n$-matrix A of* rank $p > 0$ *can be represented in the form $A = X\Lambda Y^T$, where X and Y are matrices of size $m \times p$ and $n \times p$ with orthonormal columns and Λ is a diagonal matrix of size $p \times p$.*

PROOF (Following [Schwert, 1960]). The rank of A^TA is equal to the rank of A; see Problem 8.3. Let U be an orthogonal matrix such that

$$U^TA^TAU = \operatorname{diag}(\mu_1, \dots, \mu_p, 0, \dots, 0), \quad \text{where } \mu_i > 0.$$

Further, let $y_1, \dots, y_p$ be the first p columns of U and Y the matrix formed by these columns. The columns $x_i = \lambda_i^{-1}Ay_i$, where $\lambda_i = \sqrt{\mu_i}$, constitute an orthonormal system since $(Ay_i, Ay_j) = (y_i, A^TAy_j) = \lambda_j^2(y_i, y_j)$. It is also clear that $AY = (\lambda_1x_1, \dots, \lambda_px_p) = X\Lambda$, where X is a matrix constituted from $x_1, \dots, x_p$, $\Lambda = \operatorname{diag}(\lambda_1, \dots, \lambda_p)$. Now, let us prove that $A = X\Lambda Y^T$. For this let us again consider the matrix $U = (Y, Y_0)$. Since $\operatorname{Ker} A^TA = \operatorname{Ker} A$ and $(A^TA)Y_0 = 0$, it follows that $AY_0 = 0$. Hence, $AU = (X\Lambda, 0)$ and, therefore, $A = (X\Lambda, 0)U^T = X\Lambda Y^T$. □

REMARK. Since $AU = (X\Lambda, 0)$, we have $U^TA^T = \begin{pmatrix} \Lambda X^T \\ 0 \end{pmatrix}$. Multiplying this equality by U, we get $A^T = Y\Lambda X^T$. Hence, $A^TX = Y\Lambda X^TX = Y\Lambda$ since $X^TX = I_p$. Therefore, $(X^TA)(A^TX) = (\Lambda Y^T)(Y\Lambda) = \Lambda^2$ since $Y^TY = I_p$. Thus, the columns of X are eigenvectors of AA^T.

17.3. THEOREM. *Any square matrix A can be represented in the form $A = ST$, where S and T are symmetric matrices, and if A is real, then S and T can also be considered to be real matrices.*

PROOF. First, observe that if $A = ST$, where S and T are symmetric matrices, then $A = ST = S(TS)S^{-1} = SA^TS^{-1}$, where S is a symmetric matrix. On the other hand, if $A = SA^TS^{-1}$, where S is a symmetric matrix, then $A = ST$, where $T = A^TS^{-1}$ is a symmetric matrix, since $(A^TS^{-1})^T = S^{-1}A = S^{-1}SA^TS^{-1} = A^TS^{-1}$.

If A is a cyclic block, then there exists a symmetric matrix S such that $A = SA^TS^{-1}$ (Problem 14.3). For any A there exists a matrix P such that $B = P^{-1}AP$ is in Frobenius form. For B there exists a symmetric matrix S such that $B = SB^TS^{-1}$. Hence, $A = PBP^{-1} = PSB^TS^{-1}P^{-1} = S_1AS_1^{-1}$, where $S_1 = PSP^T$ is a symmetric matrix. $\square$

To prove the theorem we could have made use of the Jordan form as well. In order to do this, it suffices to notice that, for example,

$$\begin{pmatrix} \Lambda & E & 0 \\ 0 & \Lambda & E \\ 0 & 0 & \Lambda \end{pmatrix} = \begin{pmatrix} 0 & 0 & E \\ 0 & E & 0 \\ E & 0 & 0 \end{pmatrix} \begin{pmatrix} 0 & 0 & \Lambda' \\ 0 & \Lambda' & E \\ \Lambda' & E & 0 \end{pmatrix},$$

where $\Lambda = \Lambda' = \lambda$ and $E = 1$ for the real case (or for a real λ) and for the complex case (i.e., $\lambda = a + bi$, $b \neq 0$) $\Lambda = \begin{pmatrix} a & b \\ -b & a \end{pmatrix}$, $E = \begin{pmatrix} 0 & 1 \\ 1 & 0 \end{pmatrix}$ and $\Lambda' = \begin{pmatrix} b & a \\ a & -b \end{pmatrix}$. For a Jordan block of an arbitrary size a similar decomposition also holds.

Problems

17.1 (The Gauss factorization). All minors $|a_{ij}|_1^p$, $p = 1, \dots, n$ of a matrix A of order n are nonzero. Prove that A can be represented in the form $A = T_1T_2$, where T_1 is a lower triangular and T_2 an upper triangular matrix.

17.2 (The Gram factorization). Prove that an invertible matrix X can be represented in the form $X = UT$, where U is an orthogonal matrix and T is an upper triangular matrix.

17.3 ([Ramakrishnan, 1972]). Let $B = \operatorname{diag}(1, \varepsilon, \dots, \varepsilon^{n-1})$, where $\varepsilon = \exp(\frac{2\pi i}{n})$, and $C = \|c_{ij}\|_1^n$, where $c_{ij} = \delta_{i,j-1}$ (here $j-1$ is considered modulo n). Prove that any $n \times n$-matrix M over $\mathbb{C}$ is uniquely representable in the form $M = \sum_{k,l=0}^{n-1} a_{kl}B^kC^l$.

17.4. Prove that any skew-symmetric matrix A can be represented in the form $A = S_1S_2 - S_2S_1$, where S_1 and S_2 are symmetric matrices.

18. The Smith normal form. Elementary factors of matrices

18.1. Let A be a matrix whose elements are integers or polynomials (we may assume that the elements of A belong to a commutative ring in which the notion of the greatest common divisor is defined). Further, let $f_k(A)$ be the greatest common divisor of minors of order k of A. The formula for determinant expansion with respect to a row indicates that f_k is divisible by f_{k-1}.

The formula $A^{-1} = (\operatorname{adj} A)/\det A$ shows that the elements of A^{-1} are integers (resp. polynomials) if $\det A = \pm 1$ (resp. $\det A$ is a nonzero number). The other way around, if the elements of A^{-1} are integers (resp. polynomials), then $\det A = \pm 1$ (resp. $\det A$ is a nonzero number) since $\det A \cdot \det A^{-1} = \det(AA^{-1}) = 1$. Matrices A with $\det A = \pm 1$ are called *units* (of the corresponding matrix ring). The product of units is, clearly, a unit.

18.1.1. THEOREM. *If $A' = BAC$, where B and C are unity matrices, then $f_k(A') = f_k(A)$ for all admissible k.*

PROOF. From the Binet-Cauchy formula it follows that $f_k(A')$ is divisible by $f_k(A)$. Since $A = B^{-1}A'C^{-1}$, $f_k(A)$ is divisible by $f_k(A')$. $\square$

18.1.2. THEOREM (Smith). *For any matrix A of size $m \times n$ there exist unit matrices B and C such that $BAC = \operatorname{diag}(g_1, g_2, \dots, g_p, 0, \dots, 0)$, where g_{i+1} is divisible by g_i.*

The matrix $\operatorname{diag}(g_1, g_2, \dots, g_p, 0, \dots, 0)$ is called the *Smith normal form* of A.

PROOF. The multiplication from the right (left) by the unit matrix $\left\|a_{ij}\right\|_1^n$, where $a_{ii} = 1$ for $i \neq p, q$ and $a_{pq} = a_{qp} = 1$ the other elements being zero, performs a permutation of pth column (row) with the qth one. The multiplication from the right by the unit matrix $\left\|a_{ij}\right\|_1^n$, where $a_{ii} = 1$ $(i = 1, \dots, n)$ and $a_{pq} = f$ (here p and q are fixed distinct numbers), corresponds to the addition of the pth column multiplied by f to the qth column whereas the multiplication by it from the left corresponds to the addition of the qth row multiplied by f to the pth one. It remains to verify that by such operations the matrix A can be reduced to the desired form.

Define *the norm of an integer* as its absolute value and *the norm of a polynomial* as its degree. Take a nonzero element a of the given matrix with the least norm and place it in the $(1, 1)$ position. Let us divide all elements of the first row by a with a remainder and add the multiples of the first column to the columns 2 to n so that in the first row we get the remainders after division by a.

Let us perform similar operations over columns. If, after that, in the first row and the first column there is at least one nonzero element besides a, then its norm is strictly less than that of a. Let us place this element in the position $(1, 1)$ and repeat the above operations. The norm of the upper left element strictly decreases and, therefore, at the end in the first row and in the first column we get just one nonzero element, a_{11}.

Suppose that the matrix obtained has an element a_{ij} not divisible by a_{11}. Add to the first column the column that contains a_{ij} and then add to the row that contains a_{ij} a multiple of the first row so that the element a_{ij} is replaced by the remainder after division by a_{11}. As a result we get an element whose norm is strictly less than that of a_{11}. Let us place it in position $(1, 1)$ and repeat the indicated operations. At the end we get a matrix of the form $\begin{pmatrix} g_1 & 0 \\ 0 & A' \end{pmatrix}$, where the elements of A' are divisible by g_1.

Now, we can repeat the above arguments for the matrix A'. $\square$

REMARK. Clearly, $f_k(A) = g_1 g_2 \dots g_k$.

18.2. The elements $g_1, \dots, g_p$ obtained in the Smith normal form are called *invariant factors* of A. They are expressed in terms of divisors of minors $f_k(A)$ as follows: $g_k = f_k/f_{k-1}$ if $f_{k-1} \neq 0$.

Every invariant factor g_i can be expanded in a product of powers of primes (resp. powers of irreducible polynomials). Such factors are called *elementary divisors* of A. Each factor enters the set of elementary divisors multiplicity counted.

Elementary divisors of real or complex matrix A are elementary divisors of the matrix $xI - A$. The product of all elementary divisors of a matrix A is equal, up to a sign, to its characteristic polynomial.

Problems

18.1. Compute the invariant factors of a Jordan block and of a cyclic block.

18.2. Let A be a matrix of order n, let f_{n-1} be the greatest common divisor of the $(n-1)$-minors of $xI - A$. Prove that the minimal polynomial A is equal to $\dfrac{|xI - A|}{f_{n-1}}$.

Solutions

11.1. a) The trace of $AB - BA$ is equal to 0 and, therefore, $AB - BA$ cannot be equal to I.

b) If $|A| \neq 0$ and $AB - BA = A$, then $A^{-1}AB - A^{-1}BA = I$. But $\operatorname{tr}(B - A^{-1}BA) = 0$ and $\operatorname{tr} I = n$.

11.2. Let all elements of B be equal to 1 and $\Lambda = \operatorname{diag}(\lambda_1, \dots, \lambda_n)$. Then $A = \Lambda B \Lambda^{-1}$ and if x is an eigenvector of B, then Λx is an eigenvector of A. The vector $(1, \dots, 1)$ is an eigenvector of B corresponding to the eigenvalue n and the $(n-1)$-dimensional subspace $x_1 + \dots + x_n = 0$ is the eigenspace corresponding to eigenvalue 0.

11.3. If λ is not an eigenvalue of the matrices $\pm A$, then A can be represented as half of the sum of the invertible matrices $A + \lambda I$ and $A - \lambda I$.

11.4. Obviously, the coefficients of the characteristic polynomial depend continuously on the elements of the matrix. It remains to prove that the roots of the polynomial $p(x) = x^n + a_1x^{n-1} + \dots + a_n$ depend continuously on $a_1, \dots, a_n$. It suffices to carry out the proof for the zero root (for a nonzero root x_1 we can consider the change of variables $y = x - x_1$). If $p(0) = 0$, then $a_n = 0$. Consider a polynomial $q(x) = x^n + b_1x^{n-1} + \dots + b_n$, where $|b_i - a_i| < \delta$. If $x_1, \dots, x_n$ are the roots of q, then $|x_1 \dots x_n| = |b_n| < \delta$ and, therefore, the absolute value of one of the roots of q is less than $\sqrt[n]{\delta}$. The δ required can be taken to be equal to ε^n.

11.5. If the sum of the elements of every row of A is equal to s, then $Ae = se$, where e is the column $(1, 1, \dots, 1)^T$. Therefore, $A^{-1}(Ae) = A^{-1}(se)$; hence, $A^{-1}e = (1/s)e$, i.e., the sum of the elements of every row of A^{-1} is equal to $1/s$.

11.6. Let S_1 be the first column of S. Equating the first columns of AS and $S\Lambda$, where the first column of Λ is of the form $(\lambda, 0, \dots, 0)^T$, we get $AS_1 = \lambda S_1$.

11.7. It is easy to verify that $|\lambda I - A| = \sum_{k=0}^{n} \lambda^{n-k}(-1)^k \Delta_k(A)$, where $\Delta_k(A)$ is the sum of all principal k-minors of A. It follows that

$$\sum_{i=1}^{n} |\lambda I - A_i| = \sum_{i=1}^{n} \sum_{k=0}^{n-1} \lambda^{n-k-1}(-1)^k \Delta_k(A_i).$$

It remains to notice that

$$\sum_{i=1}^{n} \Delta_k(A_i) = (n-k)\Delta_k(A),$$

since any principal k-minor of A is a principal k-minor for $n-k$ matrices A_i.

11.8. Since $\operatorname{adj}(PXP^{-1}) = P(\operatorname{adj} X)P^{-1}$, we can assume that A has the Jordan normal form. In this case $\operatorname{adj} A$ is an upper triangular matrix (by Problem 2.6) and it is easy to compute its diagonal elements.

11.9. Let $S = \left\|\delta_{i,n-j}\right\|_0^n$. Then $AS = \left\|b_{ij}\right\|_0^n$ and $SA = \left\|c_{ij}\right\|_0^n$, where $b_{ij} = a_{i,n-j}$ and $c_{ij} = a_{n-i,j}$. Therefore, the central symmetry of A means that $AS = SA$. It is also easy to see that x is a symmetric vector if $Sx = x$ and skew-symmetric if $Sx = -x$.

Let λ be an eigenvalue of A and $Ay = \lambda y$, where $y \neq 0$. Then $A(Sy) = S(Ay) = S(\lambda y) = \lambda(Sy)$. If $Sy = -y$ we can set $x = y$. If $Sy \neq -y$ we can set $x = y + Sy$ and then $Ax = \lambda x$ and $Sx = x$.

11.10. Since

$$Ae_i = a_{n-i+1,i}e_{n-i+1} = x_{n-i+1}e_{n-i+1}$$

and $Ae_{n-i+1} = x_i e_i$, the subspaces $V_i = \operatorname{Span}(e_i, e_{n-i+1})$ are invariant with respect to A. For $i \neq n-i+1$ the matrix of the restriction of A to V_i is of the form $B = \begin{pmatrix} 0 & \lambda \\ \mu & 0 \end{pmatrix}$. The eigenvalues of B are equal to $\pm\sqrt{\lambda\mu}$. If $\lambda\mu = 0$ and B is diagonalizable, then $B = 0$. Therefore, the matrix B is diagonalizable if and only if both numbers λ and μ are simultaneously equal or not equal to zero.

Thus, the matrix A is diagonalizable if and only if the both numbers x_i and x_{n-i+1} are simultaneously equal or not equal to 0 for all i.

11.11. a) Suppose the columns $x_1, \dots, x_m$ correspond to real eigenvalues $\alpha_1, \dots, \alpha_m$. Let $X = (x_1, \dots, x_m)$ and $D = \operatorname{diag}(\alpha_1, \dots, \alpha_m)$. Then $AX = XD$ and since D is a real matrix, then $AXX^* = XDX^* = X(XD)^* = X(AX)^* = XX^*A^*$. If the vectors $x_1, \dots, x_m$ are linearly independent, then $\operatorname{rank} XX^* = \operatorname{rank} X = m$ (see Problem 8.3) and, therefore, for S we can take XX^*.

Now, suppose that $AS = SA^*$ and S is a nonnegative definite matrix of rank m. Then there exists an invertible matrix P such that $S = PNP^*$, where $N = \begin{pmatrix} I_m & 0 \\ 0 & 0 \end{pmatrix}$. Let us multiply both parts of the identity $AS = SA^*$ by P^{-1} from the left and by $(P^*)^{-1}$ from the right; we get $(P^{-1}AP)N = N(P^{-1}AP)^*$. Let $P^{-1}AP = B = \begin{pmatrix} B_{11} & B_{12} \\ B_{21} & B_{22} \end{pmatrix}$, where B_{11} is a matrix of order m. Since $BN = NB^*$, we have $\begin{pmatrix} B_{11} & 0 \\ B_{21} & 0 \end{pmatrix} = \begin{pmatrix} B_{11}^* & B_{21}^* \\ 0 & 0 \end{pmatrix}$, i.e., $B = \begin{pmatrix} B_{11} & B_{12} \\ 0 & B_{22} \end{pmatrix}$, where B_{11} is an Hermitian matrix of order m. The matrix B_{11} has m linearly independent eigenvectors $z_1, \dots, z_m$ with real eigenvalues. Since $AP = PB$ and P is an invertible matrix, then the vectors $P\binom{z_1}{0}, \dots, P\binom{z_m}{0}$ are linearly independent and are eigenvectors of A corresponding to real eigenvalues.

b) The proof is essentially similar to that of a): in our case $AXX^*A^* = AX(AX)^* = XD(XD)^* = XDD^*X^* = XX^*$.

If $ASA^* = S$ and $S = PNP^*$, then $P^{-1}APN(P^{-1}AP)^* = N$, i.e.,

$$\begin{pmatrix} B_{11}B_{11}^* & B_{11}B_{21}^* \\ B_{21}B_{11}^* & B_{21}B_{21}^* \end{pmatrix} = \begin{pmatrix} I_m & 0 \\ 0 & 0 \end{pmatrix}.$$

Therefore, $B_{21} = 0$ and $P^{-1}AP = B = \begin{pmatrix} B_{11} & B_{12} \\ 0 & B_{22} \end{pmatrix}$, where B_{11} is unitary.

12.1. Let A be a Jordan block of order k. It is easy to verify that in this case $S_kA = A^TS_k$, where $S_k = \left\|\delta_{i,k+1-j}\right\|_1^k$ is an invertible matrix. If A is the direct sum of Jordan blocks, then we can take the direct sum of the matrices S_k.

12.2. The matrix P^{-1} corresponds to the permutation σ^{-1} and, therefore, $P^{-1} = \left\|q_{ij}\right\|_1^n$, where $q_{ij} = \delta_{\sigma(i)j}$. Let $P^{-1}AP = \left\|b_{ij}\right\|_1^n$. Then $b_{ij} = \sum_{s,t}\delta_{\sigma(i)s}a_{st}\delta_{t\sigma(j)} = a_{\sigma(i)\sigma(j)}$.

12.3. Let $\lambda_1, \dots, \lambda_m$ be distinct eigenvalues of A and p_i the multiplicity of the eigenvalue λ_i. Then $\operatorname{tr}(A^k) = p_1\lambda_1^k + \dots + p_m\lambda_m^k$. Therefore,

$$\left\|b_{ij}\right\|_0^{m-1} = p_1 \dots p_m \prod_{i \neq j}(\lambda_i - \lambda_j)^2 \quad \text{(See Problem 1.18).}$$

To compute $|b_{ij}|_0^m$ we can, for example, replace $p_m\lambda_m^k$ with $\lambda_m^k + (p_m - 1)\lambda_m^k$ in the expression for $\operatorname{tr}(A^k)$.

12.4. If $A' = P^{-1}AP$, then $(A' + \lambda I)^{-1}A' = P^{-1}(A + \lambda I)^{-1}AP$ and, therefore, it suffices to consider the case when A is a Jordan block. If A is invertible, then $\lim\limits_{\lambda \to 0}(A + \lambda I)^{-1} = A^{-1}$. Let $A = 0 \cdot I + N = N$ be a Jordan block with zero eigenvalue. Then

$$(N + \lambda I)^{-1}N = \lambda^{-1}(I - \lambda^{-1}N + \lambda^{-2}N^2 - \dots)N = \lambda^{-1}N - \lambda^{-2}N^2 + \dots$$

and the limit as $\lambda \to 0$ exists only if $N = 0$.

Thus, the limit indicated exists if and only if the matrix A does not have nonzero blocks with zero eigenvalues. This condition is equivalent to $\operatorname{rank} A = \operatorname{rank} A^2$.

13.1. Let $(\lambda_1, \dots, \lambda_n)$ be the diagonal of the Jordan normal form of A and $\sigma_k = \sigma_k(\lambda_1, \dots, \lambda_n)$. Then $|\lambda I - A| = \sum_{k=0}^n (-1)^k\lambda^{n-k}\sigma_k$. Therefore, it suffices to demonstrate that $f_m(A) = \sum_{k=0}^m (-1)^k A^{m-k}\sigma_k$ for all m. For $m = 1$ this equation coincides with the definition of f_1. Suppose the statement is proved for m; let us prove it for $m + 1$. Clearly,

$$f_{m+1}(A) = \sum_{k=0}^m (-1)^k A^{m-k+1}\sigma_k - \frac{1}{m+1}\operatorname{tr}\left(\sum_{k=0}^m (-1)^k A^{m-k+1}\sigma_k\right) I.$$

Since

$$\operatorname{tr}\left(\sum_{k=0}^m (-1)^k A^{m-k+1}\sigma_k\right) = \sum_{k=0}^m (-1)^k s_{m-k+1}\sigma_k,$$

where $s_p = \lambda_1^p + \dots + \lambda_n^p$, it remains to observe that

$$\sum_{k=0}^m (-1)^k s_{m-k+1}\sigma_k + (m+1)(-1)^{m+1}\sigma_{m+1} = 0 \quad \text{(see 4.1).}$$

13.2. According to the solution of Problem 13.1 the coefficients of the characteristic polynomial of X are functions of $\operatorname{tr} X, \dots, \operatorname{tr} X^n$ and, therefore, the characteristic polynomials of A and B coincide.

13.3. Let $f(\lambda)$ be an arbitrary polynomial $g(\lambda) = \lambda^n f(\lambda^{-1})$ and $B = A^{-1}$. If $0 = g(B) = B^n f(A)$, then $f(A) = 0$. Therefore, the minimal polynomial of B is proportional to $\lambda^n p(\lambda^{-1})$. It remains to observe that the highest coefficient of $\lambda^n p(\lambda^{-1})$ is equal to $\lim\limits_{\lambda\to\infty} \frac{\lambda^n p(\lambda^{-1})}{\lambda^n} = p(0)$.

13.4. As is easy to verify,

$$p\begin{pmatrix} A & I \\ 0 & A \end{pmatrix} = \begin{pmatrix} p(A) & p'(A) \\ 0 & p(A) \end{pmatrix}.$$

If $q(x) = \prod(x-\lambda_i)^{n_i}$ is the minimal polynomial of A and p is an annihilating polynomial of $\begin{pmatrix} A & I \\ 0 & A \end{pmatrix}$, then p and p' are divisible by q; among all such polynomials p the polynomial $\prod(x-\lambda_i)^{n_i+1}$ is of the minimal degree.

14.1. The minimal polynomial of a cyclic block coincides with the characteristic polynomial. The minimal polynomial of A annihilates the given cyclic blocks since it is divisible by both p and q. Since p and q are relatively prime, the minimal polynomial of A is equal to pq. Therefore, there exists a vector in V whose minimal polynomial is equal to pq.

14.2. First, let us prove that $A^k e_n = e_{n-k} + \varepsilon$, where $\varepsilon \in \operatorname{Span}(e_n, \dots, e_{n-k+1})$. We have $Ae_n = e_{n-1} + e_n$ for $k = 1$ and if the statement holds for k, then $A^{k+1} e_n = e_{n-k+1} + \lambda e_{n-k} + A\varepsilon$ and $e_{n-k}, A\varepsilon \in \operatorname{Span}(e_n, \dots, e_{n-k})$.

Therefore, expressing the coordinates of the vectors $e_n, Ae_n, \dots, A^{n-1}e_n$ with respect to the basis $e_n, e_{n-1}, \dots, e_1$ we get the matrix

$$\begin{pmatrix} 1 & \dots & \dots & * \\ 0 & 1 & & \vdots \\ \vdots & \ddots & \ddots & \vdots \\ 0 & \dots & 0 & 1 \end{pmatrix}.$$

This matrix is invertible and, therefore, the vectors $e_n, Ae_n, \dots, A^{n-1}e_n$ form a basis.

REMARK. It is possible to prove that for v we can take any vector $x_1e_1 + \dots + x_ne_n$, where $x_n \neq 0$.

14.3. Let

$$A = \begin{pmatrix} 0 & 0 & \dots & 0 & -a_n \\ 1 & 0 & \dots & 0 & -a_{n-1} \\ 0 & 1 & \dots & 0 & -a_{n-2} \\ \vdots & \vdots & \ddots & \vdots & \vdots \\ 0 & 0 & \dots & 1 & -a_1 \end{pmatrix}, \quad S = \begin{pmatrix} a_{n-1} & a_{n-2} & \dots & a_1 & 1 \\ a_{n-2} & a_{n-3} & \dots & 1 & 0 \\ \vdots & \vdots & \dots & \vdots & \vdots \\ a_1 & 1 & \dots & 0 & 0 \\ 1 & 0 & \dots & 0 & 0 \end{pmatrix}.$$

Then

$$AS = \begin{pmatrix} -a_n & 0 & 0 & \dots & 0 & 0 \\ 0 & a_{n-2} & a_{n-3} & \dots & a_1 & 1 \\ 0 & a_{n-3} & a_{n-4} & \dots & 1 & 0 \\ \vdots & \vdots & \vdots & \dots & \vdots & \vdots \\ 0 & a_1 & 1 & \dots & 0 & 0 \\ 0 & 1 & 0 & \dots & 0 & 0 \end{pmatrix}$$

is a symmetric matrix. Therefore, $AS = (AS)^T = SA^T$, i.e., $A = SA^T S^{-1}$.

15.1. By Theorem 15.3 there exists a matrix P such that the diagonal elements of $B = P^{-1}AP$ are nonzero. Consider a matrix Z whose diagonal elements are all equal to 1, the elements above the main diagonal are zeros, and the elemments below the diagonal are equal to the corresponding elements of the matrix $-B$. The eigenvalues of the lower triangular matrix Z are equal to 1 and the eigenvalues of the upper triangular matrix $B + Z$ are equal to $1 + b_{ii} \neq 1$. Therefore, for X we can take PZP^{-1}.

16.1. The operator A can be represented in the form $A = SU$, where U is an orthogonal operator and S is a positive definite symmetric operator. For a symmetric operator there exists an orthogonal basis of eigenvectors, i.e., it is a dilation along perpendicular directions.

16.2. If $A = SU$ is the polar decomposition of A, then for S there exists an orthonormal eigenbasis $e_1, \dots, e_n$ and all the eigenvalues do not exceed 1. Therefore, $Se_i = (\cos\varphi_i)e_i$. Complement the basis $e_1, \dots, e_n$ to a basis $e_1, \dots, e_n, \varepsilon_1, \dots, \varepsilon_n$ of $\mathbb{R}^{2n}$ and consider an orthogonal operator S_1 which in every plane $\operatorname{Span}(e_i, \varepsilon_i)$ acts as the rotation through an angle φ_i. The matrix of S_1 is of the form $\begin{pmatrix} S & * \\ * & * \end{pmatrix}$. Since

$$\begin{pmatrix} I & 0 \\ 0 & 0 \end{pmatrix} \begin{pmatrix} S & * \\ * & * \end{pmatrix} \begin{pmatrix} U & 0 \\ 0 & I \end{pmatrix} = \begin{pmatrix} SU & * \\ 0 & 0 \end{pmatrix},$$

it follows that $S_1 \begin{pmatrix} U & 0 \\ 0 & I \end{pmatrix}$ is the required orthogonal transformation of $\mathbb{R}^{2n}$.

17.1. Let $a_{pq} = \lambda$ be the only nonzero off-diagonal element of $X_{pq}(\lambda)$ and let the diagonal elements of $X_{pq}(\lambda)$ be equal to 1. Then $X_{pq}(\lambda)A$ is obtained from A by adding to the pth row the qth row multiplied by λ. By the hypothesis, $a_{11} \neq 0$ and, therefore, subtracting from the kth row the first row multiplied by a_{k1}/a_{11} we get a matrix with $a_{21} = \dots = a_{n1} = 0$. The hypothesis implies that $a_{22} \neq 0$. Therefore, we can subtract from the kth row ($k \geq 3$) the second row multiplied by a_{k2}/a_{22} and get a matrix with $a_{32} = \dots = a_{3n} = 0$, etc.

Therefore, by multiplying A from the right by the matrices X_{pq}, where $p > q$, we can get an upper triangular matrix T_2. Since $p > q$, the matrices X_{pq} are lower triangular and their product T is also a lower triangular matrix. The equality $TA = T_2$ implies $A = T^{-1}T_2$. It remains to observe that $T_1 = T^{-1}$ is a lower triangular matrix (see Problem 2.6); the diagonal elements of T_1 are all equal to 1.

17.2. Let $x_1, \dots, x_n$ be the columns of X. By 9.2, there exists an orthonormal set of vectors $y_1, \dots, y_n$ such that $y_i \in \operatorname{Span}(x_1, \dots, x_i)$ for $i = 1, \dots, n$. Then the matrix U whose columns are $y_1, \dots, y_n$ is orthogonal and $U = XT_1$, where T_1 is an upper triangular matrix. Therefore, $X = UT$, where $T = T_1^{-1}$ is an upper triangular matrix.

17.3. For every entry of the matrix M only one of the matrices $I, C, C^2, \dots, C^{n-1}$ has the same nonzero entry and, therefore, M is uniquely representable in the form $M = D_0 + D_1C + \dots + D_{n-1}C^{n-1}$, where the D_l are diagonal matrices. For example,

$$\begin{pmatrix} a & b \\ c & d \end{pmatrix} = \begin{pmatrix} a & 0 \\ 0 & d \end{pmatrix} + \begin{pmatrix} b & 0 \\ 0 & c \end{pmatrix} C, \quad \text{where } C = \begin{pmatrix} 0 & 1 \\ 1 & 0 \end{pmatrix}.$$

The diagonal matrices $I, B, B^2, \dots, B^{n-1}$ are linearly independent since their diagonals constitute a Vandermonde determinant. Therefore, any matrix D_l is uniquely representable as their linear combination

$$D_l = \sum_{k=0}^{n-1} a_{kl} B^k.$$

17.4. The matrix $A/2$ can be represented in the form $A/2 = S_1S_2$, where S_1 and S_2 are symmetric matrices (see 17.3). Therefore, $A = (A - A^T)/2 = S_1S_2 - S_2S_1$.

18.1. Let A be either a Jordan or cyclic block of order n. In both cases the matrix $A - xI$ has a triangular submatrix of order $n - 1$ with units 1 on the main diagonal. Therefore, $f_1 = \dots = f_{n-1} = 1$ and $f_n = p_A(x)$ is the characteristic polynomial of A. Hence, $g_1 = \dots = g_{n-1} = 1$ and $g_n = p_A(x)$.

18.2. The cyclic normal form of A is of a block diagonal form with the diagonal being formed by cyclic blocks corresponding to polynomials $p_1, p_2, \dots, p_k$, where p_1 is the minimal polynomial of A and p_i is divisible by p_{i+1}. Invariant factors of these cyclic blocks are $p_1, \dots, p_k$ (Problem 18.1), and, therefore, their Smith normal forms, are $\operatorname{diag}(1, \dots, 1, p_i)$. Hence, the Smith normal form of A is of the form $\operatorname{diag}(1, \dots, 1, p_k, \dots, p_2, p_1)$. Therefore, $f_{n-1} = p_2p_3 \dots p_k$.

CHAPTER IV

MATRICES OF SPECIAL FORM

19. Symmetric and Hermitian matrices

A real matrix A is said to be *symmetric* if $A^T = A$. In the complex case an analogue of a symmetric matrix is usually an *Hermitian matrix* for which $A^* = A$, where $A^* = \overline{A}^T$ is obtained from A by complex conjugation of its elements and transposition. (Physicists often write A^+ instead of A^*.) Sometimes, symmetric matrices with complex elements are also considered.

Let us recall the properties of Hermitian matrices proved in 11.3 and 10.3. The eigenvalues of an Hermitian matrix are real. An Hermitian matrix can be represented in the form U^*DU, where U is a unitary matrix and D is a diagonal matrix. A matrix A is Hermitian if and only if $(Ax, x) \in \mathbb{R}$ for any vector x.

19.1. To a square matrix A we can assign the *quadratic form* $q(x) = x^T Ax$, where x is the column of coordinates. Then $(x^T Ax)^T = x^T Ax$, i.e., $x^T A^T x = x^T Ax$. It follows that $2x^T Ax = x^T(A + A^T)x$, i.e., the quadratic form only depends on the symmetric part of A. Therefore, it is reasonable to assign quadratic forms to symmetric matrices only.

To a square matrix A we can also assign a *bilinear function* or a *bilinear form* $B(x, y) = x^T Ay$ (which depends on the skew-symmetric part of A, too) and if the matrix A is symmetric, then $B(x, y) = B(y, x)$, i.e., the bilinear function $B(x, y)$ is symmetric in the obvious sense. From a quadratic function $q(x) = x^T Ax$ we can recover the symmetric bilinear function $B(x, y) = x^T Ay$. Indeed,

$$2x^T Ay = (x + y)^T A(x + y) - x^T Ax - y^T Ay$$

since $y^T Ax = x^T A^T y = x^T Ay$.

In the real case a quadratic form $x^T Ax$ is said to be *positive definite* if $x^T Ax > 0$ for any nonzero x. In the complex case this definition makes no sense because any quadratic function $x^T Ax$ not only takes zero values for nonzero complex x but it takes nonreal values as well.

The notion of positive definiteness in the complex case only makes sense for Hermitian forms x^*Ax, where A is an Hermitian matrix. (Forms that are linear in one variable and antilinear in another one are sometimes called *sesquilinear* forms.) If U is a unitary matrix such that $A = U^*DU$, where D is a diagonal matrix, then $x^*Ax = (Ux)^*D(Ux)$, i.e., by the change $y = Ux$ we can represent an Hermitian form as follows

$$\sum \lambda_i y_i \overline{y}_i = \sum \lambda_i |y_i|^2.$$

An Hermitian form is positive definite if and only if all the numbers λ_i are positive.

For the matrix A of the quadratic (sesquilinear) form we write $A > 0$ and say that the *matrix* A is (positive or somehow else) definite if the corresponding quadratic (sesquilinear) form is definite in the same manner.

In particular, if A is positive definite (i.e., the Hermitian form x^*Ax is positive definite), then its trace $\lambda_1 + \cdots + \lambda_n$ and determinant $\lambda_1 \cdots \lambda_n$ are positive.

19.2.1. THEOREM (Sylvester's criterion). *Let $A = \left\|a_{ij}\right\|_1^n$ be an Hermitian matrix. Then A is positive definite if and only if all minors $|a_{ij}|_1^k$, $k = 1, \dots, n$, are positive.*

PROOF. Let the matrix A be positive definite. Then the matrix $\left\|a_{ij}\right\|_1^k$ corresponds to the restriction of a positive definite Hermitian form x^*Ax to a subspace and, therefore, $|a_{ij}|_1^k > 0$. Now, let us prove by induction on n that if $A = \left\|a_{ij}\right\|_1^n$ is an Hermitian matrix and $|a_{ij}|_1^k > 0$ for $k = 1, \dots, n$, then A is positive definite. For $n = 1$ this statement is obvious. It remains to prove that if $A' = \left\|a_{ij}\right\|_1^{n-1}$ is a positive definite matrix and $|a_{ij}|_1^n > 0$, then the eigenvalues of the Hermitian matrix $A = \left\|a_{ij}\right\|_1^n$ are all positive. There exists an orthonormal basis $e_1, \dots, e_n$ with respect to which x^*Ax is of the form $\lambda_1|y_1|^2 + \cdots + \lambda_n|y_n|^2$ and $\lambda_1 \le \lambda_2 \le \cdots \le \lambda_n$. If $y \in \operatorname{Span}(e_1, e_2)$, then $y^*Ay \le \lambda_2|y|^2$. On the other hand, if a nonzero vector y belongs to an $(n-1)$-dimensional subspace on which an Hermitian form corresponding to A' is defined, then $y^*Ay > 0$. This $(n-1)$-dimensional subspace and the two-dimensional subspace $\operatorname{Span}(e_1, e_2)$ belong to the same n-dimensional space and, therefore, they have a common nonzero vector y. It follows that $\lambda_2|y|^2 \ge y^*Ay > 0$, i.e., $\lambda_2 > 0$; hence, $\lambda_i > 0$ for $i \ge 2$. Besides, $\lambda_1 \cdots \lambda_n = |a_{ij}|_1^n > 0$ and therefore, $\lambda_1 > 0$. □

19.2.2. THEOREM (Sylvester's law of inertia). *Let an Hermitian form be reduced by a unitary transformation to the form*

$$\lambda_1|x_1|^2 + \cdots + \lambda_n|x_n|^2, \tag{1}$$

where $\lambda_i > 0$ for $i = 1, \dots, p$, $\lambda_i < 0$ for $i = p+1, \dots, p+q$, and $\lambda_i = 0$ for $i = p+q+1, \dots, n$. Then the numbers p and q do not depend on the unitary transformation.

PROOF. The expression (1) determines the decomposition of V into the direct sum of subspaces $V = V_+ \oplus V_- \oplus V_0$, where the form is positive definite, negative definite and identically zero on V_+, V_-, V_0, respectively. Let $V = W_+ \oplus W_- \oplus W_0$ be another such decomposition. Then $V_+ \cap (W_- \oplus W_0) = 0$ and, therefore, $\dim V_+ + \dim(W_- \oplus W_0) \le n$, i.e., $\dim V_+ \le \dim W_+$. Similarly, $\dim W_+ \le \dim V_+$. □

19.3. We turn to the reduction of quadratic forms to diagonal form.

THEOREM (Lagrange). *A quadratic form can always be reduced to the form*

$$q(x_1, \dots, x_n) = \lambda_1 x_1^2 + \cdots + \lambda_n x_n^2.$$

PROOF. Let $A = \left\|a_{ij}\right\|_1^n$ be the matrix of a quadratic form q. We carry out the proof by induction on n. For $n = 1$ the statement is obvious. Further, consider two cases.

a) There exists a nonzero diagonal element, say, $a_{11} \neq 0$. Then

$$q(x_1, \dots, x_n) = a_{11}y_1^2 + q'(y_2, \dots, y_n),$$

where

$$y_1 = x_1 + \frac{a_{12}x_2 + \dots + a_{1n}x_n}{a_{11}}$$

and $y_i = x_i$ for $i \geq 2$. The inductive hypothesis is applicable to q'.

b) All diagonal elements are 0. Only the case when the matrix has at least one nonzero element of interest; let, for example, $a_{12} \neq 0$. Set $x_1 = y_1 + y_2$, $x_2 = y_1 - y_2$ and $x_i = y_i$ for $i \geq 3$. Then

$$q(x_1, \dots, x_n) = 2a_{12}(y_1^2 - y_2^2) + q'(y_1, \dots, y_n),$$

where q' does not contain terms with y_1^2 and y_2^2. We can apply the change of variables from case a) to the form $q'(y_1, \dots, y_n)$. $\square$

19.4. Let the eigenvalues of an Hermitian matrix A be listed in decreasing order: $\lambda_1 \geq \dots \geq \lambda_n$. The numbers $\lambda_1, \dots, \lambda_n$ possess the following *min-max property*.

THEOREM (Courant-Fischer). *Let x run over all (admissible) unit vectors. Then*

$$\lambda_1 = \max_x (x^*Ax),$$

$$\lambda_2 = \min_{y_1} \max_{x \perp y_1} (x^*Ax),$$

$$\dots\dots\dots\dots\dots\dots\dots$$

$$\lambda_n = \min_{y_1,\dots,y_{n-1}} \max_{x \perp y_1,\dots,y_{n-1}} (x^*Ax).$$

PROOF. Let us select an orthonormal basis in which

$$x^*Ax = \lambda_1 x_1^2 + \dots + \lambda_n x_n^2.$$

Consider the subspaces $W_1 = \{x \mid x_{k+1} = \dots = x_n = 0\}$ and $W_2 = \{x \mid x \perp y_1, \dots, y_{k-1}\}$. Since $\dim W_1 = k$ and $\dim W_2 \geq n - k + 1$, we deduce that $W = W_1 \cap W_2 \neq 0$. If $x \in W$ and $|x| = 1$, then $x \in W_1$ and

$$x^*Ax = \lambda_1 x_1^2 + \dots + \lambda_k x_k^2 \geq \lambda_k (x_1^2 + \dots + x_k^2) = \lambda_k.$$

Therefore,

$$\lambda_k \leq \max_{x \in W_1 \cap W_2} (x^*Ax) \leq \max_{x \in W_2} (x^*Ax);$$

hence,

$$\lambda_k \leq \min_{y_1,\dots,y_{k-1}} \max_{x \in W_2} (x^*Ax).$$

Now, consider the vectors $y_i = (0, \dots, 0, 1, 0, \dots, 0)$ (1 stands in the ith slot). Then

$$W_2 = \{x \mid x \perp y_1, \dots, y_{k-1}\} = \{x \mid x_1 = \dots = x_{k-1} = 0\}.$$

If $x \in W_2$ and $|x| = 1$, then

$$x^*Ax = \lambda_k x_k^2 + \dots + \lambda_n x_n^2 \leq \lambda_k (x_k^2 + \dots + x_n^2) = \lambda_k.$$

Therefore,

$$\lambda_k = \max_{x \in W_2} (x^*Ax) \geq \min_{y_1,\dots,y_{k-1}} \max_{x \perp y_1,\dots,y_{k-1}} (x^*Ax). \quad \square$$

19.5. An Hermitian matrix A is called *nonnegative definite* (and we write $A \geq 0$) if $x^*Ax \geq 0$ for any column x; this condition is equivalent to the fact that all eigenvalues of A are nonnegative. In the construction of the polar decomposition (16.1) we have proved that for any nonnegative definite matrix A there exists a unique nonnegative definite matrix S such that $A = S^2$. This statement has numerous applications.

19.5.1. THEOREM. *If A is a nonnegative definite matrix and $x^*Ax = 0$ for some x, then $Ax = 0$.*

PROOF. Let $A = S^*S$. Then $0 = x^*Ax = (Sx)^*Sx$; hence, $Sx = 0$. It follows that $Ax = S^*Sx = 0$. □

Now, let us study the properties of eigenvalues of products of two Hermitian matrices, one of which is positive definite. First of all, observe that the product of two Hermitian matrices A and B is an Hermitian matrix if and only if $AB = (AB)^* = B^*A^* = BA$. Nevertheless the product of two positive definite matrices is somewhat similar to a positive definite matrix: it is a diagonalizable matrix with positive eigenvalues.

19.5.2. THEOREM. *Let A be a positive definite matrix, B an Hermitian matrix. Then AB is a diagonalizable matrix and the number of its positive, negative and zero eigenvalues is the same as that of B.*

PROOF. Let $A = S^2$, where S is an Hermitian matrix. Then the matrix AB is similar to the matrix $S^{-1}ABS = SBS$. For any invertible Hermitian matrix S if $x = Sy$, then $x^*Bx = y^*(SBS)y$ and, therefore, the matrices B and SBS correspond to the same Hermitian form only expressed in different bases. But the dimension of maximal subspaces on which an Hermitian form is positive definite, negative definite, or identically vanishes is well defined for an Hermitian form. Therefore, A is similar to an Hermitian matrix SBS which has the same number of positive, negative and zero eigenvalues as B. □

A theorem in a sense inverse to Theorem 19.5.2 is also true.

19.5.3. THEOREM. *Any diagonalizable matrix with real eigenvalues can be represented as the product of a positive definite matrix and an Hermitian matrix.*

PROOF. Let $C = PDP^{-1}$, where D is a real diagonal matrix. Then $C = AB$, where $A = PP^*$ is a positive definite matrix and $B = P^{*-1}DP^{-1}$ an Hermitian matrix. □

Problems

19.1. Prove that any Hermitian matrix of rank r can be represented as the sum of r Hermitian matrices of rank 1.

19.2. Prove that if a matrix A is positive definite, then adj A is also a positive definite matrix.

19.3. Prove that if A is a nonzero Hermitian matrix, then $\operatorname{rank} A \geq \dfrac{(\operatorname{tr} A)^2}{\operatorname{tr}(A^2)}$.

19.4. Let A be a positive definite matrix. Prove that

$$\int_{-\infty}^{\infty} e^{-x^TAx}dx = (\sqrt{\pi})^n|A|^{-1/2},$$

where n is the order of the matrix.

19.5. Prove that if the rank of a symmetric (or Hermitian) matrix A is equal to r, then it has a nonzero principal r-minor.

19.6. Let S be a symmetric invertible matrix of order n all elements of which are positive. What is the largest possible number of nonzero elements of S^{-1}?

20. Simultaneous diagonalization of a pair of Hermitian forms

20.1. THEOREM. *Let A and B be Hermitian matrices and let A be positive definite. Then there exists a matrix T such that $T^*AT = I$ and T^*BT is a diagonal matrix.*

PROOF. For A there exists a matrix Y such that $A = Y^*Y$, i.e., $Y^{*-1}AY^{-1} = I$. The matrix $C = Y^{*-1}BY^{-1}$ is Hermitian and, therefore, there exists a unitary matrix U such that U^*CU is diagonal. Since $U^*IU = I$, $T = Y^{-1}U$ is the desired matrix. □

It is not always possible to reduce simultaneously a pair of Hermitian forms to diagonal form by a change of basis. For instance, consider the Hermitian forms corresponding to matrices $\begin{pmatrix} 1 & 0 \\ 0 & 0 \end{pmatrix}$ and $\begin{pmatrix} 0 & 1 \\ 1 & 0 \end{pmatrix}$. Let $P = \begin{pmatrix} a & b \\ c & d \end{pmatrix}$ be an arbitrary invertible matrix. Then $P^*\begin{pmatrix} 1 & 0 \\ 0 & 0 \end{pmatrix}P = \begin{pmatrix} a\overline{a} & \overline{a}b \\ a\overline{b} & b\overline{b} \end{pmatrix}$ and $P^*\begin{pmatrix} 0 & 1 \\ 1 & 0 \end{pmatrix}P = \begin{pmatrix} a\overline{c} + \overline{a}c & \overline{a}d + b\overline{c} \\ a\overline{d} + \overline{b}c & b\overline{d} + \overline{b}d \end{pmatrix}$. It remains to verify that the equalities $\overline{a}b = 0$ and $\overline{a}d + b\overline{c} = 0$ cannot hold simultaneously. If $\overline{a}b = 0$ and P is invertible, then either $a = 0$ and $b \neq 0$ or $b = 0$ and $a \neq 0$. In the first case $0 = \overline{a}d + b\overline{c} = b\overline{c}$ and therefore, $c = 0$; in the second case $\overline{a}d = 0$ and, therefore, $d = 0$. In either case we get a noninvertible matrix P.

20.2. Simultaneous diagonalization. If A and B are Hermitian matrices and one of them is invertible, the following criterion for simultaneous reduction of the forms x^*Ax and x^*Bx to diagonal form is known.

20.2.1. THEOREM. *Hermitian forms x^*Ax and x^*Bx, where A is an invertible Hermitian matrix, are simultaneously reducible to diagonal form if and only if the matrix $A^{-1}B$ is diagonalizable and all its eigenvalues are real.*

PROOF. First, suppose that $A = P^*D_1P$ and $B = P^*D_2P$, where D_1 and D_2 are diagonal matrices. Then $A^{-1}B = P^{-1}D_1^{-1}D_2P$ is a diagonalizable matrix. It is also clear that the matrices D_1 and D_2 are real since $y^*D_iy \in \mathbb{R}$ for any column $y = Px$.

Now, suppose that $A^{-1}B = PDP^{-1}$, where $D = \operatorname{diag}(\lambda_1, \dots, \lambda_n)$ and $\lambda_i \in \mathbb{R}$. Then $BP = APD$ and, therefore, $P^*BP = (P^*AP)D$. Applying a permutation matrix if necessary we can assume that $D = \operatorname{diag}(\Lambda_1, \dots, \Lambda_k)$ is a block diagonal matrix, where $\Lambda_i = \lambda_i I$ and all numbers λ_i are distinct. Let us represent in the same block form the matrices $P^*BP = \left\|B_{ij}\right\|_1^k$ and $P^*AP = \left\|A_{ij}\right\|_1^k$. Since they are Hermitian, $B_{ij} = B_{ji}^*$ and $A_{ij} = A_{ji}^*$. On the other hand, $B_{ij} = \lambda_j A_{ij}$; hence, $\lambda_j A_{ij} = B_{ji}^* = \overline{\lambda}_i A_{ji}^* = \lambda_i A_{ij}$. Therefore, $A_{ij} = 0$ for $i \neq j$, i.e., $P^*AP = \operatorname{diag}(A_1, \dots, A_k)$, where $A_i^* = A_i$ and $P^*BP = \operatorname{diag}(\lambda_1 A_1, \dots, \lambda_k A_k)$. Every matrix A_i can be represented in the form $A_i = U_i D_i U_i^*$, where U_i is a unitary matrix and D_i a diagonal matrix. Let $U = \operatorname{diag}(U_1, \dots, U_k)$ and $T = PU$. Then $T^*AT = \operatorname{diag}(D_1, \dots, D_k)$ and $T^*BT = \operatorname{diag}(\lambda_1 D_1, \dots, \lambda_k D_k)$. □

There are also known certain sufficient conditions for simultaneous diagonalizability of a pair of Hermitian forms if both forms are singular.

20.2.2. THEOREM ([Newcomb, 1961]). *If Hermitian matrices A and B are nonpositive or nonnegative definite, then there exists an invertible matrix T such that T^*AT and T^*BT are diagonal.*

PROOF. Let $\operatorname{rank} A = a$, $\operatorname{rank} B = b$ and $a \le b$. There exists an invertible matrix T_1 such that $T_1^*AT_1 = \begin{pmatrix} I_a & 0 \\ 0 & 0 \end{pmatrix} = A_0$. Consider the last $n-a$ diagonal elements of $B_1 = T_1^*BT_1$. The matrix B_1 is sign-definite and, therefore, if a diagonal element of it is zero, then the whole row and column in which it is situated are zero (see Problem 20.1). Now let some of the diagonal elements considered be nonzero. It is easy to verify that

$$\begin{pmatrix} I & x^* \\ 0 & \overline{\alpha} \end{pmatrix}\begin{pmatrix} C & c^* \\ c & \gamma \end{pmatrix}\begin{pmatrix} I & 0 \\ x & \alpha \end{pmatrix} = \begin{pmatrix} * & \alpha c^* + \alpha\gamma x^* \\ \overline{\alpha}c + \overline{\alpha}\gamma x & |\alpha|^2\gamma \end{pmatrix}.$$

If $\gamma \ne 0$, then setting $\alpha = 1/\sqrt{\gamma}$ and $x = -(1/\gamma)c$ we get a matrix whose off-diagonal elements in the last row and column are zero. These transformations preserve A_0; let us prove that these transformations reduce B_1 to the form

$$B_0 = \begin{pmatrix} B_a & 0 & 0 \\ 0 & I_k & 0 \\ 0 & 0 & 0 \end{pmatrix},$$

where B_a is a matrix of size $a \times a$ and $k = b - \operatorname{rank} B_a$. Take a permutation matrix P such that the transformation $B_1 \mapsto P^*B_1P$ affects only the last $n-a$ rows and columns of B_1 and such that this transformation puts the nonzero diagonal elements (from the last $n-a$ diagonal elements) first. Then with the help of transformations indicated above we start with the last nonzero element and gradually shrinking the size of the considered matrix we eventually obtain a matrix of size $a \times a$.

Let T_2 be an invertible matrix such that $T_2^*BT_2 = B_0$ and $T_2^*AT_2 = A_0$. There exists a unitary matrix U of order a such that U^*B_aU is a diagonal matrix. Since $U^*I_aU = I_a$, it follows that $T = T_2U_1$ with $U_1 = \begin{pmatrix} U & 0 \\ 0 & I \end{pmatrix}$, is the required matrix. $\square$

20.2.3. THEOREM ([Majindar, 1963]). *Let A and B be Hermitian matrices and let there be no nonzero column x such that $x^*Ax = x^*Bx = 0$. Then there exists an invertible matrix T such that T^*AT and T^*BT are diagonal matrices.*

Since any triangular Hermitian matrix is diagonal, Theorem 20.2.3 is a particular case of the following statement.

20.2.4. THEOREM. *Let A and B be arbitrary complex square matrices and there is no nonzero column x such that $x^*Ax = x^*Bx = 0$. Then there exists an invertible matrix T such that T^*AT and T^*BT are triangular matrices.*

PROOF. If one of the matrices A and B, say B, is invertible, then $p(\lambda) = |A - \lambda B|$ is a nonconstant polynomial. If the both matrices are noninvertible, then $|A - \lambda B| = 0$ for $\lambda = 0$. In either case the equation $|A - \lambda B| = 0$ has a root λ and, therefore, there exists a column x_1 such that $Ax_1 = \lambda Bx_1$. If $\lambda \ne 0$ (resp. $\lambda = 0$) select linearly independent columns $x_2, \dots, x_n$ such that $x_i^* A x_1 = 0$ (resp. $x_i^* B x_1 = 0$) for $i = 2, \dots, n$; in either case $x_i^* A x_1 = x_i^* B x_1 = 0$ for $i = 2, \dots, n$. Indeed, if $\lambda \neq 0$, then $x_i^* A x_1 = 0$ and $x_i^* B x_1 = \lambda^{-1} x_i^* A x_1 = 0$; if $\lambda = 0$, then $x_i^* B x_1 = 0$ and $x_i^* A x_1 = 0$, since $A x_1 = 0$.

Therefore, if D is formed by columns $x_1, \dots, x_n$, then

$$D^* A D = \begin{pmatrix} x_1^* A x_1 & \dots & x_1^* A x_n \\ 0 & & \\ \vdots & A_1 & \end{pmatrix} \quad \text{and } D^* B D = \begin{pmatrix} x_1^* B x_1 & \dots & x_1^* B x_n \\ 0 & & \\ \vdots & B_1 & \end{pmatrix}.$$

Let us prove that D is invertible, i.e., that it is impossible to express the column x_1 linearly in terms of $x_2, \dots, x_n$. Indeed, assume for a moment that $x_1 = \lambda_2 x_2 + \dots + \lambda_n x_n$. Then

$$x_1^* A x_1 = (\overline{\lambda}_2 x_2^* + \dots + \overline{\lambda}_n x_n^*) A x_1 = 0.$$

Similarly, $x_1^* B x_1 = 0$; a contradiction. Hence, D is invertible.

Now, let us prove that the matrices A_1 and B_1 satisfy the hypothesis of the theorem. Suppose there exists a nonzero column $y_1 = (\alpha_2, \dots, \alpha_n)^T$ such that $y_1^* A_1 y_1 = y_1^* B_1 y_1 = 0$. As is easy to verify, $A_1 = D_1^* A D_1$ and $B_1 = D_1^* B D_1$, where D_1 is the matrix formed by the columns $x_2, \dots, x_n$. Therefore, $y^* A y = y^* B y$, where $y = D_1 y_1 = \alpha_2 x_2 + \dots + \alpha_n x_n \neq 0$, since the columns $x_2, \dots, x_n$ are linearly independent. Contradiction.

If there exists an invertible matrix T_1 such that $T_1^* A_1 T_1$ and $T_1^* B T_1$ are triangular, then the matrix $T = D \begin{pmatrix} 1 & 0 \\ 0 & T_1 \end{pmatrix}$ is a required one. For matrices of order 1 the statement is obvious and, therefore, we may use induction on the order of the matrix. □

Problems

20.1. An Hermitian matrix $A = \left\|a_{ij}\right\|_1^n$ is nonnegative definite and $a_{ii} = 0$ for some i. Prove that $a_{ij} = a_{ji} = 0$ for all j.

20.2 ([Albert, 1958]). Symmetric matrices A_i and B_i $(i = 1, 2)$ are such that the characteristic polynomials of the matrices $xA_1 + yA_2$ and $xB_1 + yB_2$ are equal for all numbers x and y. Is there necessarily an orthogonal matrix U such that $UA_iU^T = B_i$ for $i = 1, 2$?

21. Skew-symmetric matrices

A matrix A is said to be *skew-symmetric* if $A^T = -A$. In this section we consider real skew-symmetric matrices. Recall that the determinant of a skew-symmetric matrix of odd order vanishes since $|A^T| = |A|$ and $|-A| = (-1)^n|A|$, where n is the order of the matrix.

21.1.1. THEOREM. *If A is a skew-symmetric matrix, then A^2 is a symmetric nonpositive definite matrix.*

PROOF. We have $(A^2)^T = (A^T)^2 = (-A)^2 = A^2$ and $x^T A^2 x = -x^T A^T A x = -(Ax)^T Ax \leq 0$. □

COROLLARY. *The nonzero eigenvalues of a skew-symmetric matrix are purely imaginary.*

Indeed, if $Ax = \lambda x$, then $A^2x = \lambda^2 x$ and $\lambda^2 \leq 0$.

21.1.2. THEOREM. *The condition* $x^TAx = 0$ *holds for all* x *if and only if* A *is a skew-symmetric matrix.*

PROOF.

$$x^TAx = \sum_{i,j} a_{ij}x_ix_j = \sum_{i\leq j}(a_{ij}+a_{ji})x_ix_j.$$

This quadratic form vanishes for all x if and only if all its coefficients are zero, i.e., $a_{ij}+a_{ji}=0$. □

21.2. A bilinear function $B(x,y) = \sum_{i,j} a_{ij}x_iy_j$ is said to be *skew-symmetric* if $B(x,y) = -B(y,x)$. In this case

$$\sum_{i,j}(a_{ij}+a_{ji})x_iy_j = B(x,y)+B(y,x) \equiv 0,$$

i.e., $a_{ij} = -a_{ji}$.

THEOREM. *A skew-symmetric bilinear function can be reduced by a change of basis to the form*

$$\sum_{k=1}^{r}(x_{2k-1}y_{2k} - x_{2k}y_{2k-1}).$$

PROOF. Let, for instance, $a_{12} \neq 0$. Instead of x_2 and y_2, introduce variables $x_2' = a_{12}x_2 + \dots + a_{1n}x_n$ and $y_2' = a_{12}y_2 + \dots + a_{1n}y_n$. Then

$$B(x,y) = x_1y_2' - x_2'y_1 + (c_3x_3 + \dots + c_nx_n)y_2' - (c_3y_3 + \dots + c_ny_n)x_2' + \dots .$$

Instead of x_1 and y_1, introduce new variables $x_1' = x_1 + c_3x_3 + \dots + c_nx_n$ and $y_1' = y_1 + c_3y_3 + \dots + c_ny_n$. Then $B(x,y) = x_1'y_2' - x_2'y_1' + \dots$ (dots stand for the terms involving the variables x_i and y_i with $i \geq 3$). For the variables $x_3, x_4, \dots, y_3, y_4, \dots$ we can repeat the same procedure. □

COROLLARY. *The rank of a skew-symmetric matrix is an even number.*

The elements a_{ij}, where $i<j$, can be considered as independent variables. Then the proof of the theorem shows that

$$A = P^TJP, \text{ where } J = \operatorname{diag}\left(\begin{pmatrix}0 & 1\\ -1 & 0\end{pmatrix}, \dots, \begin{pmatrix}0 & 1\\ -1 & 0\end{pmatrix}\right)$$

and the elements of P are rational functions of a_{ij}. Taking into account that

$$\begin{pmatrix}0 & 1\\ -1 & 0\end{pmatrix} = \begin{pmatrix}0 & 1\\ 1 & 0\end{pmatrix}\begin{pmatrix}-1 & 0\\ 0 & 1\end{pmatrix}$$

we can represent J as the product of matrices J_1 and J_2 with equal determinants. Therefore, $A = (P^TJ_1)(J_2P) = FG$, where the elements of F and G are rational functions of the elements of A and $\det F = \det G$.

21.3. A linear operator A in Euclidean space is said to be *skew-symmetric* if its matrix is skew-symmetric with respect to an orthonormal basis.

THEOREM. *Let* $\Lambda_i = \begin{pmatrix} 0 & -\lambda_i \\ \lambda_i & 0 \end{pmatrix}$. *For a skew-symmetric operator* A *there exists an orthonormal basis with respect to which its matrix is of the form*

$$\operatorname{diag}(\Lambda_1, \dots, \Lambda_k, 0, \dots, 0).$$

PROOF. The operator A^2 is symmetric nonnegative definite. Let

$$V_\lambda = \{v \in V \mid A^2 v = -\lambda^2 v\}.$$

Then $V = \oplus V_\lambda$ and $AV_\lambda \subset V_\lambda$. If $A^2 v = 0$, then $(Av, Av) = -(A^2 v, v) = 0$, i.e., $Av = 0$. Therefore, it suffices to select an orthonormal basis in V_0.

For $\lambda > 0$ the restriction of A to V_λ has no real eigenvalues and the square of this restriction is equal to $-\lambda^2 I$. Let $x \in V_\lambda$ be a unit vector, $y = \lambda^{-1} Ax$. Then

$$(x, y) = (x, \lambda^{-1} Ax) = 0, \quad Ay = -\lambda x,$$
$$(y, y) = (\lambda^{-1} Ax, y) = \lambda^{-1}(x, -Ay) = (x, x) = 1.$$

To construct an orthonormal basis in V_λ take a unit vector $u \in V_\lambda$ orthogonal to x and y. Then $(Au, x) = (u, -Ax) = 0$ and $(Au, y) = (u, -Ay) = 0$. Further details of the construction of an orthonormal basis in V_λ are obvious. □

Problems

21.1. Prove that if A is a real skew-symmetric matrix, then $I + A$ is an invertible matrix.

21.2. An invertible matrix A is skew-symmetric. Prove that A^{-1} is also a skew-symmetric matrix.

21.3. Prove that all roots of the characteristic polynomial of AB, where A and B are skew-symmetric matrices of order $2n$, are of multiplicity greater than 1.

22. Orthogonal matrices. The Cayley transformation

A real matrix A is said to be *an orthogonal* if $AA^T = I$. This equation means that the rows of A constitute an orthonormal system. Since $A^T A = A^{-1}(AA^T)A = I$, it follows that the columns of A also constitute an orthonormal system.

A matrix A is orthogonal if and only if $(Ax, Ay) = (x, A^T Ay) = (x, y)$ for any x, y.

An orthogonal matrix is unitary and, therefore, the absolute value of its eigenvalues is equal to 1.

22.1. The eigenvalues of an orthogonal matrix belong to the unit circle centered at the origin and the eigenvalues of a skew-symmetric matrix belong to the imaginary axis. The fractional-linear transformation $f(z) = \dfrac{1-z}{1+z}$ sends the unit circle to the imaginary axis and $f(f(z)) = z$. Therefore, we may expect that the map

$$f(A) = (I - A)(I + A)^{-1}$$

sends orthogonal matrices to skew-symmetric ones and the other way round. This map is called *Cayley transformation* and our expectations are largely true. Set

$$A^{\#} = (I - A)(I + A)^{-1}.$$

We can verify the identity $(A^{\#})^{\#} = A$ in a way similar to the proof of the identity $f(f(z)) = z$; in the proof we should take into account that all matrices that we encounter in the process of this transformation commute with each other.

THEOREM. *The Cayley transformation sends any skew-symmetric matrix to an orthogonal one and any orthogonal matrix A for which $|A + I| \neq 0$ to a skew-symmetric one.*

PROOF. Since $I - A$ and $I + A$ commute, it does not matter from which side to divide and we can write the Cayley transformation as follows: $A^{\#} = \dfrac{I - A}{I + A}$. If $AA^T = I$ and $|I + A| \neq 0$, then

$$(A^{\#})^T = \frac{I - A^T}{I + A^T} = \frac{I - A^{-1}}{I + A^{-1}} = \frac{A - I}{A + I} = -A^{\#}.$$

If $A^T = -A$, then

$$(A^{\#})^T = \frac{I - A^T}{I + A^T} = \frac{I + A}{I - A} = (A^{\#})^{-1}. \quad \square$$

REMARK. The Cayley transformation can be expressed in the form

$$A^{\#} = (2I - (I + A))(I + A)^{-1} = 2(I + A)^{-1} - I.$$

22.2. If U is an orthogonal matrix and $|U + I| \neq 0$, then

$$U = (I - X)(I + X)^{-1} = 2(I + X)^{-1} - I,$$

where $X = U^{\#}$ is a skew-symmetric matrix.

If S is a symmetric matrix, then $S = U\Lambda U^T$, where Λ is a diagonal matrix and U an orthogonal matrix. If $|U + I| \neq 0$, then

$$S = (2(I + X)^{-1} - I)\Lambda(2(I + X)^{-1} - I)^T,$$

where $X = U^{\#}$.

Let us prove that similar formulas are also true when $|U + I| = 0$.

22.2.1. THEOREM ([Hsu, 1953]). *For an arbitrary square matrix A there exists a matrix $J = \operatorname{diag}(\pm 1, \dots, \pm 1)$ such that $|A + J| \neq 0$.*

PROOF. Let n be the order of A. For $n = 1$ the statement is obvious. Suppose that the statement holds for any A of order $n - 1$ and consider a matrix A of order n. Let us express A in the block form $A = \begin{pmatrix} A_1 & A_2 \\ A_3 & a \end{pmatrix}$, where A_1 is a matrix of order $n - 1$. By inductive hypothesis there exists a matrix $J_1 = \operatorname{diag}(\pm 1, \dots, \pm 1)$ such that $|A_1 + J_1| \neq 0$. Therefore,

$$\begin{vmatrix} A_1 + J_1 & A_2 \\ A_3 & a + 1 \end{vmatrix} - \begin{vmatrix} A_1 + J_1 & A_2 \\ A_3 & a - 1 \end{vmatrix} = 2|A_1 + J_1| \neq 0$$

and, therefore, at least one of the determinants on the left-hand side is nonzero. $\square$

COROLLARY. *For an orthogonal matrix U there exists a skew-symmetric matrix X and a diagonal matrix $J = \operatorname{diag}(\pm 1, \dots, \pm 1)$ such that $U = J(I - X)(I + X)^{-1}$.*

PROOF. There exists a matrix $J = \operatorname{diag}(\pm 1, \dots, \pm 1)$ such that $|U + J| \neq 0$. Clearly, $J^2 = I$. Hence, $|JU + I| \neq 0$ and, therefore, $JU = (I - X)(I + X)^{-1}$, where $X = (JU)^{\#}$. □

22.2.2. THEOREM ([Hsu, 1953]). *Any symmetric matrix S can be reduced to the diagonal form with the help of an orthogonal matrix U such that $|U + I| \neq 0$.*

PROOF. Let $S = U_1 \Lambda U_1^T$. Theorem 22.2.1 shows that there exists a matrix $J = \operatorname{diag}(\pm 1, \dots, \pm 1)$ with $|U_1 + J| \neq 0$. Then $|U_1 J + I| \neq 0$. Let $U = U_1 J$. Clearly,

$$U \Lambda U^T = U_1 J \Lambda J U_1^T = U_1 \Lambda U_1^T = S. \quad \square$$

COROLLARY. *For any symmetric matrix S there exists a skew-symmetric matrix X and a diagonal matrix Λ such that*

$$S = (2(I + X)^{-1} - I)\Lambda(2(I + X)^{-1} - I)^T.$$

Problems

22.1. Prove that if $p(\lambda)$ is the characteristic polynomial of an orthogonal matrix of order n, then $\lambda^n p(\lambda^{-1}) = \pm p(\lambda)$.

22.2. Prove that any unitary matrix of order 2 with determinant 1 is of the form $\begin{pmatrix} u & v \\ -\bar{v} & \bar{u} \end{pmatrix}$, where $|u|^2 + |v|^2 = 1$.

22.3. The determinant of an orthogonal matrix A of order 3 is equal to 1.

a) Prove that $(\operatorname{tr} A)^2 - \operatorname{tr}(A)^2 = 2 \operatorname{tr} A$.

b) Prove that $(\sum_i a_{ii} - 1)^2 + \sum_{i<j}(a_{ij} - a_{ji})^2 = 4$.

22.4. Let J be an invertible matrix. A matrix A is said to be *J-orthogonal* if $A^T J A = J$, i.e., $A^T = J A^{-1} J^{-1}$ and *J-skew-symmetric* if $A^T J = -JA$, i.e., $A^T = -J A J^{-1}$. Prove that the Cayley transformation sends J-orthogonal matrices into J-skew-symmetric ones and the other way around.

22.5 ([Djoković, 1971]). Suppose the absolute values of all eigenvalues of an operator A are equal to 1 and $|Ax| \leq |x|$ for all x. Prove that A is a unitary operator.

22.6 ([Zassenhaus, 1961]). A unitary operator U sends some nonzero vector x to a vector Ux orthogonal to x. Prove that any arc of the unit circle that contains all eigenvalues of U is of length no less than π.

23. Normal matrices

A linear operator A over $\mathbb{C}$ is said to be *normal* if $A^* A = A A^*$; the matrix of a normal operator in an orthonormal basis is called a *normal* matrix. Clearly, a matrix A is normal if and only if $A^* A = A A^*$.

The following conditions are equivalent to A being a normal operator:

1) $A = B + iC$, where B and C are commuting Hermitian operators (cf. Theorem 10.3.4);

2) $A = U \Lambda U^*$, where U is a unitary and Λ is a diagonal matrix, i.e., A has an orthonormal eigenbasis; cf. 17.1;

3) $\sum_{i=1}^n |\lambda_i|^2 = \sum_{i,j=1}^n |a_{ij}|^2$, where $\lambda_1, \dots, \lambda_n$ are eigenvalues of A, cf. Theorem 34.1.1.

23.1.1. THEOREM. *If A is a normal operator, then* $\operatorname{Ker} A^* = \operatorname{Ker} A$ *and* $\operatorname{Im} A^* = \operatorname{Im} A$.

PROOF. The conditions $A^*x = 0$ and $Ax = 0$ are equivalent, since

$$(A^*x, A^*x) = (x, AA^*x) = (x, A^*Ax) = (Ax, Ax).$$

The condition $A^*x = 0$ means that $(x, Ay) = (A^*x, y) = 0$ for all y, i.e., $x \in (\operatorname{Im} A)^\perp$. Therefore, $\operatorname{Im} A = (\operatorname{Ker} A^*)^\perp$ and $\operatorname{Im} A^* = (\operatorname{Ker} A)^\perp$. Since $\operatorname{Ker} A = \operatorname{Ker} A^*$, we have $\operatorname{Im} A = \operatorname{Im} A^*$. $\square$

COROLLARY. *If A is a normal operator, then*

$$V = \operatorname{Ker} A \oplus (\operatorname{Ker} A)^\perp = \operatorname{Ker} A \oplus \operatorname{Im} A.$$

23.1.2. THEOREM. *An operator A is normal if and only if any eigenvector of A is an eigenvector of A^*.*

PROOF. It is easy to verify that if A is a normal operator, then the operator $A - \lambda I$ is also normal and, therefore, $\operatorname{Ker}(A - \lambda I) = \operatorname{Ker}(A^* - \overline{\lambda} I)$, i.e., any eigenvector of A is an eigenvector of A^*.

Now, suppose that any eigenvector of A is an eigenvector of A^*. Let $Ax = \lambda x$ and $(y, x) = 0$. Then

$$(x, Ay) = (A^*x, y) = (\mu x, y) = \mu(x, y) = 0.$$

Take an arbitrary eigenvector e_1 of A. We can restrict A to the subspace $\operatorname{Span}(e_1)^\perp$. In this subspace take an arbitrary eigenvector e_2 of A, etc. Finally, we get an orthonormal eigenbasis of A and, therefore, A is a normal operator. $\square$

23.2. THEOREM. *If A is a normal matrix, then A^* can expressed as a polynomial of A.*

PROOF. Let $A = U\Lambda U^*$, where $\Lambda = \operatorname{diag}(\lambda_1, \dots, \lambda_n)$ and U is a unitary matrix. Then $A^* = U\Lambda^* U^*$, where $\Lambda^* = \operatorname{diag}(\overline{\lambda}_1, \dots, \overline{\lambda}_n)$. There exists an interpolation polynomial p such that $p(\lambda_i) = \overline{\lambda}_i$ for $i = 1, \dots, n$, see Appendix 3. Then

$$p(\Lambda) = \operatorname{diag}(p(\lambda_1), \dots, p(\lambda_n)) = \operatorname{diag}(\overline{\lambda}_1, \dots, \overline{\lambda}_n) = \Lambda^*.$$

Therefore, $p(A) = Up(\Lambda)U^* = U\Lambda^*U^* = A^*$. $\square$

COROLLARY. *If A and B are normal matrices and $AB = BA$, then $A^*B = BA^*$ and $AB^* = B^*A$; in particular, AB is a normal matrix.*

Problems

23.1. Let A be a normal matrix. Prove that there exists a normal matrix B such that $A = B^2$.

23.2. Let A and B be normal operators such that $\operatorname{Im} A \perp \operatorname{Im} B$. Prove that $A + B$ is a normal operator.

23.3. Prove that the matrix A is normal if and only if $A^* = AU$, where U is a unitary matrix.

23.4. Prove that if A is a normal operator and $A = SU$ is its polar decomposition, then $SU = US$.

23.5. The matrices A, B and AB are normal. Prove that BA is also normal.

24. Nilpotent matrices

24.1. A square matrix A is said to be *nilpotent* if $A^p = 0$ for some $p > 0$.

24.1.1. THEOREM. *If the order of a nilpotent matrix A is equal to n, then $A^n = 0$.*

PROOF. Select the largest positive integer p for which $A^p \neq 0$. Then $A^p x \neq 0$ for some x and $A^{p+1} = 0$. Let us prove that the vectors $x, Ax, \dots, A^p x$ are linearly independent. Suppose that $A^k x = \sum_{i>k} \lambda_i A^i x$, where $k < p$. Then $A^{p-k}(A^k x) = A^p x \neq 0$ but $A^{p-k}(\lambda_i A^i x) = 0$ since $i > k$. Contradiction. Hence, $p < n$. $\square$

24.1.2. THEOREM. *The characteristic polynomial of a nilpotent matrix A of order n is equal to λ^n.*

PROOF. The polynomial λ^n annihilates A and, therefore, the minimal polynomial of A is equal to λ^m, where $0 \leq m \leq n$, and the characteristic polynomial of A is equal to λ^n. $\square$

24.1.3. THEOREM. *Let A be a nilpotent matrix, and let k be the maximal order of Jordan blocks of A. Then $A^k = 0$ and $A^{k-1} \neq 0$.*

PROOF. Let N be the Jordan block of order m corresponding to the zero eigenvalue. Then there exists a basis $e_1, \dots, e_m$ such that $Ne_i = e_{i+1}$; hence, $N^p e_i = e_{i+p}$ (we assume that $e_{i+p} = 0$ for $i + p > m$). Thus, $N^m = 0$ and $N^{m-1} e_1 = e_m$, i.e., $N^{m-1} \neq 0$. $\square$

24.2.1. THEOREM. *Let A be a matrix of order n. The matrix A is nilpotent if and only if* $\operatorname{tr}(A^p) = 0$ *for* $p = 1, \dots, n$.

PROOF. Let us prove that the matrix A is nilpotent if and only if all its eigenvalues are zero. To this end, reduce A to the Jordan normal form. Suppose that A has nonzero eigenvalues $\lambda_1, \dots, \lambda_k$; let n_i be the sum of the orders of the Jordan blocks corresponding to the eigenvalue λ_i. Then $\operatorname{tr}(A^p) = n_1\lambda_1^p + \dots + n_k\lambda_k^p$. Since $k \leq n$, it suffices to prove that the conditions

$$n_1\lambda_1^p + \dots + n_k\lambda_k^p = 0 \quad (p = 1, \dots, k)$$

cannot hold. These conditions can be viewed as a system of equations for $n_1, \dots, n_k$. The determinant of this system is a Vandermonde determinant. It does not vanish and, therefore, $n_1 = \dots = n_k = 0$. $\square$

24.2.2. THEOREM. *Let $A : V \to V$ be a linear operator and W an invariant subspace, i.e., $AW \subset W$; let $A_1 : W \to W$ and $A_2 : V/W \to V/W$ be the operators induced by A. If operators A_1 and A_2 are nilpotent, then so is A.*

PROOF. Let $A_1^p = 0$ and $A_2^q = 0$. The condition $A_2^q = 0$ means that $A^q V \subset W$ and the condition $A_1^p = 0$ means that $A^p W = 0$. Therefore, $A^{p+q} V \subset A^p W = 0$. $\square$

24.3. The Jordan normal form of a nilpotent matrix A is a block diagonal matrix with Jordan blocks $J_{n_1}(0), \dots, J_{n_k}(0)$ on the diagonal with $n_1 + \dots + n_k = n$, where n is the order of A. We may assume that $n_1 \geq \dots \geq n_k$. The set $(n_1, \dots, n_k)$ is called a *partition of the number* n. To a partition $(n_1, \dots, n_k)$ we can assign the *Young tableau*

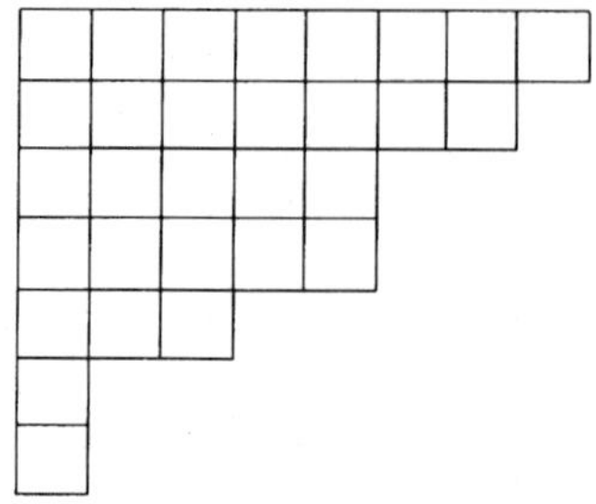

FIGURE 5

consisting of n cells, with n_i cells in the ith row and the first cells of all rows are situated in the first column, see Figure 5.

Clearly, nilpotent matrices are similar if and only if the corresponding Young tableau coincide.

The dimension of $\operatorname{Ker} A^m$ can be expressed in terms of the partition $(n_1, \dots, n_k)$. It is easy to check that

$$\begin{aligned} \dim \operatorname{Ker} A &= k = \operatorname{Card}\{j | n_j \geq 1\}, \\ \dim \operatorname{Ker} A^2 &= \dim \operatorname{Ker} A + \operatorname{Card}\{j | n_j \geq 2\}, \\ &\dots\dots\dots\dots\dots\dots\dots\dots \\ \dim \operatorname{Ker} A^m &= \dim \operatorname{Ker} A^{m-1} + \operatorname{Card}\{j | n_j \geq m\}. \end{aligned}$$

The partition $(n'_1, \dots, n'_l)$, where $n'_i = \operatorname{Card}\{j | n_j \geq i\}$, is called the *dual to the partition* $(n_1, \dots, n_k)$. Young tableaux of dual partitions of a number n are obtained from each other by transposition similar to a transposition of a matrix. If the partition $(n_1, \dots, n_k)$ corresponds to a nilpotent matrix A, then $\dim \operatorname{Ker} A^m = n'_1 + \dots + n'_m$.

Problems

24.1. Let A and B be two matrices of order n. Prove that if $A + \lambda B$ is a nilpotent matrix for $n + 1$ distinct values of λ, then A and B are nilpotent matrices.

24.2. Find matrices A and B such that $\lambda A + \mu B$ is nilpotent for any λ and μ but there exists no matrix P such that $P^{-1}AP$ and $P^{-1}BP$ are triangular matrices.

25. Projections. Idempotent matrices

25.1. An operator $P : V \to V$ is called a *projection* (or *idempotent*) if $P^2 = P$.

25.1.1. THEOREM. *In a certain basis, the matrix of a projection P is of the form* $\operatorname{diag}(1, \dots, 1, 0, \dots, 0)$.

PROOF. Any vector $v \in V$ can be represented in the form $v = Pv + (v - Pv)$, where $Pv \in \operatorname{Im} P$ and $v - Pv \in \operatorname{Ker} P$. Besides, if $x \in \operatorname{Im} P \cap \operatorname{Ker} P$, then $x = 0$. Indeed, in this case $x = Py$ and $Px = 0$ and, therefore, $0 = Px = P^2 y = Py = x$. Hence, $V = \operatorname{Im} P \oplus \operatorname{Ker} P$. For a basis of V select the union of bases of $\operatorname{Im} P$ and $\operatorname{Ker} P$. In this basis the matrix of P is of the required form. $\square$

25.1.1.1. COROLLARY. *There exists a one-to-one correspondence between projections and decompositions* $V = W_1 \oplus W_2$. *To every such decomposition there corresponds the projection* $P(w_1 + w_2) = w_1$, *where* $w_1 \in W_1$ *and* $w_2 \in W_2$, *and to every projection there corresponds a decomposition* $V = \operatorname{Im} P \oplus \operatorname{Ker} P$.

The operator P can be called the *projection onto* W_1 *parallel to* W_2.

25.1.1.2. COROLLARY. *If* P *is a projection, then* $\operatorname{rank} P = \operatorname{tr} P$.

25.1.2. THEOREM. *If* P *is a projection, then* $I - P$ *is also a projection; besides,* $\operatorname{Ker}(I - P) = \operatorname{Im} P$ *and* $\operatorname{Im}(I - P) = \operatorname{Ker} P$.

PROOF. If $P^2 = P$, then $(I-P)^2 = I - 2P + P^2 = I - P$. According to the proof of Theorem 25.1.1 $\operatorname{Ker} P$ consists of vectors $v - Pv$, i.e., $\operatorname{Ker} P = \operatorname{Im}(I - P)$. Similarly, $\operatorname{Ker}(I - P) = \operatorname{Im} P$. $\square$

COROLLARY. *If* P *is the projection onto* W_1 *parallel to* W_2, *then* $I - P$ *is the projection onto* W_2 *parallel to* W_1.

25.2. Let P be a projection and $V = \operatorname{Im} P \oplus \operatorname{Ker} P$. If $\operatorname{Im} P \perp \operatorname{Ker} P$, then Pv is an orthogonal projection of v onto $\operatorname{Im} P$; cf. 9.3.

25.2.1. THEOREM. *A projection* P *is Hermitian if and only if* $\operatorname{Im} P \perp \operatorname{Ker} P$.

PROOF. If P is Hermitian, then $\operatorname{Ker} P = (\operatorname{Im} P^*)^\perp = (\operatorname{Im} P)^\perp$. Now, suppose that P is a projection and $\operatorname{Im} P \perp \operatorname{Ker} P$. The vectors $x - Px$ and $y - Py$ belong to $\operatorname{Ker} P$; therefore, $(Px, y - Py) = 0$ and $(x - Px, Py) = 0$, i.e., $(Px, y) = (Px, Py) = (x, Py)$. $\square$

REMARK. If a projection P is Hermitian, then $(Px, y) = (Px, Py)$; in particular, $(Px, x) = |Px|^2$.

25.2.2. THEOREM. *A projection* P *is Hermitian if and only if* $|Px| \le |x|$ *for all* x.

PROOF. If the projection P is Hermitian, then $x - Px \perp x$ and, therefore, $|x|^2 = |Px|^2 + |Px - x|^2 \ge |Px|^2$. Thus, if $|Px| \le |x|$, then $\operatorname{Ker} P \perp \operatorname{Im} P$.

Now, assume that $v \in \operatorname{Im} P$ is not perpendicular to $\operatorname{Ker} P$ and v_1 is the projection of v on $\operatorname{Ker} P$. Then $|v - v_1| < |v|$ and $v = P(v - v_1)$; therefore, $|v - v_1| < |P(v - v_1)|$. Contradiction. $\square$

Hermitian projections P and Q are said to be *orthogonal* if $\operatorname{Im} P \perp \operatorname{Im} Q$, i.e., $PQ = QP = 0$.

25.2.3. THEOREM. *Let* $P_1, \dots, P_n$ *be Hermitian projections. The operator* $P = P_1 + \dots + P_n$ *is a projection if and only if* $P_i P_j = 0$ *for* $i \ne j$.

PROOF. If $P_i P_j = 0$ for $i \ne j$, then

$$P^2 = (P_1 + \dots + P_n)^2 = P_1^2 + \dots + P_n^2 = P_1 + \dots + P_n = P.$$

Now, suppose that $P = P_1 + \dots + P_n$ is a projection. This projection is Hermitian and, therefore, if $x = P_i x$, then

$$\begin{aligned} |x|^2 = |P_i x|^2 &\le |P_1 x|^2 + \dots + |P_n x|^2 \\ &= (P_1 x, x) + \dots + (P_n x, x) = (Px, x) = |Px|^2 \le |x|^2. \end{aligned}$$

Hence, $P_j x = 0$ for $i \ne j$, i.e., $P_j P_i = 0$. $\square$

25.3. Let $W \subset V$ and let $a_1, \dots, a_k$ be a basis of W. Consider the matrix A of size $n \times k$ whose columns are the coordinates of the vectors $a_1, \dots, a_k$ with respect to an orthonormal basis of V. Then $\operatorname{rank} A^*A = \operatorname{rank} A = k$, and, therefore, A^*A is invertible.

The orthogonal projection Pv of v on W can be expressed with the help of A. Indeed, on the one hand, $Pv = x_1a_1 + \dots + x_ka_k$, i.e., $Pv = Ax$, where x is the column $(x_1, \dots, x_k)^T$. On the other hand, $Pv - v \perp W$, i.e., $A^*(v - Ax) = 0$. Hence, $x = (A^*A)^{-1}A^*v$ and, therefore, $Pv = Ax = A(A^*A)^{-1}A^*v$, i.e., $P = A(A^*A)^{-1}A^*$.

If the basis $a_1, \dots, a_k$ is orthonormal, then $A^*A = I$ and, therefore, $P = AA^*$.

25.4.1. THEOREM ([Djoković, 1971]). *Let $V = V_1 \oplus \dots \oplus V_k$, where $V_i \neq 0$ for $i = 1, \dots, k$, and let $P_i : V \to V_i$ be orthogonal projections, $A = P_1 + \dots + P_k$. Then $0 < |A| \leq 1$, and $|A| = 1$ if and only if $V_i \perp V_j$ whenever $i \neq j$.*

First, let us prove two lemmas. In what follows P_i denotes the orthogonal projection to V_i and $P_{ij} : V_i \to V_j$ is the restriction of P_j onto V_i.

25.4.1.1. LEMMA. *Let $V = V_1 \oplus V_2$ and $V_i \neq 0$. Then $0 < |I - P_{12}P_{21}| \leq 1$ and the equality takes place if and only if $V_1 \perp V_2$.*

PROOF. The operators P_1 and P_2 are nonnegative definite and, therefore, the operator $A = P_1 + P_2$ is also nonnegative definite. Besides, if $Ax = P_1x + P_2x = 0$, then $P_1x = P_2x = 0$, since $P_1x \in V_1$ and $P_2x \in V_2$. Hence, $x \perp V_1$ and $x \perp V_2$ and, therefore, $x = 0$. Hence, A is positive definite and $|A| > 0$.

For a basis of V take the union of bases of V_1 and V_2. In these bases, the matrix of A is of the form $\begin{pmatrix} I & P_{21} \\ P_{12} & I \end{pmatrix}$. Consider the matrix $B = \begin{pmatrix} I & 0 \\ P_{12} & I - P_{12}P_{21} \end{pmatrix}$. As is easy to verify, $|I - P_{12}P_{21}| = |B| = |A| > 0$. Now, let us prove that the absolute value of each of the eigenvalues of $I - P_{12}P_{21}$ (i.e., of the restriction B to V_2) does not exceed 1. Indeed, if $x \in V_2$, then

$$\begin{aligned} |Bx|^2 = (Bx, Bx) &= (x - P_2P_1x, x - P_2P_1x) \\ &= |x|^2 - (P_2P_1x, x) - (x, P_2P_1x) + |P_2P_1x|^2. \end{aligned}$$

Since

$$(P_2P_1x, x) = (P_1x, P_2x) = (P_1x, x) = |P_1x|^2, \qquad (x, P_2P_1x) = |P_1x|^2$$

and $|P_1x|^2 \geq |P_2P_1x|^2$, it follows that

$$|Bx|^2 \leq |x|^2 - |P_1x|^2. \tag{1}$$

The absolute value of any eigenvalue of $I - P_{12}P_{21}$ does not exceed 1 and the determinant of this operator is positive; therefore, $0 < |I - P_{12}P_{21}| \leq 1$.

If $|I - P_{12}P_{21}| = 1$, then all eigenvalues of $I - P_{12}P_{21}$ are equal to 1 and, therefore, taking (1) into account we see that this operator is unitary; cf. Problem 22.1. Hence, $|Bx| = |x|$ for any $x \in V_2$. Taking (1) into account once again, we get $|P_1x| = 0$, i.e., $V_2 \perp V_1$. $\square$

25.4.1.2. Lemma. *Let $V = V_1 \oplus V_2$, where $V_i \neq 0$ and let H be an Hermitian operator such that* $\operatorname{Im} H = V_1$ *and* $H_1 = H|_{V_1}$ *is positive definite. Then* $0 < |H + P_2| \leq |H_1|$ *and the equality takes place if and only if* $V_1 \perp V_2$.

Proof. For a basis of V take the union of bases of V_1 and V_2. In this basis the matrix of $H + P_2$ is of the form $\begin{pmatrix} H_1 & H_1P_{21} \\ P_{12} & I \end{pmatrix}$. Indeed, since $\operatorname{Ker} H = (\operatorname{Im} H^*)^\perp = (\operatorname{Im} H)^\perp = V_1^\perp$, then $H = HP_1$; hence, $H|_{V_2} = H_1P_{21}$. It is easy to verify that

$$\begin{vmatrix} H_1 & H_1P_{21} \\ P_{12} & I \end{vmatrix} = \begin{vmatrix} H_1 & 0 \\ P_{12} & I - P_{12}P_{21} \end{vmatrix} = |H_1| \cdot |I - P_{12}P_{21}|.$$

It remains to make use of Lemma 25.4.1.1. □

Proof of Theorem 25.4.1. As in the proof of Lemma 25.4.1.1, we can show that $|A| > 0$. The proof of the inequality $|A| \leq 1$ will be carried out by induction on k. For $k = 1$ the statement is obvious. For $k > 1$ consider the space $W = V_1 \oplus \dots \oplus V_{k-1}$. Let

$$Q_i = P_i|_W \ (i = 1, \dots, k-1), \quad H_1 = Q_1 + \dots + Q_{k-1}.$$

By the inductive hypothesis $|H_1| \leq 1$; besides, $|H_1| > 0$. Applying Lemma 25.4.1.2 to $H = P_1 + \dots + P_{k-1}$ we get $0 < |A| = |H + P_k| \leq |H_1| \leq 1$.

If $|A| = 1$, then by Lemma 25.4.1.2 $V_k \perp W$. Besides, $|H_1| = 1$, hence, $V_i \perp V_j$ for $i, j \leq k - 1$. □

25.4.2. Theorem ([Djoković, 1971]). *Let N_i be a normal operator in V whose nonzero eigenvalues are equal to $\lambda_1^{(i)}, \dots, \lambda_{r_i}^{(i)}$. Further, let $r_1 + \dots + r_k \leq \dim V$. If the nonzero eigenvalues of $N = N_1 + \dots + N_k$ are equal to $\lambda_j^{(i)}$, where $j = 1, \dots, r_i$, then N is a normal operator,* $\operatorname{Im} N_i \perp \operatorname{Im} N_j$ *and $N_iN_j = 0$ for $i \neq j$.*

Proof. Let $V_i = \operatorname{Im} N_i$. Since $\operatorname{rank} N = \operatorname{rank} N_1 + \dots + \operatorname{rank} N_k$, it follows that $W = V_1 + \dots + V_k$ is the direct sum of these subspaces. For a normal operator $\operatorname{Ker} N_i = (\operatorname{Im} N_i)^\perp$, and so $\operatorname{Ker} N_i \subset W^\perp$; hence, $\operatorname{Ker} N \subset W^\perp$. It is also clear that $\dim \operatorname{Ker} N = \dim W^\perp$. Therefore, without loss of generality we may confine ourselves to a subspace W and assume that $r_1 + \dots + r_k = \dim V$, i.e., $\det N \neq 0$.

Let $M_i = N_i|_{V_i}$. For a basis of V take the union of bases of the spaces V_i. Since $N = \sum N_i = \sum N_iP_i = \sum M_iP_i$, in this basis the matrix of N takes the form

$$\begin{pmatrix} M_1P_{11} & \dots & M_1P_{k1} \\ \vdots & \vdots & \vdots \\ M_kP_{1k} & \dots & M_kP_{kk} \end{pmatrix} = \begin{pmatrix} M_1 & \dots & 0 \\ \vdots & \vdots & \vdots \\ 0 & \dots & M_k \end{pmatrix} \begin{pmatrix} P_{11} & \dots & P_{k1} \\ \vdots & \vdots & \vdots \\ P_{1k} & \dots & P_{kk} \end{pmatrix}.$$

The condition on the eigenvalues of the operators N_i and N implies $|N - \lambda I| = \prod_{i=1}^k |M_i - \lambda I|$. In particular, for $\lambda = 0$ we have $|N| = \prod_{i=1}^k |M_i|$. Hence,

$$\begin{vmatrix} P_{11} & \dots & P_{k1} \\ \vdots & \vdots & \vdots \\ P_{1k} & \dots & P_{kk} \end{vmatrix} = 1, \text{ i.e., } |P_1 + \dots + P_k| = 1.$$

Applying Theorem 25.4.1 we see that $V = V_1 \oplus \dots \oplus V_k$ is the direct sum of orthogonal subspaces. Therefore, N is a normal operator, cf. 17.1, and $N_iN_j = 0$, since $\operatorname{Im} N_j \subset (\operatorname{Im} N_i)^\perp = \operatorname{Ker} N_i$. □

Problems

25.1. Let P_1 and P_2 be projections. Prove that
a) $P_1 + P_2$ is a projection if and only if $P_1P_2 = P_2P_1 = 0$;
b) $P_1 - P_2$ is a projection if and only if $P_1P_2 = P_2P_1 = P_2$.
25.2. Find all matrices of order 2 that are projections.
25.3 (The ergodic theorem). Let A be a unitary operator. Prove that

$$\lim_{n\to\infty} \frac{1}{n}\sum_{i=0}^{n-1} A^i x = Px,$$

where P is an Hermitian projection onto $\operatorname{Ker}(A - I)$.
25.4. The operators $A_1, \dots, A_k$ in a space V of dimension n are such that $A_1 + \dots + A_k = I$. Prove that the following conditions are equivalent:
(a) the operators A_i are projections;
(b) $A_iA_j = 0$ for $i \neq j$;
(c) $\operatorname{rank} A_1 + \dots + \operatorname{rank} A_k = n$.

26. Involutions

26.1. A linear operator A is called an *involution* if $A^2 = I$. As is easy to verify, an operator P is a projection if and only if the operator $2P - I$ is an involution. Indeed, the equation

$$I = (2P - I)^2 = 4P^2 - 4P + I$$

is equivalent to the equation $P^2 = P$.

THEOREM. *Any involution takes the form* $\operatorname{diag}(\pm 1, \dots, \pm 1)$ *in some basis.*

PROOF. If A is an involution, then $P = (A + I)/2$ is a projection; this projection takes the form $\operatorname{diag}(1, \dots, 1, 0, \dots, 0)$ in a certain basis, cf. Theorem 25.1.1. In the same basis the operator $A = 2P - I$ takes the form $\operatorname{diag}(1, \dots, 1, -1, \dots, -1)$. $\square$

REMARK. Using the decomposition $x = \frac{1}{2}(x - Ax) + \frac{1}{2}(x + Ax)$ we can prove that $V = \operatorname{Ker}(A + I) \oplus \operatorname{Ker}(A - I)$.

26.2. THEOREM ([Djoković, 1967]). *A matrix A can be represented as the product of two involutions if and only if the matrices A and A^{-1} are similar.*

PROOF. If $A = ST$, where S and T are involutions, then $A^{-1} = TS = S(ST)S = SAS^{-1}$.

Now, suppose that the matrices A and A^{-1} are similar. The Jordan normal form of A is of the form $\operatorname{diag}(J_1, \dots, J_k)$ and, therefore, $\operatorname{diag}(J_1, \dots, J_k) \sim \operatorname{diag}(J_1^{-1}, \dots, J_k^{-1})$. If J is a Jordan block, then the matrix J^{-1} is similar to a Jordan block. Therefore, the matrices $J_1, \dots, J_k$ can be separated into two classes: for the matrices from the first class we have $J_\alpha^{-1} \sim J_\alpha$ and for the matrices from the second class we have $J_\alpha^{-1} \sim J_\beta$ and $J_\beta^{-1} \sim J_\alpha$. It suffices to show that a matrix J_α from the first class and the matrix $\operatorname{diag}(J_\alpha, J_\beta)$, where J_α, J_β are from the second class can be represented as products of two involutions.

The characteristic polynomial of a Jordan block coincides with the minimal polynomial and, therefore, if p and q are minimal polynomials of the matrices J_α and J_α^{-1}, respectively, then $q(\lambda) = p(0)^{-1}\lambda^n p(\lambda^{-1})$, where n is the order of J_α (see Problem 13.3).

Let $J_\alpha \sim J_\alpha^{-1}$. Then $p(\lambda) = p(0)^{-1}\lambda^n p(\lambda^{-1})$, i.e.,

$$p(\lambda) = \sum \alpha_i \lambda^{n-i},$$

where $\alpha_0 = 1$ and $\alpha_n \alpha_{n-i} = \alpha_i$. The matrix J_α is similar to a cyclic block and, therefore, there exists a basis $e_1, \dots, e_n$ such that $J_\alpha e_k = e_{k+1}$ for $k \leq n-1$ and

$$J_\alpha e_n = J_\alpha^n e_1 = -\alpha_n e_1 - \alpha_{n-1} e_2 - \dots - \alpha_1 e_n.$$

Let $Te_k = e_{n-k+1}$. Obviously, T is an involution. If $STe_k = J_\alpha e_k$, then $Se_{n-k+1} = e_{k+1}$ for $k \neq n$ and $Se_1 = -\alpha_n e_1 - \dots - \alpha_1 e_n$. Let us verify that S is an involution:

$$\begin{aligned} S^2 e_1 &= \alpha_n(\alpha_n e_1 + \dots + \alpha_1 e_n) - \alpha_{n-1} e_n - \dots - \alpha_1 e_2 \\ &= e_1 + (\alpha_n \alpha_{n-1} - \alpha_1) e_2 + \dots + (\alpha_n \alpha_1 - \alpha_{n-1}) e_n = e_1; \end{aligned}$$

clearly, $S^2 e_i = e_i$ for $i \neq 1$.

Now, consider the case $J_\alpha^{-1} \sim J_\beta$. Let $\sum \alpha_i \lambda^{n-i}$ and $\sum \beta_i \lambda^{n-i}$ be the minimal polynomials of J_α and J_β, respectively. Then

$$\sum \alpha_i \lambda^{n-i} = \beta_n^{-1} \lambda^n \sum \beta_i \lambda^{i-n} = \beta_n^{-1} \sum \beta_i \lambda^i.$$

Hence, $\alpha_{n-i}\beta_n = \beta_i$ and $\alpha_n \beta_n = \beta_0 = 1$. There exist bases $e_1, \dots, e_n$ and $\varepsilon_1, \dots, \varepsilon_n$ such that

$$J_\alpha e_k = e_{k+1}, J_\alpha e_n = -\alpha_n e_1 - \dots - \alpha_1 e_n$$

and

$$J_\beta \varepsilon_{k+1} = \varepsilon_k, J_\beta \varepsilon_1 = -\beta_1 \varepsilon_1 - \dots - \beta_n \varepsilon_n.$$

Let $Te_k = \varepsilon_k$ and $T\varepsilon_k = e_k$. If $\operatorname{diag}(J_\alpha, J_\beta) = ST$, then

$$\begin{aligned} Se_{k+1} &= ST\varepsilon_{k+1} = J_\beta \varepsilon_{k+1} = \varepsilon_k, \\ S\varepsilon_k &= e_{k+1}, \\ Se_1 &= ST\varepsilon_1 = J_\beta \varepsilon_1 = -\beta_1 \varepsilon_1 - \dots - \beta_n \varepsilon_n \end{aligned}$$

and $Se_n = -\alpha_n e_1 - \dots - \alpha_1 e_n$. Let us verify that S is an involution. The equalities $S^2 e_i = e_i$ and $S^2 \varepsilon_j = \varepsilon_j$ are obvious for $i \neq 1$ and $j \neq n$ and we have

$$\begin{aligned} S^2 e_1 &= S(-\beta_1 \varepsilon_1 - \dots - \beta_n \varepsilon_n) = -\beta_1 e_2 - \dots - \beta_{n-1} e_n + \beta_n(\alpha_n e_1 + \dots + \alpha_1 e_n) \\ &= e_1 + (\beta_n \alpha_{n-1} - \beta_1) e_2 + \dots + (\beta_n \alpha_1 - \beta_{n-1}) e_n = e_1. \end{aligned}$$

Similarly, $S^2 \varepsilon_n = \varepsilon_n$. □

COROLLARY. *If B is an invertible matrix and $X^T BX = B$, then X can be represented as the product of two involutions. In particular, any orthogonal matrix can be represented as the product of two involutions.*

PROOF. If $X^T BX = B$, then $X^T = BX^{-1}B^{-1}$, i.e., the matrices X^{-1} and X^T are similar. Besides, the matrices X and X^T are similar for any matrix X. □

Solutions

19.1. Let $S = U\Lambda U^*$, where U is a unitary matrix, $\Lambda = \operatorname{diag}(\lambda_1, \dots, \lambda_r, 0, \dots, 0)$. Then $S = S_1 + \dots + S_r$, where $S_i = U\Lambda_i U^*$, $\Lambda_i = \operatorname{diag}(0, \dots, \lambda_i, \dots, 0)$.

19.2. We can represent A in the form $U\Lambda U^{-1}$, where $\Lambda = \operatorname{diag}(\lambda_1, \dots, \lambda_n)$, $\lambda_i > 0$. Therefore, $\operatorname{adj} A = U(\operatorname{adj}\Lambda)U^{-1}$ and $\operatorname{adj}\Lambda = \operatorname{diag}(\lambda_2 \dots \lambda_n, \dots, \lambda_1 \dots \lambda_{n-1})$.

19.3. Let $\lambda_1, \dots, \lambda_r$ be the nonzero eigenvalues of A. All of them are real and, therefore, $(\operatorname{tr} A)^2 = (\lambda_1 + \dots + \lambda_r)^2 \le r(\lambda_1^2 + \dots + \lambda_r^2) = r \operatorname{tr}(A^2)$.

19.4. Let U be an orthogonal matrix such that $U^{-1}AU = \Lambda$ and $|U| = 1$. Set $x = Uy$. Then $x^T A x = y^T \Lambda y$ and $dx_1 \dots dx_n = dy_1 \dots dy_n$, since the Jacobian of this transformation is equal to $|U|$. Hence,

$$\int_{-\infty}^{\infty} e^{-x^T A x}\, dx = \int_{-\infty}^{\infty} \cdots \int_{-\infty}^{\infty} e^{-\lambda_1 y_1^2 - \dots - \lambda_n y_n^2}\, dy$$

$$= \prod_{i=1}^{n} \int_{-\infty}^{\infty} e^{-\lambda_i y_i^2}\, dy_i = \prod_{i=1}^{n} \sqrt{\frac{\pi}{\lambda_i}} = (\sqrt{\pi})^n |A|^{-\frac{1}{2}}.$$

19.5. Let the columns $i_1, \dots, i_r$ of the matrix A of rank r be linearly independent. Then all columns of A can be expressed linearly in terms of these columns, i.e.,

$$\begin{pmatrix} a_{11} & \dots & a_{1n} \\ \vdots & \vdots & \vdots \\ a_{n1} & \dots & a_{nn} \end{pmatrix} = \begin{pmatrix} a_{1i_1} & \dots & a_{1i_k} \\ \vdots & \vdots & \vdots \\ a_{ni_1} & \dots & a_{ni_k} \end{pmatrix} \begin{pmatrix} x_{11} & \dots & x_{1n} \\ \vdots & \vdots & \vdots \\ x_{i_k 1} & \dots & x_{i_k n} \end{pmatrix}.$$

In particular, for the rows $i_1, \dots, i_r$ of A we get the expression

$$\begin{pmatrix} a_{i_1 1} & \dots & a_{i_1 n} \\ \vdots & \vdots & \vdots \\ a_{i_k 1} & \dots & a_{i_k n} \end{pmatrix} = \begin{pmatrix} a_{i_1 i_1} & \dots & a_{i_1 i_k} \\ \vdots & \vdots & \vdots \\ a_{i_k i_1} & \dots & a_{i_k i_k} \end{pmatrix} \begin{pmatrix} x_{11} & \dots & x_{1n} \\ \vdots & \vdots & \vdots \\ x_{i_k 1} & \dots & x_{i_k n} \end{pmatrix}.$$

Both for a symmetric matrix and for an Hermitian matrix the linear independence of the columns $i_1, \dots, i_r$ implies the linear independence of the rows $i_1, \dots, i_k$ and, therefore,

$$\begin{vmatrix} a_{i_1 i_1} & \dots & a_{i_1 i_k} \\ \vdots & \vdots & \vdots \\ a_{i_k i_1} & \dots & a_{i_k i_k} \end{vmatrix} \neq 0.$$

19.6. The scalar product of the ith row of S by the jth column of S^{-1} vanishes for $i \neq j$. Therefore, every column of S^{-1} contains a positive and a negative element; hence, the number of nonnegative elements of S^{-1} is not less than $2n$ and the number of zero elements does not exceed $n^2 - 2n$.

An example of a matrix S^{-1} with precisely the needed number of zero elements is as follows:

$$S^{-1} = \begin{pmatrix} 1 & 1 & 1 & 1 & 1 & \dots \\ 1 & 2 & 2 & 2 & 2 & \dots \\ 1 & 2 & 1 & 1 & 1 & \dots \\ 1 & 2 & 1 & 2 & 2 & \dots \\ 1 & 2 & 1 & 2 & 1 & \dots \\ \vdots & \vdots & \vdots & \vdots & \vdots & \ddots \end{pmatrix}^{-1} = \begin{pmatrix} 2 & -1 & & & & & \\ -1 & 0 & 1 & & & & \\ & 1 & 0 & -1 & & & \\ & & -1 & 0 & & & \\ & & & & \ddots & & \\ & & & & & 0 & -s \\ & & & & & -s & s \end{pmatrix},$$

where $s = (-1)^n$.

20.1. Let $a_{ii} = 0$ and $a_{ij} \neq 0$. Take a column x such that $x_i = ta_{ij}$, $x_j = 1$, the other elements being zero. Then $x^*Ax = a_{jj} + 2t|a_{ij}|^2$. As t varies from $-\infty$ to $+\infty$ the quantity x^*Ax takes both positive and negative values.

20.2. No, not necessarily. Let $A_1 = B_1 = \operatorname{diag}(0, 1, -1)$; let

$$A_2 = \begin{pmatrix} 0 & \sqrt{2} & 2 \\ \sqrt{2} & 0 & 0 \\ 2 & 0 & 0 \end{pmatrix} \quad \text{and} \quad B_2 = \begin{pmatrix} 0 & 0 & \sqrt{2} \\ 0 & 0 & 2 \\ \sqrt{2} & 2 & 0 \end{pmatrix}.$$

It is easy to verify that

$$|xA_1 + yA_2 + \lambda I| = \lambda^3 - \lambda(x^2 + 6y^2) - 2y^2x = |xB_1 + yB_2 + \lambda I|.$$

Now, suppose there exists an orthogonal matrix U such that $UA_1U^T = B_1 = A_1$ and $UA_2U^T = B_2$. Then $UA_1 = A_1U$ and since A_1 is a diagonal matrix with distinct elements on the diagonal, U is an orthogonal diagonal matrix (see Problem 39.1 a)), i.e., $U = \operatorname{diag}(\lambda, \mu, \nu)$, where $\lambda, \mu, \nu = \pm 1$. Hence,

$$\begin{pmatrix} 0 & 0 & \sqrt{2} \\ 0 & 0 & 2 \\ \sqrt{2} & 2 & 0 \end{pmatrix} = B_2 = UA_2U^T = \begin{pmatrix} 0 & \sqrt{2}\lambda\mu & 2\lambda\mu \\ \sqrt{2}\lambda\mu & 0 & 0 \\ 2\lambda\nu & 0 & 0 \end{pmatrix}.$$

Contradiction.

21.1. The nonzero eigenvalues of A are purely imaginary and, therefore, -1 cannot be its eigenvalue.

21.2. Since $(-A)^{-1} = -A^{-1}$, it follows that $(A^{-1})^T = (A^T)^{-1} = (-A)^{-1} = -A^{-1}$.

21.3. We will repeatedly make use of the fact that for a skew-symmetric matrix A of even order $\dim \operatorname{Ker} A$ is an even number. (Indeed, the rank of a skew-symmetric matrix is an even number, see 21.2.) First, consider the case of the zero eigenvalue, i.e., let us prove that if $\dim \operatorname{Ker} AB \geq 1$, then $\dim \operatorname{Ker} AB \geq 2$. If $|B| = 0$, then $\dim \operatorname{Ker} AB \geq \dim \operatorname{Ker} B \geq 2$. If $|B| \neq 0$, then $\operatorname{Ker} AB = B^{-1} \operatorname{Ker} A$, hence, $\dim \operatorname{Ker} AB \geq 2$.

Now, suppose that $\dim \operatorname{Ker}(AB - \lambda I) \geq 1$ for $\lambda \neq 0$. We will prove that $\dim \operatorname{Ker}(AB - \lambda I) \geq 2$. If $(ABA - \lambda A)u = 0$, then $(AB - \lambda I)Au = 0$, i.e., $AU \subset \operatorname{Ker}(AB - \lambda I)$, where $U = \operatorname{Ker}(ABA - \lambda A)$. Therefore, it suffices to prove that $\dim AU \geq 2$. Since $\operatorname{Ker} A \subset U$, it follows that $\dim AU = \dim U - \dim \operatorname{Ker} A$. The matrix ABA is skew-symmetric; thus, the numbers $\dim U$ and $\dim \operatorname{Ker} A$ are even; hence, $\dim AU$ is an even number.

It remains to verify that $\operatorname{Ker} A \neq U$. Suppose that $(AB - \lambda I)Ax = 0$ implies that $Ax = 0$. Then $\operatorname{Im} A \cap \operatorname{Ker}(AB - \lambda I) = 0$. On the other hand, if $(AB - \lambda I)x = 0$, then $x = A(\lambda^{-1}Bx) \in \operatorname{Im} A$, i.e., $\operatorname{Ker}(AB - \lambda I) \subset \operatorname{Im} A$ and $\dim \operatorname{Ker}(AB - \lambda I) \geq 1$. Contradiction.

22.1. The roots of $p(\lambda)$ are such that if z is a root, then $\dfrac{1}{z} = \dfrac{\overline{z}}{z\overline{z}} = \overline{z}$ is also a root. Therefore, the polynomial $q(\lambda) = \lambda^n p(\lambda^{-1})$ has the same roots as p (with the same multiplicities). Besides, the constant term of $p(\lambda)$ is equal to ± 1 and, therefore, the leading coefficients of $p(\lambda)$ and $q(\lambda)$ can differ only in sign.

22.2. Let $\begin{pmatrix} a & b \\ c & d \end{pmatrix}$ be a unitary matrix with determinant 1. Then $\begin{pmatrix} \overline{a} & \overline{b} \\ \overline{c} & \overline{d} \end{pmatrix} = \begin{pmatrix} a & c \\ b & d \end{pmatrix}^{-1} = \begin{pmatrix} d & -c \\ -b & a \end{pmatrix}$, i.e., $\overline{a} = d$ and $\overline{b} = -c$. Besides, $ad - bc = 1$, i.e., $|a|^2 + |b|^2 = 1$.

22.3. a) A is a rotation through an angle φ and, therefore, $\operatorname{tr} A = 1 + 2\cos\varphi$ and $\operatorname{tr}(A^2) = 1 + 2\cos 2\varphi = 4\cos^2\varphi - 1$.

b) Clearly,

$$\sum_{i<j}(a_{ij} - a_{ji})^2 = \sum_{i\neq j} a_{ij}^2 - 2\sum_{i<j} a_{ij}a_{ji}$$

and

$$\operatorname{tr}(A^2) = \sum_i a_{ii}^2 + 2\sum_{i<j} a_{ij}a_{ji}.$$

On the other hand, by a)

$$\operatorname{tr}(A^2) = (\operatorname{tr} A)^2 - 2\operatorname{tr} A = (\operatorname{tr} A - 1)^2 - 1 = \left(\sum_i a_{ii} - 1\right)^2 - 1.$$

Hence, $\sum_{i<j}(a_{ij} - a_{ji})^2 + (\sum_i a_{ii} - 1)^2 - 1 = \sum_{i\neq j} a_{ij}^2 + \sum_i a_{ii}^2 = 3$.

22.4. Set $\frac{A}{B} = AB^{-1}$; then the cancellation rule takes the form: $\frac{AB}{CB} = \frac{A}{C}$. If $A^T = JA^{-1}J^{-1}$, then

$$(A^\#)^T = \frac{I - A^T}{I + A^T} = \frac{I - JA^{-1}J^{-1}}{I + JA^{-1}J^{-1}} = \frac{J(A - I)A^{-1}J^{-1}}{J(A + I)A^{-1}J^{-1}} = \frac{J(A - I)}{J(A + I)} = -JA^\# J^{-1}.$$

If $A^T = -JAJ^{-1}$, then

$$(A^\#)^T = \frac{I - A^T}{I + A^T} = \frac{I + JAJ^{-1}}{I - JAJ^{-1}} = \frac{J(I + A)J^{-1}}{J(I - A)J^{-1}} = J(A^\#)^{-1}J^{-1}.$$

22.5. Since the absolute value of each eigenvalue of A is equal to 1, it suffices to verify that A is unitarily diagonalizable. First, let us prove that A is diagonalizable. Suppose that the Jordan normal form of A has a block of order not less than 2. Then there exist vectors e_1 and e_2 such that $Ae_1 = \lambda e_1$ and $Ae_2 = \lambda e_2 + e_1$. We may assume that $|e_1| = 1$. Consider the vector $x = e_2 - (e_1, e_2)e_1$. It is easy to verify that $x \perp e_1$ and $Ax = \lambda x + e_1$. Hence, $|Ax|^2 = |\lambda x|^2 + |e_1|^2 = |x|^2 + 1$ and, therefore, $|Ax| > |x|$. Contradiction.

It remains to prove that if $Ax = \lambda x$ and $Ay = \mu y$ with $\lambda \neq \mu$, then $(x, y) = 0$. Suppose that $(x, y) \neq 0$. Replacing x by αx, where α is an appropriate complex number, we can assume that $\operatorname{Re}[(\lambda\overline{\mu} - 1)(x, y)] > 0$. Then

$$|A(x + y)|^2 - |x + y|^2 = |\lambda x + \mu y|^2 - |x + y|^2 = 2\operatorname{Re}[(\lambda\overline{\mu} - 1)(x, y)] > 0,$$

i.e., $|Az| > |z|$, where $z = x + y$. Contradiction.

22.6. Let $\lambda_1, \dots, \lambda_n$ be the eigenvalues of an operator U and $e_1, \dots, e_n$ the corresponding pairwise orthogonal eigenvectors. Then $x = \sum x_i e_i$ and $Ux = \sum \lambda_i x_i e_i$; hence, $0 = (Ux, x) = \sum \lambda_i |x_i|^2$. Let $t_i = |x_i|^2 |x|^{-2}$. Since $t_i \geq 0$, $\sum t_i = 1$ and $\sum t_i \lambda_i = 0$, the origin belongs to the interior of the convex hull of $\lambda_1, \dots, \lambda_n$.

23.1. Let $A = U\Lambda U^*$, where U is a unitary matrix, $\Lambda = \operatorname{diag}(\lambda_1, \dots, \lambda_n)$. Set $B = UDU^*$, where $D = \operatorname{diag}(\sqrt{\lambda_1}, \dots, \sqrt{\lambda_n})$.

23.2. By assumption $\operatorname{Im} B \subset (\operatorname{Im} A)^\perp = \operatorname{Ker} A^*$, i.e., $A^*B = 0$. Similarly, $B^*A = 0$. Hence, $(A^* + B^*)(A + B) = A^*A + B^*B$. Since $\operatorname{Ker} A = \operatorname{Ker} A^*$ and $\operatorname{Im} A = \operatorname{Im} A^*$ for a normal operator A, we similarly deduce that $(A + B)(A^* + B^*) = AA^* + BB^*$.

23.3. If $A^* = AU$, where U is a unitary matrix, then $A = U^*A^*$ and, therefore, $UA = UU^*A^* = A^*$. Hence, $AU = UA$ and $A^*A = AUA = AAU = AA^*$.

If A is a normal operator, then there exists an orthonormal eigenbasis $e_1, \dots, e_n$ for A such that $Ae_i = \lambda_i e_i$ and $A^*e_i = \overline{\lambda}_i e_i$. Let $U = \operatorname{diag}(d_1, \dots, d_n)$, where $d_i = \overline{\lambda}_i / \lambda_i$ for $\lambda_i \neq 0$ and $d_i = 1$ for $\lambda_i = 0$. Then $A^* = AU$.

23.4. Consider an orthonormal basis in which A is a diagonal operator. We can assume that $A = \operatorname{diag}(d_1, \dots, d_k, 0, \dots, 0)$, where $d_i \neq 0$. Then

$$S = \operatorname{diag}(|d_1|, \dots, |d_k|, 0, \dots, 0).$$

Let $D = \operatorname{diag}(d_1, \dots, d_k)$ and $D_+ = \operatorname{diag}(|d_1|, \dots, |d_k|)$. The equalities

$$\begin{pmatrix} D & 0 \\ 0 & 0 \end{pmatrix} = \begin{pmatrix} D_+ & 0 \\ 0 & 0 \end{pmatrix} \begin{pmatrix} U_1 & U_2 \\ U_3 & U_4 \end{pmatrix} = \begin{pmatrix} D_+U_1 & D_+U_2 \\ 0 & 0 \end{pmatrix}$$

hold only if $U_1 = D_+^{-1}D = \operatorname{diag}(e^{i\varphi_1}, \dots, e^{i\varphi_k})$ and, therefore, $\begin{pmatrix} U_1 & U_2 \\ U_3 & U_4 \end{pmatrix}$ is a unitary matrix only if $U_2 = 0$ and $U_3 = 0$. Clearly,

$$\begin{pmatrix} D_+ & 0 \\ 0 & 0 \end{pmatrix} \begin{pmatrix} U_1 & 0 \\ 0 & U_4 \end{pmatrix} = \begin{pmatrix} U_1 & 0 \\ 0 & U_4 \end{pmatrix} \begin{pmatrix} D_+ & 0 \\ 0 & 0 \end{pmatrix}.$$

23.5. A matrix X is normal if and only if $\operatorname{tr}(X^*X) = \sum |\lambda_i|^2$, where λ_i are eigenvalues of X; cf. 34.1. Besides, the eigenvalues of $X = AB$ and $Y = BA$ coincide; cf. 11.7. It remains to verify that $\operatorname{tr}(X^*X) = \operatorname{tr}(Y^*Y)$. This is easy to do if we take into account that $A^*A = AA^*$ and $B^*B = BB^*$.

24.1. The matrix $(A + \lambda B)^n$ can be represented in the form

$$(A + \lambda B)^n = A^n + \lambda C_1 + \dots + \lambda^{n-1} C_{n-1} + \lambda^n B^n,$$

where matrices $C_1, \dots, C_{n-1}$ do not depend on λ. Let $a, c_1, \dots, c_{n-1}, b$ be the elements of the matrices $A^n, C_1, \dots, C_{n-1}, B$ occupying the (i, j)th positions. Then $a + \lambda c_1 + \dots + \lambda^{n-1} c_{n-1} + \lambda^n b = 0$ for $n + 1$ distinct values of λ. We have obtained a system of $n + 1$ equations for $n + 1$ unknowns $a, c_1, \dots, c_{n-1}, b$. The determinant of this system is a Vandermonde determinant and, therefore, it is nonzero. Hence, the system obtained has only the zero solution. In particular, $a = b = 0$ and, therefore, $A^n = B^n = 0$.

24.2. Let $A = \begin{pmatrix} 0 & 1 & 0 \\ 0 & 0 & -1 \\ 0 & 0 & 0 \end{pmatrix}$, $B = \begin{pmatrix} 0 & 0 & 0 \\ 1 & 0 & 0 \\ 0 & 1 & 0 \end{pmatrix}$ and $C = \lambda A + \mu B$. As is easy to verify, $C^3 = 0$.

It is impossible to reduce A and B to triangular form simultaneously since AB is not nilpotent.

25.1. a) Since $P_1^2 = P_1$ and $P_2^2 = P_2$, the equality $(P_1 + P_2)^2 = P_1 + P_2$ is equivalent to $P_1P_2 = -P_2P_1$. Multiplying this by P_1 once from the right and once from the left we get $P_1P_2P_1 = -P_2P_1$ and $P_1P_2 = -P_1P_2P_1$, respectively; therefore, $P_1P_2 = P_2P_1 = 0$.

b) Since $I - (P_1 - P_2) = (I - P_1) + P_2$, we deduce that $P_1 - P_2$ is a projection if and only if $(I - P_1)P_2 = P_2(I - P_1) = 0$, i.e., $P_1P_2 = P_2P_1 = P_2$.

25.2. If P is a matrix of order 2 and $\operatorname{rank} P = 1$, then $\operatorname{tr} P = 1$ and $\det P = 0$ (if $\operatorname{rank} P \neq 1$, then $P = I$ or $P = 0$). Hence, $P = \frac{1}{2}\begin{pmatrix} 1+a & b \\ c & 1-a \end{pmatrix}$, where $a^2 + bc = 1$.

It is also clear that if $\operatorname{tr} P = 1$ and $\det P = 0$, then by the Cayley-Hamilton theorem $P^2 - P = P^2 - (\operatorname{tr} P)P + \det P = 0$.

25.3. Since $\operatorname{Im}(I - A) = \operatorname{Ker}((I - A)^*)^\perp$, any vector x can be represented in the form $x = x_1 + x_2$, where $x_1 \in \operatorname{Im}(I - A)$ and $x_2 \in \operatorname{Ker}(I - A^*)$. It suffices to consider, separately, x_1 and x_2. The vector x_1 is of the form $y - Ay$ and, therefore,

$$\left| \frac{1}{n} \sum_{i=0}^{n-1} A^i x_i \right| = \left| \frac{1}{n}(y - A^n y) \right| \leq \frac{2|y|}{n} \to 0 \text{ as } n \to \infty.$$

Since $x_2 \in \operatorname{Ker}(I - A^*)$, it follows that $x_2 = A^*x_2 = A^{-1}x_2$, i.e., $Ax_2 = x_2$. Hence,

$$\lim_{n\to\infty} \frac{1}{n} \sum_{i=0}^{n-1} A^i x_2 = \lim_{n\to\infty} \frac{1}{n} \sum_{i=0}^{n-1} x_2 = x_2.$$

25.4. $(b) \Rightarrow (a)$. It suffices to multiply the identity $A_1 + \dots + A_k = I$ by A_i.

$(a) \Rightarrow (c)$. Since A_i is a projection, $\operatorname{rank} A_i = \operatorname{tr} A_i$. Hence, $\sum \operatorname{rank} A_i = \sum \operatorname{tr} A_i = \operatorname{tr}(\sum A_i) = n$.

$(c) \Rightarrow (b)$. Since $\sum A_i = I$, we have $\operatorname{Im} A_1 + \dots + \operatorname{Im} A_k = V$. But $\operatorname{rank} A_1 + \dots + \operatorname{rank} A_k = \dim V$ and, therefore, $V = \operatorname{Im} A_1 \oplus \dots \oplus \operatorname{Im} A_k$.

For any $x \in V$ we have

$$A_j x = (A_1 + \dots + A_k)A_j x = A_1 A_j x + \dots + A_k A_j x,$$

where $A_i A_j x \in \operatorname{Im} A_i$ and $A_j x \in \operatorname{Im} A_j$. Hence, $A_i A_j = 0$ for $i \neq j$ and $A_j^2 = A_j$.

CHAPTER V

MULTILINEAR ALGEBRA

27. Multilinear maps and tensor products

27.1. Let $V, V_1, \dots, V_k$ be linear spaces; $\dim V_i = n_i$. A map

$$f : V_1 \times \dots \times V_k \longrightarrow V$$

is said to be *multilinear* (or *k-linear*) if it is linear in every of its k variables when the other variables are fixed.

In the spaces $V_1, \dots, V_k$, select bases $\{e_{1i}\}, \dots, \{e_{kj}\}$. If f is a multilinear map, then

$$f\left(\sum x_{1i}e_{1i}, \dots, \sum x_{kj}e_{kj}\right) = \sum x_{1i}\dots x_{kj} f(e_{1i}, \dots, e_{kj}).$$

The map f is determined by the $n_1 \dots n_k$ values $f(e_{1i}, \dots, e_{kj}) \in V$ and these values can be arbitrary. Consider the space $V_1 \otimes \dots \otimes V_k$ of dimension $n_1 \dots n_k$ and a certain basis in it whose elements we will denote by $e_{1i} \otimes \dots \otimes e_{kj}$. Further, consider a map $p : V_1 \times \dots \times V_k \mapsto V_1 \otimes \dots \otimes V_k$ given by the formula

$$p\left(\sum x_{1i}e_{1i}, \dots, \sum x_{kj}e_{kj}\right) = \sum x_{1i}\dots x_{kj} e_{1i} \otimes \dots \otimes e_{kj}$$

and denote the element $p(v_1, \dots, v_k)$ by $v_1 \otimes \dots \otimes v_k$. To every multilinear map f there corresponds a linear map

$$\tilde{f} : V_1 \otimes \dots \otimes V_k \longrightarrow V, \text{ where } \tilde{f}(e_{1i} \otimes \dots \otimes e_{kj}) = f(e_{1i}, \dots, e_{kj}),$$

and this correspondence between multilinear maps f and linear maps $\tilde{f}$ is one-to-one. It is also easy to verify that $\tilde{f}(v_1 \otimes \dots \otimes v_k) = f(v_1, \dots, v_k)$ for any vectors $v_i \in V_i$.

To an element $v_1 \otimes \dots \otimes v_k$ we can assign a multilinear function on $V_1^* \times \dots \times V_k^*$ defined by the formula

$$f(w_1, \dots, w_k) = w_1(v_1) \dots w_k(v_k).$$

If we extend this map by linearity we get an isomorphism of the space $V_1 \otimes \dots \otimes V_k$ with the space of linear functions on $V_1^* \times \dots \times V_k^*$. This gives us an invariant definition of the space $V_1 \otimes \dots \otimes V_k$ and this space is called the *tensor product of the spaces* $V_1, \dots, V_k$.

A linear map $V_1^* \otimes \dots \otimes V_k^* \longrightarrow (V_1 \otimes \dots \otimes V_k)^*$ that sends $f_1 \otimes \dots \otimes f_k \in V_1^* \otimes \dots \otimes V_k^*$ to a multilinear function $f(v_1, \dots, v_k) = f_1(v_1) \dots f_k(v_k)$ is a canonical isomorphism.

27.2.1. THEOREM. *Let* Hom(V, W) *be the space of linear maps* $V \longrightarrow W$. *Then there exists a canonical isomorphism* $\alpha : V^* \otimes W \longrightarrow \text{Hom}(V, W)$.

PROOF. Let $\{e_i\}$ and $\{\varepsilon_j\}$ be bases of V and W. Set

$$\alpha(e_i^* \otimes \varepsilon_j)v = e_i^*(v)\varepsilon_j = v_i\varepsilon_j$$

and extend α by linearity to the whole space. If $v \in V$, $f \in V^*$ and $w \in W$, then $\alpha(f \otimes w)v = f(v)w$ and, therefore, α can be invariantly defined.

Let $Ae_p = \sum_q a_{qp}\varepsilon_q$; then $A(\sum_p v_p e_p) = \sum_{p,q} a_{qp}v_p\varepsilon_q$. Hence, the matrix (a_{qp}), where $a_{qp} = \delta_{qj}\delta_{pi}$, corresponds to the map $\alpha(e_i^* \otimes \varepsilon_j)$. Such matrices constitute a basis of Hom(V, W). It is also clear that the dimensions of $V^* \otimes W$ and Hom(V, W) are equal. $\square$

27.2.2. THEOREM. *Let V be a linear space over a field K. Consider the convolution* $\varepsilon : V^* \otimes V \longrightarrow K$ *given by the formula* $\varepsilon(x^* \otimes y) = x^*(y)$ *and extended to the whole space by linearity. Then* $\operatorname{tr} A = \varepsilon\alpha^{-1}(A)$ *for any linear operator A in V.*

PROOF. Select a basis in V. It suffices to carry out the proof for the matrices $E_{ij} = (a_{pq})$, where $a_{qp} = \delta_{qj}\delta_{pi}$. Clearly, $\operatorname{tr} E_{ij} = \delta_{ij}$ and

$$\varepsilon\alpha^{-1}(E_{ij}) = \varepsilon(e_i^* \otimes e_j) = e_i^*(e_j) = \delta_{ij}. \quad \square$$

REMARK. The space $V^* \otimes V$ and the maps α and ε are invariantly defined. Therefore, Theorem 27.2.2 gives an invariant definition of the trace of a matrix.

27.3. A *tensor of type* (p, q) on V is an element of the space

$$T_p^q(V) = \underbrace{V^* \otimes \cdots \otimes V^*}_{p \text{ factors}} \otimes \underbrace{V \otimes \cdots \otimes V}_{q \text{ factors}}$$

isomorphic to the space of linear functions on $V \times \cdots \times V \times V^* \times \cdots \times V^*$ (with p factors V and q factors V^*). The number p is called the *covariant valency* of the tensor, q its *contravariant valency*, and $p+q$ its *total valency*. *Vectors* are tensors of type $(0, 1)$ and *covectors* are tensors of type $(1, 0)$.

Let a basis $e_1, \dots, e_n$ be selected in V and let $e_1^*, \dots, e_n^*$ be the dual basis of V^*. Each tensor T of type (p, q) is of the form

$$T = \sum T_{i_1 \dots i_p}^{j_1 \dots j_q} e_{i_1}^* \otimes \cdots \otimes e_{i_p}^* \otimes e_{j_1} \otimes \cdots \otimes e_{j_q}; \tag{1}$$

the numbers $T_{i_1 \dots i_p}^{j_1 \dots j_q}$ are called the *coordinates of the tensor T* in the basis $e_1, \dots, e_n$.

Let us establish how coordinates of a tensor change under the passage to another basis. Let $\varepsilon_j = Ae_j = \sum a_{ij}e_i$ and $\varepsilon_j^* = \sum b_{ij}e_i^*$. It is easy to see that $B = (A^T)^{-1}$, cf. 5.3.

Introduce notations $a_j^i = a_{ij}$, $b_i^j = b_{ij}$ and denote the tensor (1) by $\sum T_\alpha^\beta e_\alpha^* \otimes e_\beta$ for brevity. Then

$$\sum T_\alpha^\beta e_\alpha^* \otimes e_\beta = \sum S_\mu^\nu \varepsilon_\mu^* \otimes \varepsilon_\nu = \sum S_\mu^\nu b_\alpha^\mu a_\nu^\beta e_\alpha^* \otimes e_\beta,$$

i.e.,

$$T^{j_1\dots j_q}_{i_1\dots i_p} = b^{l_1}_{i_1}\dots b^{l_p}_{i_p} a^{j_1}_{k_1}\dots a^{j_q}_{k_q} S^{k_1\dots k_q}_{l_1\dots l_p} \tag{2}$$

(here summation over repeated indices is assumed). Formula (2) relates the coordinates S of the tensor in the basis $\{\varepsilon_i\}$ with the coordinates T in the basis $\{e_i\}$.

On tensors of type (1, 1) (which can be identified with linear operators) a convolution is defined; it sends $v^* \otimes w$ to $v^*(w)$. The convolution maps an operator to its trace; cf. Theorem 27.2.2.

Let $1 \le i \le p$ and $1 \le j \le q$. Consider a linear map $T^q_p(V) \longrightarrow T^{q-1}_{p-1}(V)$:

$$f_1 \otimes \dots \otimes f_p \otimes v_1 \otimes \dots \otimes v_q \mapsto f_i(v_j) f_{\hat{\imath}} \otimes v_{\hat{\jmath}},$$

where $f_{\hat{\imath}}$ and $v_{\hat{\jmath}}$ are tensor products of $f_1, \dots, f_p$ and $v_1, \dots, v_q$ with f_i and v_j, respectively, omitted. This map is called the *convolution of a tensor with respect to its ith lower index and jth upper index.*

27.4. Linear maps $A_i : V_i \longrightarrow W_i$, $(i = 1, \dots, k)$ induce a linear map

$$A_1 \otimes \dots \otimes A_k : V_1 \otimes \dots \otimes V_k \longrightarrow W_1 \otimes \dots \otimes W_k,$$
$$e_{1i} \otimes \dots \otimes e_{kj} \mapsto A_1 e_{1i} \otimes \dots \otimes A_k e_{kj}.$$

As is easy to verify, this map sends $v_1 \otimes \dots \otimes v_k$ to $A_1 v_1 \otimes \dots \otimes A_k v_k$. The map $A_1 \otimes \dots \otimes A_k$ is called the *tensor product of operators* $A_1, \dots, A_k$.

If $Ae_j = \sum a_{ij}\varepsilon_i$ and $Be'_q = \sum b_{pq}\varepsilon'_p$, then $A \otimes B(e_j \otimes e'_q) = \sum a_{ij} b_{pq} \varepsilon_i \otimes \varepsilon'_p$. Hence, by appropriately ordering the basis $e_i \otimes e'_q$ and $\varepsilon_i \otimes \varepsilon'_p$ we can express the matrix $A \otimes B$ in either of the forms

$$\begin{pmatrix} a_{11}B & \dots & a_{1n}B \\ \vdots & \ddots & \vdots \\ a_{m1}B & \dots & a_{mn}B \end{pmatrix} \text{ or } \begin{pmatrix} b_{11}A & \dots & b_{1l}A \\ \vdots & \ddots & \vdots \\ b_{k1}A & \dots & b_{kl}A \end{pmatrix}.$$

The matrix $A \otimes B$ is called the *Kronecker product* †of matrices A and B.

The following properties of the Kronecker product are easy to verify:

1) $(A \otimes B)^T = A^T \otimes B^T$;

2) $(A \otimes B)(C \otimes D) = AC \otimes BD$ provided all products are well defined, i.e., the matrices are of agreeble sizes;

3) if A and B are orthogonal matrices, then $A \otimes B$ is an orthogonal matrix;

4) if A and B are invertible matrices, then $(A \otimes B)^{-1} = A^{-1} \otimes B^{-1}$.

Note that properties 3) and 4) follow from properties 1) and 2).

THEOREM. *Let the eigenvalues of matrices A and B be equal to $\alpha_1, \dots, \alpha_m$ and $\beta_1, \dots, \beta_n$, respectively. Then the eigenvalues of $A \otimes B$ are equal to $\alpha_i\beta_j$ and the eigenvalues of $A \otimes I_n + I_m \otimes B$ are equal to $\alpha_i + \beta_j$.*

PROOF. Let us reduce the matrices A and B to their Jordan normal forms (it suffices to reduce them to a triangular form, actually). For a basis in the tensor product of the spaces take the product of the bases which normalize A and B. It remains to notice that $J_p(\alpha) \otimes J_q(\beta)$ is an upper triangular matrix with diagonal $(\alpha\beta, \dots, \alpha\beta)$ and $J_p(\alpha) \otimes I_q$ and $I_p \otimes J_q(\beta)$ are upper triangular matrices whose diagonals are $(\alpha, \dots, \alpha)$ and $(\beta, \dots, \beta)$, respectively. □

COROLLARY. $\det(A \otimes B) = (\det A)^n (\det B)^m$.

27.5. The tensor product of operators can be used to solve matrix equations of the form

$$A_1 X B_1 + \dots + A_s X B_s = C, \tag{1}$$

where

$$V^k \xrightarrow{B_i} V^l \xrightarrow{X} V^m \xrightarrow{A_i} V^n.$$

Let us prove that the natural identifications

$$\mathrm{Hom}(V^l, V^m) = (V^l)^* \otimes V^m \text{ and } \mathrm{Hom}(V^k, V^n) = (V^k)^* \otimes V^n$$

send the map $X \mapsto A_i X B_i$ to $B_i^T \otimes A_i$, i.e., equation (1) takes the form

$$(B_1^T \otimes A_1 + \dots + B_s^T \otimes A_s) X = C,$$

where $X \in (V^l)^* \otimes V^m$ and $C \in (V^k)^* \otimes V^n$. Indeed, if $f \otimes v \in (V^l)^* \otimes V^m$ corresponds to the map $Xx = (f \otimes v)x = f(x)v$, then $B^T f \otimes Av \in (V^k)^* \otimes V^n$ corresponds to the map $(B^T f \otimes Av)y = f(By)Av = AXBy$.

27.5.1. THEOREM. *Let A and B be square matrices. If they have no common eigenvalues, then the equation $AX - XB = C$ has a unique solution for any C. If A and B do have a common eigenvalue, then depending on C this equation either has no solutions or has infinitely many of them.*

PROOF. The equation $AX - XB = C$ can be rewritten as $(I \otimes A - B^T \otimes I)X = C$. The eigenvalues of the operator $I \otimes A - B^T \otimes I$ are equal to $\alpha_i - \beta_j$, where α_i are eigenvalues of A and β_j are eigenvalues of B^T, i.e., eigenvalues of B. The operator $I \otimes A - B^T \otimes I$ is invertible if and only if $\alpha_i - \beta_j \neq 0$ for all i and j. □

27.5.2. THEOREM. *Let A and B be square matrices of the same order. The equation $AX - XB = \lambda X$ has a nonzero solution if and only if $\lambda = \alpha_i - \beta_j$, where α_i and β_j are certain eigenvalues of A and B, respectively.*

PROOF. The equation $(I \otimes A - B^T \otimes I)X = \lambda X$ has a nonzero solution if λ is an eigenvalue of $I \otimes A - B^T \otimes I$, i.e., $\lambda = \alpha_i - \beta_j$. □

27.6. To a multilinear function $f \in \mathrm{Hom}(V \times \dots \times V, K) \cong \otimes^p V^*$ we can assign a subspace $W_f \subset V^*$ spanned by covectors ξ of the form

$$\xi(x) = f(a_1, \dots, a_{i-1}, x, a_i, \dots, a_{p-1}),$$

where the vectors $a_1, \dots, a_{p-1}$ and the number i are fixed.

27.6.1. THEOREM. $f \in \otimes^p W_f$.

PROOF. Let $\varepsilon_1, \dots, \varepsilon_r$ be a basis of W_f. Let us complement it to a basis $\varepsilon_1, \dots, \varepsilon_n$ of V^*. We have to prove that $f = \sum f_{i_1 \dots i_p} \varepsilon_{i_1} \otimes \dots \otimes \varepsilon_{i_p}$, where $f_{i_1 \dots i_p} = 0$ when one of the indices $i_1, \dots, i_p$ is greater than r. Let $e_1, \dots, e_n$ be the basis dual to $\varepsilon_1, \dots, \varepsilon_n$. Then $f(e_{j_1}, \dots, e_{j_p}) = f_{j_1 \dots j_p}$. On the other hand, if $j_k > r$, then

$$f(\dots e_{j_k} \dots) = \lambda_1 \varepsilon_1(e_{j_k}) + \dots + \lambda_r \varepsilon_r(e_{j_k}) = 0. \quad \square$$

27.6.2. THEOREM. *Let* $f = \sum f_{i_1\dots i_p}\varepsilon_{i_1}\otimes\dots\otimes\varepsilon_{i_p}$, *where* $\varepsilon_1,\dots,\varepsilon_r \in V^*$. *Then* $W_f \in \operatorname{Span}(\varepsilon_1,\dots,\varepsilon_r)$.

PROOF. Clearly,

$$f(a_1,\dots,a_{k-1},x,a_k,\dots,a_{p-1})$$
$$= \sum f_{i_1\dots i_p}\varepsilon_{i_1}(a_1)\dots\varepsilon_{i_k}(x)\dots\varepsilon_{i_p}(a_{p-1}) = \sum c_s\varepsilon_s(x). \quad \square$$

Problems

27.1. Prove that $v\otimes w = v'\otimes w' \neq 0$ if and only if $v = \lambda v'$ and $w' = \lambda w$.
27.2. Let $A_i : V_i \longrightarrow W_i$ $(i = 1, 2)$ be linear maps. Prove that
a) $\operatorname{Im}(A_1\otimes A_2) = (\operatorname{Im} A_1)\otimes(\operatorname{Im} A_2)$;
b) $\operatorname{Im}(A_1\otimes A_2) = (\operatorname{Im} A_1\otimes W_2)\cap(W_1\otimes\operatorname{Im} A_2)$;
c) $\operatorname{Ker}(A_1\otimes A_2) = \operatorname{Ker} A_1\otimes W_2 + W_1\otimes\operatorname{Ker} A_2$.
27.3. Let $V_1, V_2 \subset V$ and $W_1, W_2 \subset W$. Prove that

$$(V_1\otimes W_1)\cap(V_2\otimes W_2) = (V_1\cap V_2)\otimes(W_1\cap W_2).$$

27.4. Let V be an Euclidean space and let V^* be canonically identified with V. Prove that the operator $A = I - 2a\otimes a$ is a symmetry with respect to $a^\perp$.
27.5. Let $A(x, y)$ be a bilinear function on a Euclidean space such that if $x \perp y$, then $A(x, y) = 0$. Prove that $A(x, y)$ is proportional to the inner product (x, y).

28. Symmetric and skew-symmetric tensors

28.1. To every permutation $\sigma \in S_q$ we can assign a linear operator

$$f_\sigma\colon T_0^q(V) \longrightarrow T_0^q(V)$$
$$v_1\otimes\dots\otimes v_q \mapsto v_{\sigma(1)}\otimes\dots\otimes v_{\sigma(q)}.$$

A tensor $T \in T_0^q(V)$ said to be *symmetric* (resp. *skew-symmetric*) if $f_\sigma(T) = T$ (resp. $f_\sigma(T) = (-1)^\sigma T$) for any σ. The symmetric tensors constitute a subspace $S^q(V)$ and the skew-symmetric tensors constitute a subspace $\Lambda^q(V)$ in $T_0^q(V)$. Clearly, $S^q(V)\cap\Lambda^q(V) = 0$ for $q \geq 2$.

The operator $S = \frac{1}{q!}\sum_\sigma f_\sigma$ is called the *symmetrization* and $A = \frac{1}{q!}\sum_\sigma(-1)^\sigma f_\sigma$ the *skew-symmetrization* or *alternation*.

28.1.1. THEOREM. *S is the projection of $T_0^q(V)$ onto $S^q(V)$ and A is the projection onto $\Lambda^q(V)$.*

PROOF. Obviously, the symmetrization of any tensor is a symmetric tensor and on symmetric tensors S is the identity operator.

Since

$$f_\sigma(AT) = \frac{1}{q!}\sum_\tau(-1)^\tau f_\sigma f_\tau(T) = (-1)^\sigma\frac{1}{q!}\sum_{\rho=\sigma\tau}(-1)^\rho f_\rho(T) = (-1)^\sigma AT,$$

it follows that $\operatorname{Im} A \subset \Lambda^q(V)$. If T is skew-symmetric, then

$$AT = \frac{1}{q!}\sum_\sigma (-1)^\sigma f_\sigma(T) = \frac{1}{q!}\sum_\sigma (-1)^\sigma(-1)^\sigma T = T. \qquad \square$$

We introduce notations:

$$S(e_{i_1}\otimes\dots\otimes e_{i_q}) = e_{i_1}\dots e_{i_q} \text{ and } A(e_{i_1}\otimes\dots\otimes e_{i_q}) = e_{i_1}\wedge\dots\wedge e_{i_q}.$$

For example, $e_ie_j = \frac{1}{2}(e_i\otimes e_j + e_j\otimes e_i)$ and $e_i\wedge e_j = \frac{1}{2}(e_i\otimes e_j - e_j\otimes e_i)$. If $e_1,\dots,e_n$ is a basis of V, then the tensors $e_{i_1}\dots e_{i_q}$ span $S^q(V)$ and the tensors $e_{i_1}\wedge\dots\wedge e_{i_q}$ span $\Lambda^q(V)$. The tensor $e_{i_1}\dots e_{i_q}$ depends only on the number of times each e_i enters this product and, therefore, we can set $e_{i_1}\dots e_{i_q} = e_1^{k_1}\dots e_n^{k_n}$, where k_i is the multiplicity of occurrence of e_i in $e_{i_1}\dots e_{i_q}$. The tensor $e_{i_1}\wedge\dots\wedge e_{i_q}$ changes sign under the permutation of any two factors e_{i_α} and e_{i_β} and, therefore, $e_{i_1}\wedge\dots\wedge e_{i_q} = 0$ if $e_{i_\alpha} = e_{i_\beta}$; hence, the tensors $e_{i_1}\wedge\dots\wedge e_{i_q}$, where $1\le i_1<\dots<i_q\le n$, span the space $\Lambda^q(V)$. In particular, $\Lambda^q(V) = 0$ for $q > n$.

28.1.2. THEOREM. *The elements* $e_1^{k_1}\dots e_n^{k_n}$, *where* $k_1+\dots+k_n = q$, *form a basis of* $S^q(V)$ *and the elements* $e_{i_1}\wedge\dots\wedge e_{i_q}$, *where* $1\le i_1<\dots<i_q\le n$, *form a basis of* $\Lambda^q(V)$.

PROOF. It suffices to verify that these vectors are linearly independent. If the sets $(k_1,\dots,k_n)$ and $(l_1,\dots,l_n)$ are distinct, then the tensors $e_1^{k_1}\dots e_n^{k_n}$ and $e_1^{l_1}\dots e_n^{l_n}$ are linear combinations of two nonintersecting subsets of basis elements of $T_0^q(V)$. For tensors of the form $e_{i_1}\wedge\dots\wedge e_{i_q}$ the proof is similar. $\square$

COROLLARY. $\dim\Lambda^q(V) = \binom{n}{q}$ *and* $\dim S^q(V) = \binom{n+q-1}{q}$.

PROOF. Clearly, the number of ordered tuples $i_1,\dots,i_n$ such that $1\le i_1<\dots< i_q\le n$ is equal to $\binom{n}{q}$. To compute the number of of ordered tuples $k_1,\dots,k_n$ such that such that $k_1+\dots+k_n = q$, we proceed as follows. To each such set assign a sequence of $q+n-1$ balls among which there are q white and $n-1$ black ones. In this sequence, let k_1 white balls come first, then one black ball followed by k_2 white balls, next one black ball, etc. From $n+q-1$ balls we can select q white balls in $\binom{n+q-1}{q}$ ways. $\square$

28.2. In $\Lambda(V) = \oplus_{q=0}^n\Lambda^q(V)$, we can introduce the *wedge product* setting $T_1\wedge T_2 = A(T_1\otimes T_2)$ for $T_1\in\Lambda^p(V)$ and $T_2\in\Lambda^q(V)$ and extending the operation to $\Lambda(V)$ by linearity. The obtained algebra $\Lambda(V)$ is called the *exterior* or *Grassmann algebra of* V.

THEOREM. *The algebra* $\Lambda(V)$ *is associative and skew-commutative, i.e.,* $T_1\wedge T_2 = (-1)^{pq}T_2\wedge T_1$ *for* $T_1\in\Lambda^p(V)$ *and* $T_2\in\Lambda^q(V)$.

PROOF. Instead of tensors T_1 and T_2 from $\Lambda(V)$ it suffices to consider tensors from the tensor algebra (we will denote them by the same letters) $T_1 = x_1\otimes\dots\otimes x_p$ and $T_2 = x_{p+1}\otimes\dots\otimes x_{p+q}$. First, let us prove that $A(T_1\otimes T_2) = A(A(T_1)\otimes T_2)$. Since

$$A(x_1\otimes\dots\otimes x_p) = \frac{1}{p!}\sum_{\sigma\in S_p}(-1)^\sigma x_{\sigma(1)}\otimes\dots\otimes x_{\sigma(p)},$$

it follows that

$$A(A(T_1)\otimes T_2) = A\left(\frac{1}{p!}\sum_{\sigma\in S_p}(-1)^\sigma x_{\sigma(1)}\otimes\dots\otimes x_{\sigma(p)}\otimes x_{p+1}\otimes\dots\otimes x_{p+q}\right)$$
$$= \frac{1}{p!(p+q)!}\sum_{\sigma\in S_p}\sum_{\tau\in S_{p+q}}(-1)^{\sigma\tau}x_{\tau(\sigma(1))}\otimes\dots\otimes x_{\tau(p+q)}.$$

It remains to notice that

$$\sum_{\sigma\in S_p}(-1)^{\sigma\tau}x_{\tau(\sigma(1))}\otimes\dots\otimes x_{\tau(p+q)} = p!\sum(-1)^{\tau_1}x_{\tau_1(1)}\otimes\dots\otimes x_{\tau_1(p+q)},$$

where $\tau_1 = (\tau(\sigma(1)),\dots,\tau(\sigma(p)),\tau(p+1),\dots,\tau(p+q))$.

We similarly prove that $A(T_1\otimes T_2) = A(T_1\otimes A(T_2))$ and, therefore,

$$(T_1\wedge T_2)\wedge T_3 = A(A(T_1\otimes T_2)\otimes T_3) = A(T_1\otimes T_2\otimes T_3)$$
$$= A(T_1\otimes A(T_2\otimes T_3)) = T_1\wedge(T_2\wedge T_3).$$

Clearly,

$$x_{p+1}\otimes\dots\otimes x_{p+q}\otimes x_1\otimes\dots\otimes x_p = x_{\sigma(1)}\otimes\dots\otimes x_{\sigma(p+q)},$$

where $\sigma = (p+1,\dots,p+q,1,\dots,p)$. To place 1 in the first position, etc. p in the pth position in σ we have to perform pq transpositions. Hence, $(-1)^\sigma = (-1)^{pq}$ and $A(T_1\otimes T_2) = (-1)^{pq}A(T_2\otimes T_1)$. $\square$

In $\Lambda(V)$, the kth power of ω, i.e., $\underbrace{\omega\wedge\dots\wedge\omega}_{k\text{ times}}$ is denoted by $\Lambda^k\omega$; in particular, $\Lambda^0\omega = 1$.

28.3. By a *skew-symmetric function* on $V\times\dots\times V$ we mean a multilinear function $f(v_1,\dots,v_q)$ such that $f(v_{\sigma(1)},\dots,v_{\sigma(q)}) = (-1)^\sigma f(v_1,\dots,v_q)$ for any permutation σ.

THEOREM. *The space $\Lambda^q(V^*)$ is canonically isomorphic to the space $(\Lambda^q V)^*$ and also to the space of skew-symmetric functions on $V\times\dots\times V$.*

PROOF. As is easy to verify

$$(f_1\wedge\dots\wedge f_q)(v_1,\dots,v_q) = A(f_1\otimes\dots\otimes f_q)(v_1,\dots,v_q)$$
$$= \frac{1}{q!}\sum_\sigma(-1)^\sigma f_1(v_{\sigma(1)}),\dots,f_q(v_{\sigma(q)})$$

is a skew-symmetric function. If $e_1,\dots,e_n$ is a basis of V, then a skew-symmetric function f is given by its values $f(e_{i_1},\dots,e_{i_q})$, where $1\le i_1<\dots<i_q\le n$, and each such set of values corresponds to a skew-symmetric function. Therefore, the dimension

of the space of skew-symmetric functions is equal to the dimension of $\Lambda^q(V^*)$; hence, these spaces are isomorphic.

Now, let us construct the canonical isomorphism $\Lambda^q(V^*) \longrightarrow (\Lambda^q V)^*$. A linear map $V^* \otimes \cdots \otimes V^* \longrightarrow (V \otimes \cdots \otimes V)^*$ which sends $(f_1, \dots, f_q) \in V^* \otimes \cdots \otimes V^*$ to a multilinear function $f(v_1, \dots, v_q) = f_1(v_1) \dots f_q(v_q)$ is a canonical isomorphism. Consider the restriction of this map onto $\Lambda^q(V^*)$. The element $f_1 \wedge \cdots \wedge f_q = A(f_1 \otimes \cdots \otimes f_q) \in \Lambda^q(V^*)$ turns into the multilinear function $f(v_1, \dots, v_q) = \frac{1}{q!}\sum_\sigma (-1)^\sigma f_1(v_{\sigma(1)}) \dots f_q(v_{\sigma(q)})$. The function f is skew-symmetric; therefore, we get a map $\Lambda^q(V^*) \longrightarrow (\Lambda^q V)^*$. Let us verify that this map is an isomorphism. To a multilinear function f on $V \times \cdots \times V$ there corresponds, by 27.1, a linear function $\widetilde{f}$ on $V \otimes \cdots \otimes V$. Clearly,

$$\begin{aligned}\widetilde{f}(A(v_1 \otimes \cdots \otimes v_q)) &= \left(\frac{1}{q!}\right)^2 \sum_{\sigma,\tau} (-1)^{\sigma\tau} f_1(v_{\sigma\tau(1)}) \dots f_q(v_{\sigma\tau(q)}) \\ &= \frac{1}{q!} \sum_\sigma (-1)^\sigma f_1(v_{\sigma(1)}) \dots f_q(v_{\sigma(q)}) = \frac{1}{q!} \begin{vmatrix} f_1(v_1) & \dots & f_1(v_q) \\ \vdots & \vdots & \vdots \\ f_q(v_1) & \dots & f_q(v_q) \end{vmatrix}.\end{aligned}$$

Let $e_1, \dots, e_n$ and $\varepsilon_1, \dots, \varepsilon_n$ be dual bases of V and V^*. The elements $e_{i_1} \wedge \cdots \wedge e_{i_q}$ form a basis of $\Lambda^q V$. Consider the dual basis of $(\Lambda^q V)^*$. The above arguments imply that under our restrictions the element $\varepsilon_{i_1} \wedge \cdots \wedge \varepsilon_{i_q}$ turns into a basis element dual to $e_{i_1} \wedge \cdots \wedge e_{i_q}$, multiplied by $(q!)^{-1}$. $\square$

REMARK. As a byproduct we have proved that

$$\widetilde{f}(A(v_1 \otimes \cdots \otimes v_q)) = \frac{1}{q!}\widetilde{f}(v_1 \otimes \cdots \otimes v_q) \quad \text{for } f \in \Lambda^q(V^*).$$

28.4.1. THEOREM. $T_0^2(V) = \Lambda^2(V) \oplus S^2(V)$.

PROOF. It suffices to notice that

$$a \otimes b = \frac{1}{2}(a \otimes b - b \otimes a) + \frac{1}{2}(a \otimes b + b \otimes a). \quad \square$$

28.4.2. THEOREM. *The following canonical isomorphisms take place:*
a) $\Lambda^q(V \oplus W) \cong \oplus_{i=0}^q (\Lambda^i V \otimes \Lambda^{q-i} W)$;
b) $S^q(V \oplus W) \cong \oplus_{i=0}^q (S^i V \otimes S^{q-i} W)$.

PROOF. Clearly, $\Lambda^i V \subset T_0^i(V \oplus W)$ and $\Lambda^{q-i} W \subset T_0^{q-i}(V \oplus W)$. Therefore, there exists a canonical embedding $\Lambda^i V \otimes \Lambda^{q-i} W \subset T_0^q(V \oplus W)$. Let us project $T_0^q(V \oplus W)$ to $\Lambda^q(V \oplus W)$ using alternation. As a result, we get a canonical map

$$\Lambda^i V \otimes \Lambda^{q-i} W \longrightarrow \Lambda^q(V \oplus W)$$

that acts as follows:

$$(v_1 \wedge \cdots \wedge v_i) \otimes (w_1 \wedge \cdots \wedge w_{q-i}) \mapsto v_1 \wedge \cdots \wedge v_i \wedge w_1 \wedge \cdots \wedge w_{q-i}.$$

Selecting bases in V and W, it is easy to verify that the resulting map

$$\bigoplus_{i=0}^{q}(\Lambda^i V \otimes \Lambda^{q-i} W) \longrightarrow \Lambda^q(V \oplus W)$$

is an isomorphism.

For $S^q(V \oplus W)$ the proof is similar. $\square$

28.4.3. THEOREM. *If* $\dim V = n$, *then there exists a canonical isomorphism* $\Lambda^p V \cong (\Lambda^{n-p} V)^* \otimes \Lambda^n V$.

PROOF. The exterior product is a map $\Lambda^p V \times \Lambda^{n-p} V \longrightarrow \Lambda^n V$ and therefore to every element of $\Lambda^p V$ there corresponds a map $\Lambda^{n-p} V \longrightarrow \Lambda^n V$. As a result we get a map

$$\Lambda^p V \longrightarrow \operatorname{Hom}(\Lambda^{n-p} V, \Lambda^n V) \cong (\Lambda^{n-p} V)^* \otimes \Lambda^n V.$$

Let us prove that this map is an isomorphism. Select a basis $e_1, \dots, e_n$ in V. To $e_{i_1} \wedge \dots \wedge e_{i_p}$ there corresponds a map which sends $e_{j_1} \wedge \dots \wedge e_{j_{n-p}}$ to 0 or $\pm e_1 \wedge \dots \wedge e_n$, depending on whether the sets $\{i_1, \dots, i_p\}$ and $\{j_1, \dots, j_{n-p}\}$ intersect or are complementary in $\{1, \dots, n\}$. Such maps constitute a basis in $\operatorname{Hom}(\Lambda^{n-p} V, \Lambda^n V)$. $\square$

28.5. A linear operator $B : V \longrightarrow V$ induces a linear operator $B_q : T_0^q(V) \longrightarrow T_0^q(V)$ which maps $v_1 \otimes \dots \otimes v_q$ to $Bv_1 \otimes \dots \otimes Bv_q$. If $T = v_1 \otimes \dots \otimes v_q$, then $B_q f_\sigma(T) = f_\sigma B_q(T)$ and, therefore,

$$B_q f_\sigma(T) = f_\sigma B_q(T) \text{ for any } T \in T_0^q(V). \tag{1}$$

Consequently, B_q sends symmetric tensors to symmetric ones and skew-symmetric tensors to skew-symmetric ones. The restrictions of B_q to $S^q(V)$ and $\Lambda^q(V)$ will be denoted by $S^q B$ and $\Lambda^q B$, respectively. Let S and A be symmetrization and alternation, respectively. The equality (1) implies that $B_q S = S B_q$ and $B_q A = A B_q$. Hence,

$$B_q(e_1^{k_1} \dots e_n^{k_n}) = (Be_1)^{k_1} \dots (Be_n)^{k_n} \text{ and } B_q(e_{i_1} \wedge \dots \wedge e_{i_q}) = (Be_{i_1}) \wedge \dots \wedge (Be_{i_q}).$$

Introduce the *lexicographic order* on the set of indices $(i_1, \dots, i_q)$, i.e., assume that

$$(i_1, \dots, i_q) < (j_1, \dots, j_q) \text{ if } i_1 = j_1, \dots, i_r = j_r \text{ and } i_{r+1} < j_{r+1} \text{ (or } i_1 < j_1).$$

Let us lexicographically order the basis vectors $e_1^{k_1} \dots e_n^{k_n}$ and $e_{i_1} \wedge \dots \wedge e_{i_q}$.

28.5.1. THEOREM. *Let* $B_q(e_{j_1} \wedge \dots \wedge e_{j_q}) = \sum_{1 \le i_1 < \dots < i_q \le n} b_{j_1 \dots j_q}^{i_1 \dots i_q} e_{i_1} \wedge \dots \wedge e_{i_q}$. *Then* $b_{j_1 \dots j_q}^{i_1 \dots i_q}$ *is equal to the minor* $B\binom{i_1 \dots i_q}{j_1 \dots j_q}$ *of* B.

PROOF. Clearly,

$$\begin{aligned} Be_{j_1} \wedge \dots \wedge Be_{j_q} &= \left(\sum_{i_1} b_{i_1 j_1} e_{i_1}\right) \wedge \dots \wedge \left(\sum_{i_q} b_{i_q j_q} e_{i_q}\right) \\ &= \sum_{i_1, \dots, i_q} b_{i_1 j_1} \dots b_{i_q j_q} e_{i_1} \wedge \dots \wedge e_{i_q} \\ &= \sum_{1 \le i_1 < \dots < i_q \le n} \left(\sum_\sigma (-1)^\sigma b_{i_{\sigma(1)} j_1} \dots b_{i_{\sigma(q)} j_q}\right) e_{i_1} \wedge \dots \wedge e_{i_q}. \quad \square \end{aligned}$$

COROLLARY. *The matrix of operator $\Lambda^q B$ with respect to the lexicographically ordered basis $e_{i_1} \wedge \dots \wedge e_{i_q}$ is the compound matrix $C_q(B)$* (see 2.6).

28.5.2. THEOREM. *If the matrix of an operator B is triangular in the basis $e_1, \dots, e_n$, then the matrices of $S^q B$ and $\Lambda^q B$ are triangular in the lexicographically ordered bases $e_1^{k_1} \dots e_n^{k_n}$ (for $k_1 + \dots + k_n = q$) and $e_{i_1} \wedge \dots \wedge e_{i_q}$ (for $1 \le i_1 < \dots < i_q \le n$).*

PROOF. Let $Be_i \in \operatorname{Span}(e_1, \dots, e_i)$, i.e., $Be_i \le e_i$ with respect to our order. If $i_1 \le j_1, \dots, i_q \le j_q$, then $e_{i_1} \wedge \dots \wedge e_{i_q} \le e_{j_1} \wedge \dots \wedge e_{j_q}$ and $e_{i_1} \dots e_{i_q} = e_1^{k_1} \dots e_n^{k_n} \le e_1^{l_1} \dots e_n^{l_n} = e_{j_1} \dots e_{j_q}$. Hence,

$$\Lambda^q B(e_{i_1} \wedge \dots \wedge e_{i_q}) \le e_{i_1} \wedge \dots \wedge e_{i_q} \text{ and } S^q B(e_1^{k_1} \dots e_n^{k_n}) \le e_1^{k_1} \dots e_n^{k_n}. \quad \square$$

28.5.3. THEOREM. $\det(\Lambda^q B) = (\det B)^p$, *where* $p = \binom{n-1}{q-1}$ *and* $\det(S^q B) = (\det B)^r$, *where* $r = \frac{q}{n}\binom{n+q-1}{q}$.

PROOF. We may assume that B is an operator over $\mathbb{C}$. Let $e_1, \dots, e_n$ be the Jordan basis for B. By Theorem 28.5.2, the matrices of $\Lambda^q B$ and $S^q B$ are triangular in the lexicographically ordered bases $e_{i_1} \wedge \dots \wedge e_{i_q}$ and $e_1^{k_1} \dots e_n^{k_n}$. If a diagonal element λ_i corresponds to e_i, then the diagonal elements $\lambda_{i_1} \dots \lambda_{i_q}$ and $\lambda_1^{k_1} \dots \lambda_n^{k_n}$, where $k_1 + \dots + k_n = q$, correspond to $e_{i_1} \wedge \dots \wedge e_{i_q}$ and $e_1^{k_1} \dots e_n^{k_n}$. Hence, the product of all diagonal elements of the matrices $\Lambda^q B$ and $S^q B$ is a polynomial in λ of total degree $q \dim \Lambda^q(V)$ and $q \dim S^q(V)$, respectively. Hence, $|\Lambda^q B| = |B|^p$ and $|S^q B| = |B|^r$, where $p = \frac{q}{n}\binom{n}{q}$ and $r = \frac{q}{n}\binom{n+q-1}{q}$. $\square$

COROLLARY (Sylvester's identity). *Since $\Lambda^q B = C_q(B)$ is the compound matrix* (see Corollary 28.5.1), $\det(C_q(B)) = (\det B)^p$, *where* $p = \binom{n-1}{q-1}$.

To a matrix B of order n we can assign a polynomial

$$\Lambda_B(t) = 1 + \sum_{q=1}^{n} \operatorname{tr}(\Lambda^q B) t^q$$

and a series

$$S_B(t) = 1 + \sum_{q=1}^{\infty} \operatorname{tr}(S^q B) t^q.$$

28.5.4. THEOREM. $S_B(t) = (\Lambda_B(-t))^{-1}$.

PROOF. As in the proof of Theorem 28.5.3 we see that if B is a triangular matrix with diagonal $(\lambda_1, \dots, \lambda_n)$, then $\Lambda^q B$ and $S^q B$ are triangular matrices with diagonal elements $\lambda_{i_1} \dots \lambda_{i_q}$ and $\lambda_1^{k_1} \dots \lambda_n^{k_n}$, where $k_1 + \dots + k_n = q$. Hence,

$$\Lambda_B(-t) = (1 - t\lambda_1) \dots (1 - t\lambda_n)$$

and

$$S_B(t) = (1 + t\lambda_1 + t^2\lambda_1^2 + \dots) \dots (1 + t\lambda_n + t^2\lambda_n^2 + \dots).$$

It remains to notice that

$$(1 - t\lambda_i)^{-1} = 1 + t\lambda_i + t^2\lambda_i^2 + \dots \quad \square$$

Problems

28.1. A trilinear function f is symmetric with respect to the first two arguments and skew-symmetric with respect to the last two arguments. Prove that $f = 0$.

28.2. Let $f : \mathbb{R}^m \times \mathbb{R}^m \longrightarrow \mathbb{R}^n$ be a symmetric bilinear map such that $f(x, x) \neq 0$ for $x \neq 0$ and $(f(x,x), f(y,y)) \leq |f(x,y)|^2$. Prove that $m \leq n$.

28.3. Let $\omega = e_1 \wedge e_2 + e_3 \wedge e_4 + \dots + e_{2n-1} \wedge e_{2n}$, where $e_1, \dots, e_{2n}$ is a basis of a vector space. Prove that $\Lambda^n \omega = n! e_1 \wedge \dots \wedge e_{2n}$.

28.4. Let A be a matrix of order n. Prove that $\det(A + I) = 1 + \sum_{q=1}^n \operatorname{tr}(\Lambda^q A)$.

28.5. Let d be the determinant of a system of linear equations

$$\left(\sum_{j=1}^n a_{ij} x_j\right)\left(\sum_{q=1}^n a_{pq} x_q\right) = 0, \quad (i, p = 1, \dots, n),$$

where the unknowns are the $(n+1)n/2$ lexicographically ordered quantities $x_i x_j$. Prove that $d = (\det(a_{ij}))^{n+1}$.

28.6. Let $s_k = \operatorname{tr} A^k$ and let σ_k be the sum of the principal minors of order k of the matrix A. Prove that for any positive integer m we have

$$s_m - s_{m-1}\sigma_1 + s_{m-2}\sigma_2 - \dots + (-1)^m m \sigma_m = 0.$$

28.7. Prove the Binet-Cauchy formula with the help of the wedge product.

29. The Pfaffian

29.1. If $A = \left\| a_{ij} \right\|_1^n$ is a skew-symmetric matrix, then $\det A$ is a polynomial in the indeterminates a_{ij}, where $i < j$; let us denote this polynomial by $P(a_{ij})$. For n odd we have $P \equiv 0$ (see Problem 1.1), and if n is even, then A can be represented as $A = XJX^T$, where the elements of X are rational functions in a_{ij} and $J = \operatorname{diag}\left(\begin{pmatrix} 0 & 1 \\ -1 & 0 \end{pmatrix}, \dots, \begin{pmatrix} 0 & 1 \\ -1 & 0 \end{pmatrix}\right)$ (see 21.2). Since $\det X = f(a_{ij})/g(a_{ij})$, where f and g are polynomials, it follows that

$$P = \det(XJX^T) = (f/g)^2.$$

Therefore, $f^2 = Pg^2$, i.e., f^2 is divisible by g^2; hence, f is divisible by g, i.e., $f/g = Q$ is a polynomial. As a result we get $P = Q^2$, where Q is a polynomial in variables a_{ij}, i.e., the determinant of a skew-symmetric matrix considered as a polynomial in a_{ij}, where $i < j$, is a perfect square.

This result can be also obtained by another method which also gives an explicit expression for Q. Let $e_1, \dots, e_{2n}$ be a basis in V. First, let us assign to a skew-symmetric matrix $A = \left\| a_{ij} \right\|_1^{2n}$ the element $\omega = \sum_{i<j} a_{ij} e_i \wedge e_j \in \Lambda^2(V)$ and then to ω assign $\Lambda^n \omega = f(A) e_1 \wedge \dots \wedge e_{2n} \in \Lambda^{2n}(V)$. The function $f(A)$ can be easily expressed in terms of the elements of A and it does not depend on the choice of a basis.

Now, let us express the elements ω and $\Lambda^n \omega$ with respect to a new basis $\varepsilon_j = \sum x_{ij} e_i$. We can verify that $\sum_{i<j} a_{ij} e_i \wedge e_j = \sum_{i<j} b_{ij} \varepsilon_i \wedge \varepsilon_j$, where $A = XBX^T$ and $\varepsilon_1 \wedge \dots \wedge \varepsilon_{2n} = (\det X) e_1 \wedge \dots \wedge e_{2n}$. Hence,

$$f(A) e_1 \wedge \dots \wedge e_{2n} = f(B) \varepsilon_1 \wedge \dots \wedge \varepsilon_{2n} = (\det X) f(B) e_1 \wedge \dots \wedge e_{2n},$$

i.e., $f(XBX^T) = (\det X)f(B)$. If A is an invertible skew-symmetric matrix, then it can be represented in the form

$$A = XJX^T, \quad \text{where} \quad J = \operatorname{diag}\left(\begin{pmatrix} 0 & 1 \\ -1 & 0 \end{pmatrix}, \dots, \begin{pmatrix} 0 & 1 \\ -1 & 0 \end{pmatrix}\right).$$

Hence, $f(A) = f(XJX^T) = (\det X)f(J)$ and $\det A = (\det X)^2 = (f(A)/f(J))^2$.

Let us prove that

$$f(A) = n! \sum_{\sigma} (-1)^{\sigma} a_{i_1 i_2} a_{i_3 i_4} \dots a_{i_{2n-1} i_{2n}},$$

where $\sigma = \binom{1 \ \dots \ 2n}{i_1 \ \dots \ i_{2n}}$ and the summation is taken over all partitions of $\{1, \dots, 2n\}$ into pairs $\{i_k, i_{k+1}\}$, where $i_k < i_{k+1}$ (note that the summation is taken not over all permutations σ, but over partitions!). Let $\omega_{ij} = a_{ij} e_i \wedge e_j$; then $\omega_{ij} \wedge \omega_{kl} = \omega_{kl} \wedge \omega_{ij}$ and $\omega_{ij} \wedge \omega_{kl} = 0$ if some of the indices i, j, k, l coincide. Hence,

$$\begin{aligned} \Lambda^n \left(\sum \omega_{ij}\right) &= \sum \omega_{i_1 i_2} \wedge \dots \wedge \omega_{i_{2n-1} i_{2n}} \\ &= \sum a_{i_1 i_2} \dots a_{i_{2n-1} i_{2n}} e_{i_1} \wedge \dots \wedge e_{i_{2n}} \\ &= \sum (-1)^{\sigma} a_{i_1 i_2} \dots a_{i_{2n-1} i_{2n}} e_1 \wedge \dots \wedge e_{2n} \end{aligned}$$

and precisely $n!$ summands have $a_{i_1 i_2} \dots a_{i_{2n-1} i_{2n}}$ as the coefficient. Indeed, each of the n elements $\omega_{i_1 i_2}, \dots, \omega_{i_{2n-1} i_{2n}}$ can be selected in any of the n factors in $\Lambda^n(\sum \omega_{ij})$ and in each factor we select exactly one such element. In particular, $f(J) = n!$.

The polynomial $\operatorname{Pf}(A) = f(A)/f(J) = \pm\sqrt{\det A}$ considered as a polynomial in the variables a_{ij}, $i < j$, is called the *Pfaffian*. It is easy to verify that for matrices of order 2 and 4, respectively, the Pfaffian is equal to a_{12} and $a_{12}a_{34} - a_{13}a_{24} + a_{14}a_{23}$.

29.2. Let $1 \le \sigma_1 < \dots < \sigma_{2k} \le 2n$. The set $\{\sigma_1, \dots, \sigma_{2k}\}$ can be complemented to the set $\{1, 2, \dots, 2n\}$ by the set $\{\overline{\sigma}_1, \dots, \overline{\sigma}_{2(n-k)}\}$, where $\overline{\sigma}_1 < \dots < \overline{\sigma}_{2(n-k)}$. As a result to the set $\{\sigma_1, \dots, \sigma_{2k}\}$ we have assigned the permutation $\sigma = (\sigma_1 \dots \sigma_{2k} \overline{\sigma}_1 \dots \overline{\sigma}_{2(n-k)})$. It is easy to verify that $(-1)^{\sigma} = (-1)^a$, where $a = (\sigma_1 - 1) + (\sigma_2 - 2) + \dots + (\sigma_{2k} - 2k)$.

The Pfaffian of a submatrix of a skew-symmetric matrix $M = \|m_{ij}\|_1^{2n}$, where $m_{ij} = (-1)^{i+j-1}$ for $i < j$, possesses the following property.

29.2.1. THEOREM. *Let $P_{\sigma_1 \dots \sigma_{2k}} = \operatorname{Pf}(M')$, where $M' = \|m_{\sigma_i \sigma_j}\|_1^{2k}$. Then $P_{\sigma_1 \dots \sigma_{2k}} = (-1)^{\sigma}$, where $\sigma = (\sigma_1 \dots \sigma_{2k} \overline{\sigma}_1 \dots \overline{\sigma}_{2(n-k)})$ (see above).*

PROOF. Let us apply induction on k. Clearly, $P_{\sigma_1 \sigma_2} = m_{\sigma_1 \sigma_2} = (-1)^{\sigma_1 + \sigma_2 + 1}$. The sign of the permutation corresponding to $\{\sigma_1, \sigma_2\}$ is equal to $(-1)^a$, where $a = (\sigma_1 - 1) + (\sigma_2 - 2) \equiv (\sigma_1 + \sigma_2 + 1) \mod 2$.

Using Problem 29.1, it is easy to verify that

$$P_{\sigma_1 \dots \sigma_{2k}} = \sum_{i=2}^{2k} (-1)^i P_{\sigma_1 \sigma_i} P_{\sigma_2 \dots \hat{\sigma}_i \dots \sigma_{2k}}.$$

By inductive hypothesis, $P_{\sigma_1\dots\hat{\sigma}_i\dots\sigma_{2k}} = (-1)^\tau$, where $\tau = (\sigma_2\dots\hat{\sigma}_i\dots\sigma_{2k}12\dots 2n)$. The signs of permutations σ and τ are equal to $(-1)^a$ and $(-1)^b$, respectively, where $a = (\sigma_1 - 1) + \dots + (\sigma_{2k} - 2k)$ and

$$b = (\sigma_2 - 1) + (\sigma_3 - 2) + \dots + (\sigma_{i-1} - i + 2) + (\sigma_{i+1} - i + 1) + \dots + (\sigma_{2k} - 2k + 2).$$

Hence, $(-1)^\tau = (-1)^\sigma(-1)^{\sigma_1+\sigma_2+1}$. Therefore,

$$P_{\sigma_1\dots\sigma_{2k}} = \sum_{i=2}^{2k}(-1)^i(-1)^{\sigma_1+\sigma_2+1}(-1)^\sigma(-1)^{\sigma_1+\sigma_i+1} = (-1)^\sigma\sum_{i=2}^{2k}(-1)^i = (-1)^\sigma. \quad \square$$

29.2.2. THEOREM (Lieb). *Let A be a skew-symmetric matrix of order $2n$. Then*

$$\operatorname{Pf}(A + \lambda^2 M) = \sum_{k=0}^{n}\lambda^{2k}P_k,\ \textit{where}\ P_k = \sum_\sigma A\begin{pmatrix}\sigma_1 & \dots & \sigma_{2(n-k)}\\ \sigma_1 & \dots & \sigma_{2(n-k)}\end{pmatrix}.$$

PROOF ([Kahane, 1971]). The matrices A and M will be considered as elements $\sum_{i<j} a_{ij}e_i \wedge e_j$ and $\sum_{i<j} m_{ij}e_i \wedge e_j$, respectively, in $\Lambda^2 V$. Since $A \wedge M = M \wedge A$, the Newton binomial formula holds:

$$\begin{aligned}\Lambda^n(A + \lambda^2 M) &= \sum_{k=0}^{n}\binom{n}{k}\lambda^{2k}(\Lambda^k M)\wedge(\Lambda^{n-k}A)\\ &= \sum_{k=0}^{n}\binom{n}{k}\lambda^{2k}\sum(k!P_{\sigma_1\dots\sigma_{2k}})((n-k)!P_k)e_{\sigma_1}\wedge\dots\wedge e_{\sigma_k}\wedge\cdots.\end{aligned}$$

By Theorem 29.2.1, $P_{\sigma_1\dots\sigma_{2k}} = (-1)^\sigma$. It is also clear that $e_{\sigma_1}\wedge\dots\wedge e_{\sigma_k}\wedge\dots = (-1)^\sigma e_1\wedge\dots\wedge e_{2n}$. Hence,

$$\Lambda^n(A + \lambda^2 M) = n!\sum_{k=0}^{n}\lambda^{2k}P_k e_1\wedge\dots\wedge e_n$$

and, therefore, $\operatorname{Pf}(A + \lambda^2 M) = \sum_{k=0}^{n}\lambda^{2k}P_k$. $\square$

Problems

29.1. Let $\operatorname{Pf}(A) = a_{pq}C_{pq} + f$, where f does not depend on a_{pq} and let A_{pq} be the matrix obtained from A by deleting its pth and qth columns and rows. Prove that $C_{pq} = (-1)^{p+q+1}\operatorname{Pf}(A_{pq})$.

29.2. Let X be a matrix of order $2n$ whose rows are the coordinates of vectors $x_1,\dots,x_{2n}$ and $g_{ij} = \langle x_i, x_j\rangle$, where $\langle a, b\rangle = \sum_{k=1}^{n}(a_{2k-1}b_{2k} - a_{2k}b_{2k-1})$ for vectors $a = (a_1,\dots,a_{2n})$ and $b = (b_1,\dots,b_{2n})$. Prove that $\det X = \operatorname{Pf}(G)$, where $G = \|g_{ij}\|_1^{2n}$.

30. Decomposable skew-symmetric and symmetric tensors

30.1. A skew-symmetric tensor $\omega \in \Lambda^k(V)$ is said to be *decomposable* (or *simple* or *split*) if it can be represented in the form $\omega = x_1 \wedge \cdots \wedge x_k$, where $x_i \in V$.

A symmetric tensor $T \in S^k(V)$ said to be *decomposable* (or *simple* or *split*) if it can be represented in the form $T = S(x_1 \otimes \cdots \otimes x_k)$, where $x_i \in V$.

30.1.1. THEOREM. *If* $x_1 \wedge \cdots \wedge x_k = y_1 \wedge \cdots \wedge y_k \neq 0$, *then* $\operatorname{Span}(x_1, \dots, x_k) = \operatorname{Span}(y_1, \dots, y_k)$.

PROOF. Suppose for instance, that $y_1 \notin \operatorname{Span}(x_1, \dots, x_k)$. Then the vectors $e_1 = x_1, \dots, e_k = x_k$ and $e_{k+1} = y_1$ can be complemented to a basis. Expanding the vectors $y_2, \dots, y_k$ with respect to this basis we get

$$e_1 \wedge \cdots \wedge e_k = e_{k+1} \wedge \left(\sum a_{i_2 \dots i_k} e_{i_2} \wedge \cdots \wedge e_{i_k}\right).$$

This equality contradicts the linear independence of the vectors $e_{i_1} \wedge \cdots \wedge e_{i_k}$. $\square$

COROLLARY. *To any decomposable skew-symmetric tensor* $\omega = x_1 \wedge \cdots \wedge x_k$ *a* k*-dimensional subspace* $\operatorname{Span}(x_1, \dots, x_k)$ *can be assigned; this subspace does not depend on the expansion of* ω, *but only on the tensor* ω *itself.*

30.1.2. THEOREM ([Merris, 1975]). *If* $S(x_1 \otimes \cdots \otimes x_k) = S(y_1 \otimes \cdots \otimes y_k) \neq 0$, *then* $\operatorname{Span}(x_1, \dots, x_k) = \operatorname{Span}(y_1, \dots, y_k)$.

PROOF. Suppose, for instance, that $y_1 \notin \operatorname{Span}(x_1, \dots, x_k)$. Let $T = S(x_1 \otimes \cdots \otimes x_k)$ be a nonzero tensor. To any multilinear function $f : V \times \cdots \times V \longrightarrow K$ there corresponds a linear function $\tilde{f} : V \otimes \cdots \otimes V \longrightarrow K$. The tensor T is nonzero and, therefore, there exists a linear function $\tilde{f}$ such that $\tilde{f}(T) \neq 0$. A multilinear function f is a linear combination of products of linear functions and, therefore, there exist linear functions $g_1, \dots, g_k$ such that $\tilde{g}(T) \neq 0$, where $g = g_1 \dots g_k$.

Consider linear functions $h_1, \dots, h_k$ that coincide with $g_1 \dots g_k$ on the subspace $\operatorname{Span}(x_1, \dots, x_k)$ and vanish on y_1. Let $h = h_1 \dots h_k$. Then $\tilde{h}(T) = \tilde{g}(T) \neq 0$. On the other hand, $T = S(y_1 \otimes \cdots \otimes y_k)$ and, therefore,

$$\tilde{h}(T) = \sum_{\sigma} h_1(y_{\sigma(1)}) \dots h_k(y_{\sigma(k)}) = 0,$$

since $h_i(y_1) = 0$ is present in every summand. We obtained a contradiction and, therefore, $y_1 \in \operatorname{Span}(x_1, \dots, x_k)$.

Similar arguments prove that

$$\operatorname{Span}(y_1, \dots, y_k) \subset \operatorname{Span}(x_1, \dots, x_k) \text{ and } \operatorname{Span}(x_1, \dots, x_k) \subset \operatorname{Span}(y_1, \dots, y_k). \quad \square$$

30.2. From the definition of decomposability alone it is impossible to determine after a finite number of operations if a skew-symmetric tensor $\sum_{i_1 < \cdots < i_k} a_{i_1 \dots i_k} e_{i_1} \wedge \cdots \wedge e_{i_k}$ is decomposable. We will show that the decomposability condition for a skew-symmetric tensor is equivalent to a system of equations for its coordinates $a_{i_1 \dots i_k}$ (*Plücker relations*, see Corollary 30.2.2). Let us make several preliminary remarks.

For any $v^* \in V^*$ let us consider a map $i(v^*) : \Lambda^k V \longrightarrow \Lambda^{k-1} V$ given by the formula

$$\langle i(v^*)T, f\rangle = \langle T, v^* \wedge f\rangle \text{ for any } f \in (\Lambda^{k-1} V)^* \text{ and } T \in \Lambda^k V.$$

For a given $v \in V$ we define a similar map $i(v) : \Lambda^k V^* \longrightarrow \Lambda^{k-1} V^*$.

To a subspace $\Lambda \subset \Lambda^k V$ assign the subspace $\Lambda^\perp = \{v^* | i(v^*)\Lambda = 0\} \subset V^*$; clearly, if $\Lambda \subset \Lambda^1 V = V$, then $\Lambda^\perp$ is the annihilator of Λ (see 5.5).

30.2.1. THEOREM. *$W = (\Lambda^\perp)^\perp$ is the minimal subspace of V for which Λ belongs to the subspace $\Lambda^k W \subset \Lambda^k V$.*

PROOF. If $\Lambda \subset \Lambda^k W_1$ and $v^* \in W_1^\perp$, then $i(v^*)\Lambda = 0$; hence, $W_1^\perp \subset \Lambda^\perp$; therefore, $W_1 \supset (\Lambda^\perp)^\perp$. It remains to demonstrate that $\Lambda \subset \Lambda^k W$. Let $V = W \oplus U$, $u_1, \dots, u_a$ a basis of U, $w_1, \dots, w_{n-a}$ a basis of W, $u_1^*, \dots, w_{n-a}^*$ the dual basis of V^*. Then $u_j^* \in W^\perp = \Lambda^\perp$, i.e., $i(u_j^*)\Lambda = 0$. If $j \in \{j_1, \dots, j_b\}$ and $\{j'_1, \dots, j'_{b-1}) \cup \{j\} = \{j_1, \dots, j_b\}$, then the map $i(u_j^*)$ sends $w_{i_1} \wedge \dots \wedge w_{i_{k-b}} \wedge u_{j_1} \wedge \dots \wedge u_{j_b}$ to

$$\lambda w_{i_1} \wedge \dots \wedge w_{i_{k-b}} \wedge u_{j'_1} \wedge \dots \wedge u_{j'_{b-1}};$$

otherwise $i(u_j^*)$ sends this tensor to 0. Therefore,

$$i(u_j^*)(\Lambda^{k-b} W \otimes \Lambda^b U) \subset \Lambda^{k-b} W \otimes \Lambda^{b-1} U.$$

Let $\sum_{\alpha=1}^a \Lambda_\alpha \otimes u_\alpha$ be the component of an element from the space Λ which belongs to $\Lambda^{k-1} W \otimes \Lambda^1 U$. Then $i(u_\beta^*)(\sum_\alpha \Lambda_\alpha \otimes u_\alpha) = 0$ and, therefore, for all f we have

$$0 = \left\langle i(u_\beta^*) \sum_\alpha \Lambda_\alpha \otimes u_\alpha, f \right\rangle = \left\langle \sum_\alpha \Lambda_\alpha \otimes u_\alpha, u_\beta^* \wedge f \right\rangle = \langle \Lambda_\beta \otimes u_\beta, u_\beta^* \wedge f\rangle.$$

But if $\Lambda_\beta \otimes u_\beta \neq 0$, then it is possible to choose f so that $\langle \Lambda_\beta \otimes u_\beta, u_\beta^* \wedge f\rangle \neq 0$. We similarly prove that the components of any element of $\Lambda^{k-i} W \otimes \Lambda^i U$ in Λ are zero for $i > 0$, i.e., $\Lambda \subset \Lambda^k W$.

Let $\omega \in \Lambda^k V$. Let us apply Theorem 30.2.1 to $\Lambda = \operatorname{Span}(\omega)$. If $w_1, \dots, w_m$ is a basis of W, then $\omega = \sum a_{i_1 \dots i_k} w_{i_1} \wedge \dots \wedge w_{i_k}$. Therefore, the skew-symmetric tensor ω is decomposable if and only if $m = k$, i.e., $\dim W = k$. If ω is not decomposable, then $\dim W > k$. □

30.2.2. THEOREM. *Let $W = (\operatorname{Span}(\omega)^\perp)^\perp$. Let $\omega \in \Lambda^k V$ and $W' = \{w \in W \mid w \wedge \omega = 0\}$. The skew-symmetric tensor ω is decomposable if and only if $W' = W$.*

PROOF. If $\omega = v_1 \wedge \dots \wedge v_k \neq 0$, then $W = \operatorname{Span}(v_1, \dots, v_k)$; hence, $w \wedge \omega = 0$ for any $w \in W$.

Now, suppose that ω is not decomposable, i.e., $\dim W = m > k$. Let $w_1, \dots, w_m$ be a basis of W. Then $\omega = \sum a_{i_1 \dots i_k} w_{i_1} \wedge \dots \wedge w_{i_k}$. We may assume that $a_{1\dots k} \neq 0$. Let $\alpha = w_{k+1} \wedge \dots \wedge w_m \in \Lambda^{m-k} W$. Then $\omega \wedge \alpha = a_{1\dots k} w_1 \wedge \dots \wedge w_m \neq 0$. In particular, $\omega \wedge w_m \neq 0$, i.e., $w_m \notin W'$. □

COROLLARY (Plücker relations). *Let* $\omega = \sum_{i_1<\dots<i_k} a_{i_1\dots i_k} e_{i_1} \wedge \dots \wedge e_{i_k}$ *be a skew-symmetric tensor. It is decomposable if and only if*

$$\left(\sum_{i_1<\dots<i_k} a_{i_1\dots i_k} e_{i_1} \wedge \dots \wedge e_{i_k} \right) \wedge \left(\sum_j a_{j_1\dots j_{k-1}j} e_j \right) = 0$$

for any $j_1 < \dots < j_{k-1}$. (To determine the coefficient $a_{j_1\dots j_{k-1}j}$ for $j_{k-1} > j$ we assume that $a_{\dots ij\dots} = -a_{\dots ji\dots}$).

PROOF. In our case

$$\Lambda^\perp = \{v^* | \langle \omega, f \wedge v^* \rangle = 0 \text{ for any } f \in \Lambda^{k-1}(V^*)\}.$$

Let $\varepsilon_1, \dots, \varepsilon_n$ be the basis dual to $e_1, \dots, e_n$, $f = \varepsilon_{j_1} \wedge \dots \wedge \varepsilon_{j_{k-1}}$, and $v^* = \sum v_i \varepsilon_i$. Then

$$\langle \omega, f \wedge v^* \rangle = \left\langle \sum_{i_1<\dots<i_k} a_{i_1\dots i_k} e_{i_1} \wedge \dots \wedge e_{i_k}, \sum_j v_j \varepsilon_{j_1} \wedge \dots \wedge \varepsilon_{j_{k-1}} \wedge \varepsilon_j \right\rangle$$
$$= \frac{1}{n!} \sum a_{j_1\dots j_{k-1}j} v_j.$$

Therefore,

$$\Lambda^\perp = \{v^* = \sum v_j \varepsilon_j \mid \sum a_{j_1\dots j_{k-1}j} v_j = 0 \text{ for any } j_1, \dots, j_{k-1}\};$$

hence, $W = (\Lambda^\perp)^\perp = \{w = \sum_j a_{j_1\dots j_{k-1}j} e_j\}$. By Theorem 30.2.2, ω is decomposable if and only if $\omega \wedge w = 0$ for all $w \in W$. $\square$

EXAMPLE. For $k = 2$ for every fixed p we get a relation

$$\left(\sum_{i<j} a_{ij} e_i \wedge e_j \right) \wedge \left(\sum_q a_{pq} e_p \right) = 0.$$

In this relation the coefficient at $e_i \wedge e_j \wedge e_q$ is equal to $a_{ij}a_{pq} - a_{ip}a_{pj} + a_{jp}a_{pi}$ and the relation

$$a_{ij}a_{pq} - a_{iq}a_{pj} + a_{jq}a_{pi} = 0$$

is nontrivial only if the numbers i, j, p, q are distinct.

Problems

30.1. Let $\omega \in \Lambda^k V$ and $e_1 \wedge \dots \wedge e_r \neq 0$ for some $e_i \in V$. Prove that $\omega = \omega_1 \wedge e_1 \wedge \dots \wedge e_r$ if and only if $\omega \wedge e_i = 0$ for $i = 1, \dots, r$.

30.2. Let $\dim V = n$ and $\omega \in \Lambda^{n-1} V$. Prove that ω is a decomposable skew-symmetric tensor.

30.3. Let $e_1, \dots, e_{2n}$ be linearly independent, $\omega = \sum_{i=1}^n e_{2i-1} \wedge e_{2i}$, and $\Lambda = \operatorname{Span}(\omega)$. Find the dimension of $W = (\Lambda^\perp)^\perp$.

30.4. Let tensors $z_1 = x_1 \wedge \dots \wedge x_r$ and $z_2 = y_1 \wedge \dots \wedge y_r$ be nonproportional; $X = \operatorname{Span}(x_1, \dots, x_r)$ and $Y = \operatorname{Span}(y_1, \dots, y_r)$. Prove that $\operatorname{Span}(z_1, z_2)$ consists of decomposable skew-symmetric tensors if and only if $\dim(X \cap Y) = r - 1$.

30.5. Let $W \subset \Lambda^k V$ consist of decomposable skew-symmetric tensors. To every $\omega = x_1 \wedge \dots \wedge x_k \in W$ assign the subspace $[\omega] = \operatorname{Span}(x_1, \dots, x_k) \subset V$. Prove that either all subspaces $[\omega]$ have a common $(k-1)$-dimensional subspace or all of them belong to one $(k+1)$-dimensional subspace.

31. The tensor rank

31.1. The space $V \otimes W$ consists of linear combinations of elements of the form $v \otimes w$. Not every element of this space, however, can be represented in the form $v \otimes w$. The *rank* of an element $T \in V \otimes W$ is the least number k such that $T = v_1 \otimes w_1 + \dots + v_k \otimes w_k$.

31.1.1. THEOREM. *If* $T = \sum a_{ij} e_i \otimes \varepsilon_j$, *where* $\{e_i\}$ *and* $\{\varepsilon_j\}$ *are bases of* V *and* W, *then* $\operatorname{rank} T = \operatorname{rank}\|a_{ij}\|$.

PROOF. Let $v_p = \sum \alpha_i^p e_i$, $w_p = \sum \beta_j^p \varepsilon_j$, α^p a column $(\alpha_1^p, \dots, \alpha_n^p)^T$ and β^p a row $(\beta_1^p, \dots, \beta_m^p)$. If $T = v_1 \otimes w_1 + \dots + v_k \otimes w_k$, then $\|a_{ij}\| = \alpha^1\beta^1 + \dots + \alpha^k\beta^k$. The least number k for which such a decomposition of $\|a_{ij}\|$ is possible is equal to the rank of this matrix (see 8.2). $\square$

31.1.1.1. COROLLARY. *The set* $\{T \in V \otimes W \mid \operatorname{rank} T \leq k\}$ *is given by algebraic equations and, therefore, is closed; in particular, if* $\lim_{i\to\infty} T_i = T$ *and* $\operatorname{rank} T_i \leq k$, *then* $\operatorname{rank} T \leq k$.

31.1.1.2. COROLLARY. *The rank of an element of a real subspace* $V \otimes W$ *does not change under complexifications.*

For an element $T \in V_1 \otimes \dots \otimes V_p$ its *rank* can be similarly defined as the least number k for which $T = v_1^1 \otimes \dots \otimes v_p^1 + \dots + v_1^k \otimes \dots \otimes v_p^k$. It turns out that for $p \geq 3$ the properties formulated in Corollaries 31.1.1.1 and 31.1.1.2 do not hold. Before we start studying the properties of the tensor rank let us explain why the interest in it.

31.2. In the space of matrices of order n select the basis $e_{\alpha\beta} = \|\delta_{i\alpha}\delta_{j\beta}\|_1^n$ and let $\varepsilon_{\alpha\beta}$ be the dual basis. Then $A = \sum_{i,j} a_{ij}e_{ij}$, $B = \sum_{i,j} b_{ij}e_{ij}$ and

$$AB = \sum_{i,j,k} a_{ik}b_{kj}e_{ij} = \sum_{i,j,k} \varepsilon_{ik}(A)\varepsilon_{kj}(B)e_{ij}.$$

Thus, the calculation of the product of two matrices of order n reduces to calculation of n^3 products $\varepsilon_{ik}(A)\varepsilon_{kj}(B)$ of linear functions. Is the number n^3 the least possible one?

It turns out that no, it is not. For example, for matrices of order 2 we can indicate 7 pairs of linear functions f_p and g_p and 7 matrices E_p such that $AB = \sum_{p=1}^7 f_p(A)g_p(B)E_p$. This decomposition was constructed in [Strassen, 1969]. The computation of the least number of such triples is equivalent to the computation of the rank of the tensor

$$\sum_{i,j,k} \varepsilon_{ik} \otimes \varepsilon_{kj} \otimes e_{ij} = \sum_p f_p \otimes g_p \otimes E_p.$$

Identify the space of vectors with the space of covectors, and introduce, for brevity, the notation $a = e_{11}$, $b = e_{12}$, $c = e_{21}$ and $d = e_{22}$. It is easy to verify that for matrices of order 2

$$\sum_{i,j,k} \varepsilon_{ik} \otimes \varepsilon_{kj} \otimes e_{ij} = (a \otimes a + b \otimes c) \otimes a + (a \otimes b + b \otimes d) \otimes b$$
$$+ (c \otimes a + d \otimes c) \otimes c + (c \otimes b + d \otimes d) \otimes d.$$

Strassen's decomposition is of the form $\sum \varepsilon_{ik} \otimes \varepsilon_{kj} \otimes e_{ij} = \sum_{p=1}^{7} T_p$, where

$$T_1 = (a-d) \otimes (a-d) \otimes (a+d), \qquad T_2 = d \otimes (a+c) \otimes (a+c),$$
$$T_3 = (a-b) \otimes d \otimes (a-b), \qquad T_4 = a \otimes (b+d) \otimes (b+d),$$
$$T_5 = (c-d) \otimes a \otimes (c-d), \qquad T_6 = (b-d) \otimes (c+d) \otimes a,$$
$$T_7 = (c-a) \otimes (a+b) \otimes d.$$

This decomposition leads to the following algorithm for computing the product of matrices $A = \begin{pmatrix} a_1 & b_1 \\ c_1 & d_1 \end{pmatrix}$ and $B = \begin{pmatrix} a_2 & b_2 \\ c_2 & d_2 \end{pmatrix}$. Let

$$\begin{array}{lllll} S_1 = a_1 - d_1, & S_2 = a_2 - d_2, & S_3 = a_1 - b_1, & S_4 = b_1 - d_1, & S_5 = c_2 + d_2, \\ S_6 = a_2 + c_2, & S_7 = b_2 + d_2, & S_8 = c_1 - d_1, & S_9 = c_1 - a_1, & S_{10} = a_2 + b_2; \\ P_1 = S_1 S_2, & P_2 = S_3 d_2, & P_3 = S_4 S_5, & P_4 = d_1 S_6, & P_5 = a_1 S_7, \\ P_6 = S_8 a_2, & P_7 = S_9 S_{10}; & S_{11} = P_1 + P_2, & S_{12} = S_{11} + P_3, & S_{13} = S_{12} + P_4, \\ S_{14} = P_5 - P_2, & S_{15} = P_4 + P_6, & S_{16} = P_1 + P_5, & S_{17} = S_{16} - P_6, & S_{18} = S_{17} + P_7. \end{array}$$

Then $AB = \begin{pmatrix} S_{13} & S_{14} \\ S_{15} & S_{18} \end{pmatrix}$. Strassen's algorithm for computing AB requires just 7 multiplications and 18 additions (or subtractions)[1].

31.3. Let V be a two-dimensional space with basis $\{e_1, e_2\}$. Consider the tensor

$$T = e_1 \otimes e_1 \otimes e_1 + e_1 \otimes e_2 \otimes e_2 + e_2 \otimes e_1 \otimes e_2.$$

31.3.1. THEOREM. *The rank of T is equal to 3, but there exists a sequence of tensors of rank ≤ 2 which converges to T.*

PROOF. Let

$$T_\lambda = \lambda^{-1}[e_1 \otimes e_1 \otimes (-e_2 + \lambda e_1) + (e_1 + \lambda e_2) \otimes (e_1 + \lambda e_2) \otimes e_2].$$

Then $T_\lambda - T = \lambda e_2 \otimes e_2 \otimes e_2$ and, therefore, $\lim_{\lambda \to 0} |T_\lambda - T| = 0$.

Suppose that

$$\begin{aligned} T &= a \otimes b \otimes c + u \otimes v \otimes w = (\alpha_1 e_1 + \alpha_2 e_2) \otimes b \otimes c + (\lambda_1 e_1 + \lambda_2 e_2) \otimes v \otimes w \\ &= e_1 \otimes (\alpha_1 b \otimes c + \lambda_1 v \otimes w) + e_2 \otimes (\alpha_2 b \otimes c + \lambda_2 v \otimes w). \end{aligned}$$

Then

$$e_1 \otimes e_1 + e_2 \otimes e_2 = \alpha_1 b \otimes c + \lambda_1 v \otimes w \text{ and } e_1 \otimes e_2 = \alpha_2 b \otimes c + \lambda_2 v \otimes w.$$

Hence, linearly independent tensors $b \otimes c$ and $v \otimes w$ of rank 1 belong to the space $\operatorname{Span}(e_1 \otimes e_1 + e_2 \otimes e_2, e_1 \otimes e_2)$. The latter space can be identified with the space of matrices of the form $\begin{pmatrix} x & y \\ 0 & x \end{pmatrix}$. But all such matrices of rank 1 are linearly dependent. Contradiction. $\square$

---

[1]Strassen's algorithm is of importance nowadays since modern computers add (subtract) much faster than multiply.

COROLLARY. *The subset of tensors of rank ≤ 2 in $T_0^3(V)$ is not closed, i.e., it cannot be defined by a system of algebraic equations.*

31.3.2. Let us consider the tensor

$$T_1 = e_1 \otimes e_1 \otimes e_1 - e_2 \otimes e_2 \otimes e_1 + e_1 \otimes e_2 \otimes e_2 + e_2 \otimes e_1 \otimes e_2.$$

Let $\operatorname{rank}_{\mathbb{R}} T_1$ denote the rank of T_1 over $\mathbb{R}$ and $\operatorname{rank}_{\mathbb{C}} T_1$ be the rank of T_1 over $\mathbb{C}$.

THEOREM. $\operatorname{rank}_{\mathbb{R}} T_1 \neq \operatorname{rank}_{\mathbb{C}} T_1$.

PROOF. It is easy to verify that $T_1 = (a_1 \otimes a_1 \otimes a_2 + a_2 \otimes a_2 \otimes a_1)/2$, where $a_1 = e_1 + ie_2$ and $a_2 = e_1 - ie_2$. Hence, $\operatorname{rank}_{\mathbb{C}} T_1 \leq 2$. Now, suppose that $\operatorname{rank}_{\mathbb{R}} T_1 \leq 2$. Then as in the proof of Theorem 31.3.1 we see that linearly independent tensors $b \otimes c$ and $v \otimes w$ of rank 1 belong to $\operatorname{Span}(e_1 \otimes e_1 + e_2 \otimes e_2, e_1 \otimes e_2 - e_2 \otimes e_1)$, which can be identified with the space of matrices of the form $\begin{pmatrix} x & y \\ -y & x \end{pmatrix}$. But over $\mathbb{R}$ among such matrices there is no matrix of rank 1. $\square$

Problems

31.1. Let $U \subset V$ and $T \in T_0^p(U) \subset T_0^p(V)$. Prove that the rank of T does not depend on whether T is considered as an element of $T_0^p(U)$ or as an element of $T_0^p(V)$.

31.2. Let $e_1, \dots, e_k$ be linearly independent vectors, $e_i^{\otimes p} = e_i \otimes \dots \otimes e_i \in T_0^p(V)$, where $p \geq 2$. Prove that the rank of $e_1^{\otimes p} + \dots + e_k^{\otimes p}$ is equal to k.

32. Linear transformations of tensor products

The tensor product $V_1 \otimes \dots \otimes V_p$ is a linear space; this space has an additional structure — the rank function on its elements. Therefore, we can, for instance, consider linear transformations that send tensors of rank k to tensors of rank k. The most interesting case is that of maps of $\operatorname{Hom}(V_1, V_2) = V_1^* \otimes V_2$ into itself. Observe also that if $\dim V_1 = \dim V_2 = n$, then to invertible maps from $\operatorname{Hom}(V_1, V_2)$ there correspond tensors of rank n, i.e., the condition $\det A = 0$ can be interpreted in terms of the tensor rank.

32.1. If $A : U \longrightarrow U$ and $B : V \longrightarrow V$ are invertible linear operators, then the linear operator $T = A \otimes B : U \otimes V \longrightarrow U \otimes V$ preserves the rank of elements of $U \otimes V$.

If $\dim U = \dim V$, there is one more type of transformations that preserve the rank of elements. Take an arbitrary isomorphism $\varphi : U \longrightarrow V$ and define a map

$$S : U \otimes V \longrightarrow U \otimes V, \quad S(u \otimes v) = \varphi^{-1} v \otimes \varphi u.$$

Then any transformation of the form TS, where $T = A \otimes B$ is a transformation of the first type, preserves the rank of the elements from $U \otimes V$.

REMARK. It is easy to verify that S is an involution.

In terms of matrices the first type transformations are of the form $X \mapsto AXB$ and the second type transformations are of the form $X \mapsto AX^T B$. The second type transformations do not reduce to the first type transformations (see Problem 32.1).

THEOREM ([Marcus, Moyls, 1959 (b)]). *Let a linear map $T : U \otimes V \longrightarrow U \otimes V$ send any element of rank* 1 *to an element of rank* 1. *Then either $T = A \otimes B$ or $T = (A \otimes B)S$ and the second case is possible only if* $\dim U = \dim V$.

PROOF (Following [Grigoriev, 1979]). We need the following statement.

LEMMA. *Let elements $\alpha_1, \alpha_2 \in U \otimes V$ be such that* $\operatorname{rank}(t_1\alpha_1 + t_2\alpha_2) \leq 1$ *for any numbers t_1 and t_2. Then α_i can be represented in the form $\alpha_i = u_i \otimes v_i$, where $u_1 = u_2$ or $v_1 = v_2$.*

PROOF. Suppose that $\alpha_i = u_i \otimes v_i$ and $\alpha_1 + \alpha_2 = u \otimes v$ and also that $\operatorname{Span}(u_1) \neq \operatorname{Span}(u_2)$ and $\operatorname{Span}(v_1) \neq \operatorname{Span}(v_2)$. Then without loss of generality we may assume that $\operatorname{Span}(u) \neq \operatorname{Span}(u_1)$. On one hand,

$$(f \otimes g)(u \otimes v) = f(u)g(v),$$

and on the other hand,

$$(f \otimes g)(u \otimes v) = (f \otimes g)(u_1 \otimes v_1 + u_2 \otimes v_2) = f(u_1)g(v_1) + f(u_2)g(u_2).$$

Therefore, selecting $f \in U^*$ and $g \in V^*$ so as $f(u) = 0$, $f(u_1) \neq 0$ and $g(u_2) = 0$, $g(u_1) \neq 0$ we get a contradiction. □

In what follows we will assume that $\dim V \geq \dim U \geq 2$. Besides, for convenience we will denote fixed vectors by a and b, while variable vectors from U and V will be denoted by u and v, respectively. Applying the above lemma to $T(a \otimes b_1)$ and $T(a \otimes b_2)$, where $\operatorname{Span}(b_1) \neq \operatorname{Span}(b_2)$, we get $T(a \otimes b_i) = a' \otimes b_i'$ or $T(a \otimes b_i) = a_i' \otimes b'$. Since $\operatorname{Ker} T = 0$, it follows that $\operatorname{Span}(b_1') \neq \operatorname{Span}(b_2')$ (resp. $\operatorname{Span}(a_1') \neq \operatorname{Span}(a_2')$).

It is easy to verify that in the first case $T(a \otimes v) = a' \otimes v'$ for any $v \in V$. For the proof, it suffices to apply the lemma to $T(a \otimes b_1)$ and $T(a \otimes v)$ and also to $T(a \otimes b_2)$ and $T(a \otimes v)$. Indeed, the case $T(a \otimes v) = c' \otimes b_1'$, where $\operatorname{Span}(c') \neq \operatorname{Span}(a')$, is impossible. Similarly, in the second case $T(a \otimes v) = f(v) \otimes b'$, where $f : V \longrightarrow U$ is a map (obviously a linear one). In the second case the map of the subspace $a \otimes V$ to $U \otimes b'$ is a monomorphisms. Hence, $\dim V \leq \dim U$ and, therefore, $\dim U = \dim V$.

Consider the map T_1 which is equal to T in the first case and to TS in the second case. Then for a fixed a we have $T_1(a \otimes v) = a' \otimes Bv$, where $B : V \longrightarrow V$ is an invertible operator. Let $\operatorname{Span}(a_1) \neq \operatorname{Span}(a)$. Then either $T_1(a_1 \otimes v) = a_1' \otimes Bv$ or $T_1(a_1 \otimes v) = a' \otimes B_1 v$, where $\operatorname{Span}(B_1) \neq \operatorname{Span}(B)$. Applying the lemma to $T(a \otimes v)$ and $T(u \otimes v)$ and also to $T(a_1 \otimes v)$ and $T(u \otimes v)$ we see that in the first case $T_1(u \otimes v) = Au \otimes Bv$, and in the second case $T_1(u \otimes v) = a' \otimes f(u, v)$. In the second case the space $U \otimes V$ is monomorphically mapped into $a' \otimes V$ which is impossible. □

COROLLARY. *If a linear map $T : U \otimes V \longrightarrow U \otimes V$ sends any rank* 1 *element into a rank* 1 *element, then it sends any rank k element into a rank k element.*

32.2. Let $M_{n,n}$ be the space of matrices of order n and $T : M_{n,n} \longrightarrow M_{n,n}$ a linear map.

32.2.1. THEOREM ([Marcus, Moyls, 1959 (a)]). *If T preserves the determinant, then T preserves the rank as well.*

PROOF. For convenience, denote by I_r and 0_r the unit and the zero matrix of order r, respectively, and set $A \oplus B = \begin{pmatrix} A & 0 \\ 0 & B \end{pmatrix}$.

First, let us prove that if T preserves the determinant, then T is invertible. Suppose that $T(A) = 0$, where $A \neq 0$. Then $0 < \operatorname{rank} A < n$. There exist invertible matrices M and N such that $MAN = I_r \oplus 0_{n-r}$, where $r = \operatorname{rank} A$ (cf. Theorem 6.3.2). For any matrix X of order n we have

$$\begin{aligned}|MAN + X| \cdot |MN|^{-1} &= |A + M^{-1}XN^{-1}| \\ &= |T(A + M^{-1}XN^{-1})| = |T(M^{-1}XN^{-1})| = |X| \cdot |MN|^{-1}.\end{aligned}$$

Therefore, $|MAN + X| = |X|$. Setting $X = 0_r \oplus I_{n-r}$ we get a contradiction.

Let $\operatorname{rank} A = r$ and $\operatorname{rank} T(A) = s$. Then there exist invertible matrices M_1, N_1 and M_2, N_2 such that $M_1AN_1 = I_r \oplus 0_{n-r} = Y_1$ and $M_2T(A)N_2 = I_s \oplus 0_{n-s} = Y_2$. Consider a map $f : M_{n,n} \longrightarrow M_{n,n}$ given by the formula $f(X) = M_2T(M_1^{-1}XN_1^{-1})N_2$. This map is linear and $|f(X)| = k|X|$, where $k = |M_2M_1^{-1}N_1^{-1}N_2|$. Besides, $f(Y_1) = M_2T(A)N_2 = Y_2$. Consider a matrix $Y_3 = 0_r \oplus I_{n-r}$. Then $|\lambda Y_1 + Y_3| = \lambda^r$ for all λ. On the other hand,

$$|f(\lambda Y_1 + Y_3)| = |\lambda Y_2 + f(Y_3)| = p(\lambda),$$

where p is a polynomial of degree not greater than s. It follows that $r \leq s$. Since $|B| = |TT^{-1}(B)| = |T^{-1}(B)|$, the map T^{-1} also preserves the determinant. Hence, $s \leq r$. $\square$

Let us say that a linear map $T : M_{n,n} \longrightarrow M_{n,n}$ *preserves eigenvalues* if the sets of eigenvalues (multiplicities counted) of X and $T(X)$ coincide for any X.

32.2.2. THEOREM ([Marcus, Moyls, 1959 (a)]). a) *If T preserves eigenvalues, then either $T(X) = AXA^{-1}$ or $T(X) = AX^TA^{-1}$.*

b) *If T, given over $\mathbb{C}$, preserves eigenvalues of Hermitian matrices, then either $T(X) = AXA^{-1}$ or $T(X) = AX^TA^{-1}$.*

PROOF. a) If T preserves eigenvalues, then T preserves the rank as well and, therefore, either $T(X) = AXB$ or $T(X) = AX^TB$ (see 32.1). It remains to prove that $T(I) = I$. The determinant of a matrix is equal to the product of its eigenvalues and, therefore, T preserves the determinant. Hence,

$$|X - \lambda I| = |T(X) - \lambda T(I)| = |CT(X) - \lambda I|,$$

where $C = T(I)^{-1}$ and, therefore, the eigenvalues of X and $CT(X)$ coincide; besides, the eigenvalues of X and $T(X)$ coincide by the assumption. The map T is invertible (see the proof of Theorem 32.2.1) and, therefore, any matrix Y can be represented in the form $T(X)$ which means that the eigenvalues of Y and CY coincide.

The matrix C can be represented in the form $C = SU$, where U is a unitary matrix and S an Hermitian positive definite matrix. The eigenvalues of U^{-1} and $CU^{-1} = S$ coincide, but the eigenvalues of U^{-1} are of the form $e^{i\varphi}$, whereas the eigenvalues of S are positive. It follows that $S = U = I$ and $C = I$, i.e., $T(I) = I$.

b) It suffices to prove that if T preserves eigenvalues of Hermitian matrices, then T preserves eigenvalues of all matrices. Any matrix X can be represented in the form $X = P + iQ$, where P and Q are Hermitian matrices. For any real x the matrix

$A = P + xQ$ is Hermitian. If the eigenvalues of A are equal to $\lambda_1, \dots, \lambda_n$, then the eigenvalues of A^m are equal to $\lambda_1^m, \dots, \lambda_n^m$ and, therefore, $\operatorname{tr}(A^m) = \operatorname{tr}(T(A)^m)$. Both sides of this identity are polynomials in x of degree not exceeding m. Two polynomials whose values are equal for all real x coincide and, therefore, their values at $x = i$ are also equal. Hence, $\operatorname{tr}(X^m) = \operatorname{tr}(T(X)^m)$ for any X. It remains to use the result of Problem 13.2. □

32.3. Theorem ([Marcus, Purves, 1959]). a) *Let $T : M_{n,n} \longrightarrow M_{n,n}$ be a linear map that sends invertible matrices into invertible ones. Then T is an invertible map.*

b) *If, in addition, $T(I) = I$, then T preserves eigenvalues.*

Proof. a) If $|T(X)| = 0$, then $|X| = 0$. For $X = A - \lambda I$ we see that if

$$|T(A - \lambda I)| = |T(I)| \cdot |T(I)^{-1}T(A) - \lambda I| = 0,$$

then $|A - \lambda I| = 0$. Therefore, the eigenvalues of $T(I)^{-1}T(A)$ are eigenvalues of A.

Suppose that $A \neq 0$ and $T(A) = 0$. For a matrix A we can find a matrix X such that the matrices X and $X + A$ have no common eigenvalues (see Problem 15.1); hence, the matrices $T(I)^{-1}T(A + X)$ and $T(I)^{-1}T(X)$ have no common eigenvalues. On the other hand, these matrices coincide since $T(A + X) = T(X)$. Contradiction.

b) If $T(I) = I$, then the proof of a) implies that the eigenvalues of $T(A)$ are eigenvalues of A. Hence, if the eigenvalues of $B = T(A)$ are simple (nonmultiple), then the eigenvalues of B coincide with the eigenvalues of $A = T^{-1}(B)$. For a matrix B with multiple eigenvalues we consider a sequence of matrices B_i with simple eigenvalues that converges to it (see Theorem 43.5.2) and observe that the eigenvalues of the matrices B_i tend to eigenvalues of B (see Problem 11.6). □

Problems

32.1. Let X be a matrix of size $m \times n$, where $mn > 1$. Prove that the map $X \mapsto X^T$ cannot be represented in the form $X \mapsto AXB$ and the map $X \mapsto X$ cannot be represented in the form $X \mapsto AX^TB$.

32.2. Let $f : M_{n,n} \longrightarrow M_{n,n}$ be an invertible map and $f(XY) = f(X)f(Y)$ for any matrices X and Y. Prove that $f(X) = AXA^{-1}$, where A is a fixed matrix.

Solutions

27.1. Complement vectors v and w to bases of V and W, respectively. If $v' \otimes w' = v \otimes w$, then the decompositions of v' and w' with respect to these bases are of the form λv and μw, respectively. It is also clear that $\lambda v \otimes \mu w = \lambda\mu(v \otimes w)$, i.e., $\mu = 1/\lambda$.

27.2. a) The statement obviously follows from the definition.

b) Take bases of the spaces $\operatorname{Im} A_1$ and $\operatorname{Im} A_2$ and complement them to bases $\{e_i\}$ and $\{\varepsilon_j\}$ of the spaces W_1 and W_2, respectively. The space $\operatorname{Im} A_1 \otimes W_2$ is spanned by the vectors $e_i \otimes \varepsilon_j$, where $e_i \in \operatorname{Im} A_1$, and the space $W_1 \otimes \operatorname{Im} A_2$ is spanned by the vectors $e_i \otimes \varepsilon_j$, where $\varepsilon_j \in \operatorname{Im} A_2$. Therefore, the space $(\operatorname{Im} A_1 \otimes W_2) \cap (W_1 \otimes \operatorname{Im} A_2)$ is spanned by the vectors $e_i \otimes \varepsilon_j$, where $e_i \in \operatorname{Im} A_1$ and $\varepsilon_j \in \operatorname{Im} A_2$, i.e., this space coincides with $\operatorname{Im} A_1 \otimes \operatorname{Im} A_2$.

c) Take bases in $\operatorname{Ker} A_1$ and $\operatorname{Ker} A_2$ and complement them to bases $\{e_i\}$ and $\{\varepsilon_j\}$ in V_1 and V_2, respectively. The map $A_1 \otimes A_2$ sends $e_i \otimes \varepsilon_j$ to 0 if either $e_i \in \operatorname{Ker} A_1$

or $\varepsilon_j \in \operatorname{Ker} A_2$; the set of other elements of the form $e_i \otimes \varepsilon_j$ is mapped into a basis of the space $\operatorname{Im} A_1 \otimes \operatorname{Im} A_2$, i.e., into linearly independent elements.

27.3. Select a basis $\{v_i\}$ in $V_1 \cap V_2$ and complement it to bases $\{v_j^1\}$ and $\{v_k^2\}$ of V_1 and V_2, respectively. The set $\{v_i, v_j^1, v_k^2\}$ is a basis of $V_1 + V_2$. Similarly, construct a basis $\{w_\alpha, w_\beta^1, w_\gamma^2\}$ of $W_1 + W_2$. Then $\{v_i \otimes w_\alpha, v_i \otimes w_\beta^1, v_j^1 \otimes w_\alpha, v_j^1 \otimes w_\beta^1\}$ and $\{v_i \otimes w_\alpha, v_i \otimes w_\gamma^2, v_k^2 \otimes w_\alpha, v_k^2 \otimes w_\gamma^2\}$ are bases of $V_1 \otimes W_1$ and $V_2 \otimes W_2$, respectively, and the elements of these bases are also elements of a basis for $(V_1 + V_2) \otimes (W_1 + W_2)$, i.e., they are linearly independent. Hence, $\{v_i \otimes w_\alpha\}$ is a basis of $(V_1 \otimes W_1) \cap (V_2 \otimes W_2)$.

27.4. Clearly, $Ax = x - 2(a, x)a$, i.e., $Aa = -a$ and $Ax = x$ for $x \in a^\perp$.

27.5. Fix $a \neq 0$; then $A(a, x)$ is a linear function; hence, $A(a, x) = (b, x)$, where $b = B(a)$ for some linear map B. If $x \perp a$, then $A(a, x) = 0$, i.e., $(b, x) = 0$. Hence, $a^\perp \subset b^\perp$ and, therefore, $B(a) = b = \lambda(a)a$. Since $A(u + v, x) = A(u, x) + A(v, x)$, it follows that

$$\lambda(u + v)(u + v) = \lambda(u)u + \lambda(v)v.$$

If the vectors u and v are linearly independent, then $\lambda(u) = \lambda(v) = \lambda$ and any other vector w is linearly independent of one of the vectors u or v; hence, $\lambda(w) = \lambda$. For a one-dimensional space the statement is obvious.

28.1. Let us successively change places of the first two arguments and the second two arguments:

$$\begin{aligned} f(x, y, z) = f(y, x, z) = -f(y, z, x) &= -f(z, y, x) \\ &= f(z, x, y) = f(x, z, y) = -f(x, y, z); \end{aligned}$$

hence, $2f(x, y, z) = 0$.

28.2. Let us extend f to a bilinear map $\mathbb{C}^m \times \mathbb{C}^m \longrightarrow \mathbb{C}^n$. Consider the equation $f(z, z) = 0$, i.e., the system of quadratic equations

$$f_1(z, z) = 0, \quad \ldots, \quad f_n(z, z) = 0.$$

Suppose $n < m$. Then this system has a nonzero solution $z = x + iy$. The second condition implies that $y \neq 0$. It is also clear that

$$0 = f(z, z) = f(x + iy, x + iy) = f(x, x) - f(y, y) + 2if(x, y).$$

Hence, $f(x, x) = f(y, y) \neq 0$ and $f(x, y) = 0$. This contradicts the first condition.

28.3. The elements $\alpha_i = e_{2i-1} \wedge e_{2i}$ belong to $\Lambda^2(V)$; hence, $\alpha_i \wedge \alpha_j = \alpha_j \wedge \alpha_i$ and $\alpha_i \wedge \alpha_i = 0$. Thus,

$$\Lambda^n \omega = \sum_{i_1, \ldots, i_n} \alpha_{i_1} \wedge \cdots \wedge \alpha_{i_n} = n!\,\alpha_1 \wedge \cdots \wedge \alpha_n = n!\,e_1 \wedge \cdots \wedge e_{2n}.$$

28.4. Let the diagonal of the Jordan normal form of A be occupied by numbers $\lambda_1, \ldots, \lambda_n$. Then $\det(A + I) = (1 + \lambda_1) \ldots (1 + \lambda_n)$ and $\operatorname{tr}(\Lambda^q A) = \sum_{i_1 < \cdots < i_q} \lambda_{i_1} \ldots \lambda_{i_q}$; see the proof of Theorem 28.5.3.

28.5. If $A = \|a_{ij}\|_1^n$, then the matrix of the system of equations under consideration is equal to $S^2(A)$. Besides, $\det S^2(A) = (\det A)^r$, where $r = \frac{2}{n}\binom{n+2-1}{2} = n+1$ (see Theorem 28.5.3).

28.6. It is easy to verify that $\sigma_k = \operatorname{tr}(\Lambda^k A)$. If in a Jordan basis the diagonal of A is of the form $(\lambda_1, \dots, \lambda_n)$, then $s_k = \lambda_1^k + \dots + \lambda_n^k$ and $\sigma_k = \sum \lambda_{i_1} \dots \lambda_{i_k}$. The required identity for the functions s_k and σ_k was proved in 4.1.

28.7. Let e_j and ε_j, where $1 \le j \le m$, be dual bases. Let $v_i = \sum a_{ij} e_j$ and $f_i = \sum b_{ji} \varepsilon_j$. The quantity $n!\langle v_1 \wedge \dots \wedge v_n, f_1 \wedge \dots \wedge f_n \rangle$ can be computed in two ways. On the one hand, it is equal to

$$\begin{vmatrix} f_1(v_1) & \dots & f_1(v_n) \\ \vdots & \ddots & \vdots \\ f_n(v_1) & \dots & f_n(v_n) \end{vmatrix} = \begin{vmatrix} \sum a_{1j} b_{j1} & \dots & \sum a_{nj} b_{j1} \\ \vdots & \ddots & \vdots \\ \sum a_{1j} b_{jn} & \dots & \sum a_{nj} b_{jn} \end{vmatrix} = \det AB.$$

On the other hand, it is equal to

$$\begin{aligned} n! \left\langle \sum_{k_1,\dots,k_n} a_{1k_1} \dots a_{nk_n} e_{k_1} \wedge \dots \wedge e_{k_n}, \sum_{l_1,\dots,l_n} b_{l_1 1} \dots b_{l_n n} \varepsilon_{l_1} \wedge \dots \wedge \varepsilon_{l_n} \right\rangle \\ = \sum_{k_1 \le \dots \le k_n} A_{k_1 \dots k_n} B^{l_1 \dots l_n} n! \langle e_{k_1} \wedge \dots \wedge e_{k_n}, \varepsilon_{l_1} \wedge \dots \wedge \varepsilon_{l_n} \rangle \\ = \sum_{k_1 < \dots < k_n} A_{k_1 \dots k_n} B^{k_1 \dots k_n}. \end{aligned}$$

29.1. Since $\operatorname{Pf}(A) = \sum (-1)^\sigma a_{i_1 i_2} \dots a_{i_{2n-1} i_{2n}}$, where the sum runs over all partitions of $\{1, \dots, 2n\}$ into pairs $\{i_{2k-1}, i_{2k}\}$ with $i_{2k-1} < i_{2k}$, we have

$$a_{i_1 i_2} C_{i_1 i_2} = a_{i_1 i_2} \sum (-1)^\sigma a_{i_3 i_4} \dots a_{i_{2n-1} i_{2n}}.$$

It remains to observe that the signs of the permutations

$$\sigma = \begin{pmatrix} 1 & 2 & \dots & 2n \\ i_1 & i_2 & \dots & i_{2n} \end{pmatrix}$$

and

$$\tau = \begin{pmatrix} i_1 & i_2 & 1 & 2 & \dots & \bar{i}_1 & \dots & \bar{i}_2 & \dots & 2n \\ i_1 & i_2 & i_3 & i_4 & \dots & \dots & \dots & \dots & \dots & i_{2n} \end{pmatrix}$$

differ by the factor of $(-1)^{i_1+i_2+1}$.

29.2. Let $J = \operatorname{diag}\left(\begin{pmatrix} 0 & 1 \\ -1 & 0 \end{pmatrix}, \dots, \begin{pmatrix} 0 & 1 \\ -1 & 0 \end{pmatrix} \right)$. It is easy to verify that $G = XJX^T$. Hence, $\operatorname{Pf}(G) = \det X$.

30.1. Clearly, if $\omega = \omega_1 \wedge e_1 \wedge \dots \wedge e_r$, then $\omega \wedge e_i = 0$. Now, suppose that $\omega \wedge e_i = 0$ for $i = 1, \dots, r$ and $e_1 \wedge \dots \wedge e_r \ne 0$. Let us complement vectors $e_1, \dots, e_r$ to a basis $e_1, \dots, e_n$ of V. Then

$$\omega = \sum a_{i_1} \dots a_{i_k} e_{i_1} \wedge \dots \wedge e_{i_k},$$

where

$$\sum a_{i_1\dots i_k} e_{i_1} \wedge \dots \wedge e_{i_k} \wedge e_i = \omega \wedge e_i = 0 \quad \text{for } i = 1, \dots, r.$$

If the nonzero tensors $e_{i_1} \wedge \dots \wedge e_{i_k} \wedge e_i$ are linearly dependent, then the tensors $e_{i_1} \wedge \dots \wedge e_{i_k}$ are also linearly dependent. Hence, $a_{i_1\dots i_k} = 0$ for $i \notin \{i_1, \dots, i_k\}$. It follows that $a_{i_1\dots i_k} \neq 0$ only if $\{1, \dots, r\} \subset \{i_1, \dots, i_k\}$ and, therefore,

$$\omega = \left(\sum b_{i_1\dots i_{k-r}} e_{i_1} \wedge \dots \wedge e_{i_{k-r}}\right) \wedge e_1 \wedge \dots \wedge e_r.$$

30.2. Consider the linear map $f : V \longrightarrow \Lambda^n V$ given by the formula $f(v) = v \wedge \omega$. Since $\dim \Lambda^n V = 1$, it follows that $\dim \operatorname{Ker} f \geq n - 1$. Hence, there exist linearly independent vectors $e_1, \dots, e_{n-1}$ belonging to $\operatorname{Ker} f$, i.e., $e_i \wedge \omega = 0$ for $i = 1, \dots, n-1$. By Problem 30.1 $\omega = \lambda e_1 \wedge \dots \wedge e_{n-1}$.

30.3. Let $W_1 = \operatorname{Span}(e_1, \dots, e_{2n})$. Let us prove that $W = W_1$. The space W is the minimal space for which $\Lambda \subset \Lambda^2 W$ (see Theorem 30.2.1). Clearly, $\Lambda \subset \Lambda^2 W_1$; hence, $W \subset W_1$ and $\dim W \leq \dim W_1 = 2n$. On the other hand, $\Lambda^n \omega \in \Lambda^{2n} W$ and $\Lambda^n \omega = n! e_1 \wedge \dots \wedge e_{2n}$ (see Problem 28.3). Hence, $\Lambda^{2n} W \neq 0$, i.e., $\dim W \geq 2n$.

30.4. Under the change of bases of X and Y the tensors z_1 and z_2 are replaced by proportional tensors. Therefore, we can assume that

$$z_1 + z_2 = (v_1 \wedge \dots \wedge v_k) \wedge (x_1 \wedge \dots \wedge x_{r-k} + y_1 \wedge \dots \wedge y_{r-k}),$$

where $v_1, \dots, v_k$ is a basis of $X \cap Y$, and the vectors $x_1, \dots, x_{r-k}$ and $y_1, \dots, y_{r-k}$ complement it to bases of X and Y. Suppose that $z_1 + z_2 = u_1 \wedge \dots \wedge u_2$. Let $u = v + x + y$, where $v \in \operatorname{Span}(v_1, \dots, v_k)$, $x \in \operatorname{Span}(x_1, \dots, x_{r-k})$ and $y \in \operatorname{Span}(y_1, \dots, y_{r-k})$. Then

$$(z_1 + z_2) \wedge u = (v_1 \wedge \dots \wedge v_k) \wedge (x_1 \wedge \dots \wedge x_{r-k} \wedge y + y_1 \wedge \dots \wedge y_{r-k} \wedge x).$$

If $r - k > 1$, then the nonzero tensors $x_1 \wedge \dots \wedge x_{r-k} \wedge y$ and $y_1 \wedge \dots \wedge y_{r-k} \wedge x$ are linearly independent. This means that in this case the equality $(z_1 + z_2) \wedge u = 0$ implies that $u \in \operatorname{Span}(v_1, \dots, v_k)$, i.e., $\operatorname{Span}(u_1, \dots, u_r) \subset \operatorname{Span}(v_1, \dots, v_k)$ and $r \leq k$. We get a contradiction; hence, $r - k = 1$.

30.5. Any two subspaces $[\omega_1]$ and $[\omega_2]$ have a common $(k-1)$-dimensional subspace (see Problem 30.4). It remains to use Theorem 9.6.1.

31.1. Let us complement the basis $e_1, \dots, e_k$ of U to a basis $e_1, \dots, e_n$ of V. Let $T = \sum \alpha_i v_1^i \otimes \dots \otimes v_p^i$. Each element $v_j^i \in V$ can be represented in the form $v_j^i = u_j^i + w_j^i$, where $u_j^i \in U$ and $w_j^i \in \operatorname{Span}(e_{k+1}, \dots, e_n)$. Hence, $T = \sum \alpha_i u_1^i \otimes \dots \otimes u_p^i + \cdots$. Expanding the elements denoted by dots with respect to the basis $e_1, \dots, e_n$, we can easily verify that no nonzero linear combination of them can belong to $T_0^p(U)$. Since $T \in T_0^p(U)$, it follows that $T = \sum \alpha_i u_1^i \otimes \dots \otimes u_p^i$, i.e., the rank of T in $T_0^p(U)$ is not greater than its rank in $T_0^p(V)$. The converse inequality is obvious.

31.2. Let

$$e_1^{\otimes p} + \dots + e_k^{\otimes p} = u_1^1 \otimes \dots \otimes u_p^1 + \dots + u_1^r \otimes \dots \otimes u_p^r.$$

By Problem 31.1, we may assume that $u_j^i \in \operatorname{Span}(e_1, \dots, e_k)$. Then $u_1^i = \sum_j \alpha_{ij} e_j$, i.e.,

$$\sum_i u_1^i \otimes \dots \otimes u_p^i = \sum_j e_j \otimes \left(\sum_i \alpha_{ij} u_2^i \right) \otimes \dots \otimes u_p^i.$$

Hence

$$\sum_i \alpha_{ij} u_2^i \otimes \dots \otimes u_p^i = e_j^{\otimes p-1}$$

and, therefore, k linearly independent vectors $e_1^{\otimes p-1}, \dots, e_k^{\otimes p-1}$ belong to the space

$$\operatorname{Span}(u_2^1 \otimes \dots \otimes u_p^1, \quad \dots, \quad u_2^r \otimes \dots \otimes u_p^r)$$

whose dimension does not exceed r. Hence, $r \geq k$.

32.1. Suppose that $AXB = X^T$ for all matrices X of size $m \times n$. Then the matrices A and B are of size $n \times m$ and $\sum_{k,s} a_{ik} x_{ks} b_{sj} = x_{ji}$. Hence, $a_{ij} b_{ij} = 1$ and $a_{ik} b_{sj} = 0$ if $k \neq j$ or $s \neq i$. The first set of equalities implies that all elements of A and B are nonzero, but then the second set of equalities cannot hold.

The equality $AX^T B = X$ cannot hold for all matrices X either, because it implies $B^T X A^T = X^T$.

32.2. Let $B, X \in M_{n,n}$. The equation $BX = \lambda X$ has a nonzero solution X if and only if λ is an eigenvalue of B. If λ is an eigenvalue of B, then $BX = \lambda X$ for a nonzero matrix X. Hence, $f(B) f(X) = \lambda f(X)$ and, therefore, λ is an eigenvalue of $f(B)$. Let $B = \operatorname{diag}(\beta_1, \dots, \beta_n)$, where β_i are distinct nonzero numbers. Then the matrix $f(B)$ is similar to B, i.e., $f(B) = A_1 B A_1^{-1}$.

Let $g(X) = A_1^{-1} f(X) A_1$. Then $g(B) = B$. If $X = \left\| x_{ij} \right\|_1^n$, then $BX = \left\| \beta_i x_{ij} \right\|_1^n$ and $XB = \left\| x_{ij} \beta_j \right\|_1^n$. Hence, $BX = \beta_i X$ only if all rows of X except the ith one are zero and $XB = \beta_j X$ only if all columns of X except the jth are zero. Let E_{ij} be given by $E_{ij} = \left\| a_{pq} \right\|_1^n$, where $a_{pq} = \delta_{pi} \delta_{qj}$. Then $Bg(E_{ij}) = \beta_i g(E_{ij})$ and $g(E_{ij}) B = \beta_j g(E_{ij})$ and, therefore, $g(E_{ij}) = \alpha_{ij} E_{ij}$. As is easy to see, $E_{ij} = E_{i1} E_{1j}$. Hence, $\alpha_{ij} = \alpha_{i1} \alpha_{1j}$. Besides, $E_{ii}^2 = E_{ii}$; hence, $\alpha_{ii}^2 = \alpha_{ii}$ and, therefore, $\alpha_{i1} \alpha_{1i} = \alpha_{ii} = 1$, i.e., $\alpha_{1i} = \alpha_{i1}^{-1}$. It follows that $\alpha_{ij} = \alpha_{i1} \cdot \alpha_{j1}^{-1}$. Hence, $g(X) = A_2 X A_2^{-1}$, where $A_2 = \operatorname{diag}(\alpha_{11}, \dots, \alpha_{n1})$, and, therefore, $f(X) = AXA^{-1}$, where $A = A_1 A_2$.

CHAPTER VI

MATRIX INEQUALITIES

33. Inequalities for symmetric and Hermitian matrices

33.1. Let A and B be Hermitian matrices. We will write that $A > B$ (resp. $A \geq B$) if $A - B$ is a positive (resp. nonnegative) definite matrix. The inequality $A > 0$ means that A is positive definite.

33.1.1. THEOREM. *If $A > B > 0$, then $A^{-1} < B^{-1}$.*

PROOF. By Theorem 20.1, there exists a matrix P such that $A = P^*P$ and $B = P^*DP$, where $D = \operatorname{diag}(d_1, \dots, d_n)$. The inequality $x^*Ax > x^*Bx$ is equivalent to the inequality $y^*y > y^*Dy$, where $y = Px$. Hence, $A > B$ if and only if $d_i > 1$. Therefore, $A^{-1} = Q^*Q$ and $B^{-1} = Q^*D_1Q$, where $D_1 = \operatorname{diag}(d_1^{-1}, \dots, d_n^{-1})$ and $d_i^{-1} < 1$ for all i; thus, $A^{-1} < B^{-1}$. □

33.1.2. THEOREM. *If $A > 0$, then $A + A^{-1} \geq 2I$.*

PROOF. Let us express A in the form $A = U^*DU$, where U is a unitary matrix and $D = \operatorname{diag}(d_1, \dots, d_n)$, where $d_i > 0$. Then

$$x^*(A + A^{-1})x = x^*U^*(D + D^{-1})Ux \geq 2x^*U^*Ux = 2x^*x,$$

since $d_i + d_i^{-1} \geq 2$. □

33.1.3. THEOREM. *If A is a real matrix and $A > 0$, then*

$$(A^{-1}x, x) = \max_y (2(x, y) - (Ay, y)).$$

PROOF. There exists for a matrix A an orthonormal basis such that $(Ax, x) = \sum \alpha_i x_i^2$. Since

$$2x_iy_i - \alpha_i y_i^2 = -\alpha_i(y_i - \alpha_i^{-1}x_i)^2 + \alpha_i^{-1}x_i^2,$$

it follows that

$$\max_y (2(x, y) - (Ay, y)) = \sum \alpha_i^{-1}x_i^2 = (A^{-1}x, x)$$

and the maximum is attained at $y = (y_1, \dots, y_n)$, where $y_i = \alpha_i^{-1}x_i$. □

33.2.1. THEOREM. *Let* $A = \begin{pmatrix} A_1 & B \\ B^* & A_2 \end{pmatrix} > 0$. *Then* $\det A \le \det A_1 \det A_2$.

PROOF. The matrices A_1 and A_2 are positive definite. It is easy to verify (see 3.1) that

$$\det A = \det A_1 \det(A_2 - B^* A_1^{-1} B).$$

The matrix $B^* A_1^{-1} B$ is positive definite; hence, $\det(A_2 - B^* A_1^{-1} B) \le \det A_2$ (see Problem 33.1). Thus, $\det A \le \det A_1 \det A_2$ and the equality is only attained if $B^* A_1^{-1} B = 0$, i.e., $B = 0$. □

33.2.1.1. COROLLARY (Hadamard's inequality). *If a matrix* $A = \|a_{ij}\|_1^n$ *is positive definite, then* $\det A \le a_{11}a_{22}\dots a_{nn}$ *and the equality is only attained if* A *is a diagonal matrix.*

33.2.1.2. COROLLARY. *If* X *is an arbitrary matrix, then*

$$|\det X|^2 \le \sum_i |x_{1i}|^2 \cdots \sum_i |x_{ni}|^2.$$

To prove Corollary 33.2.1.2 it suffices to apply Corollary 33.2.1.1 to the matrix $A = XX^*$.

33.2.2. THEOREM. *Let* $A = \begin{pmatrix} A_1 & B \\ B^* & A_2 \end{pmatrix}$ *be a positive definite matrix, where* B *is a square matrix. Then*

$$|\det B|^2 \le \det A_1 \det A_2.$$

PROOF ([Everitt, 1958]). Since

$$T^*AT = \begin{pmatrix} A_1 & 0 \\ 0 & A_2 - B^* A_1^{-1} B \end{pmatrix} > 0 \text{ for } T = \begin{pmatrix} I & -A_1^{-1}B \\ 0 & I \end{pmatrix},$$

we directly deduce that $A_2 - B^* A_1^{-1} B > 0$. Hence,

$$\det(B^* A_1^{-1} B) \le \det(B^* A_1^{-1} B) + \det(A_2 - B^* A_1^{-1} B) \le \det A_2$$

(see Problem 33.1), i.e.,

$$|\det B|^2 = \det(BB^*) \le \det A_1 \det A_2. \quad \square$$

33.2.3 THEOREM (Szasz's inequality). *Let* A *be a positive definite nondiagonal matrix of order* n; *let* P_k *be the product of all principal* k*-minors of* A. *Then*

$$P_1 > P_2^{a_2} > \dots > P_{n-1}^{a_{n-1}} > P_n, \text{ where } a_k = \binom{n-1}{k-1}^{-1}.$$

PROOF ([Mirsky, 1957]). The required inequality can be rewritten in the form $P_k^{n-k} > P_{k+1}^k$ $(1 \le k \le n-1)$. For $n = 2$ the proof is obvious. For a diagonal matrix we have $P_k^{n-k} = P_{k+1}^k$. Suppose that $P_k^{n-k} > P_{k+1}^k$ $(1 \le k \le n-1)$ for some $n \ge 2$. Consider a matrix A of order $n+1$. Let A_r be the matrix obtained from A by deleting the rth row and the rth column; let $P_{k,r}$ be the product of all principal k-minors of A_r. By the inductive hypothesis

$$(1) \qquad P_{k,r}^{n-k} \ge P_{k+1,r}^k \quad \text{for } 1 \le k \le n-1 \text{ and } 1 \le r \le n+1,$$

where at least one of the matrices A_r is not a diagonal one and, therefore, at least one of the inequalities (1) is strict. Hence,

$$\prod_{r=1}^{n+1} P_{k,r}^{n-k} > \prod_{r=1}^{n+1} P_{k+1,r}^k \quad \text{for } 1 \le k \le n-1,$$

i.e., $P_k^{(n-k)(n+1-k)} > P_{k+1}^{(n-k)k}$. Extracting the $(n-k)$th root for $n \ne k$ we get the required conclusion.

For $n = k$ consider the matrix $\operatorname{adj} A = B = \left\|b_{ij}\right\|_1^{n+1}$. Since $A > 0$, it follows that $B > 0$ (see Problem 19.4). By Hadamard's inequality

$$b_{11} \dots b_{n+1,n+1} > \det B = (\det A)^n,$$

i.e., $P_n > P_{n+1}^n$. $\square$

REMARK. The inequality $P_1 > P_n$ coincides with Hadamard's inequality.

33.3.1. THEOREM. *Let $\alpha_i > 0$, $\sum \alpha_i = 1$ and $A_i > 0$. Then*

$$|\alpha_1 A_1 + \dots + \alpha_k A_k| \ge |A_1|^{\alpha_1} \dots |A_k|^{\alpha_k}.$$

PROOF ([Mirsky, 1955]). First, consider the case $k = 2$. Let $A, B > 0$. Then $A = P^*\Lambda P$ and $B = P^*P$, where $\Lambda = \operatorname{diag}(\lambda_1, \dots, \lambda_n)$. Hence,

$$|\alpha A + (1-\alpha)B| = |P^*P| \cdot |\alpha\Lambda + (1-\alpha)I| = |B| \prod_{i=1}^n (\alpha\lambda_i + 1 - \alpha).$$

If $f(t) = \lambda^t$, where $\lambda > 0$, then $f''(t) = (\ln\lambda)^2\lambda^t \ge 0$ and, therefore,

$$f(\alpha x + (1-\alpha)y) \le \alpha f(x) + (1-\alpha)f(y) \quad \text{for } 0 < \alpha < 1.$$

For $x = 1$ and $y = 0$ we get $\lambda^\alpha \le \alpha\lambda + 1 - \alpha$. Hence,

$$\prod(\alpha\lambda_i + 1 - \alpha) \ge \prod \lambda_i^\alpha = |\Lambda|^\alpha = |A|^\alpha |B|^{-\alpha}.$$

The rest of the proof will be carried out by induction on k; we will assume that $k \ge 3$. Since

$$\alpha_1 A_1 + \dots + \alpha_k A_k = (1-\alpha_k)B + \alpha_k A_k$$

and the matrix

$$B = \frac{\alpha_1}{1-\alpha_k}A_1 + \cdots + \frac{\alpha_{k-1}}{1-\alpha_k}A_{k-1}$$

is positive definite, it follows that

$$|\alpha_1 A_1 + \cdots + \alpha_k A_k| \geq \left|\frac{\alpha_1}{1-\alpha_k}A_1 + \cdots + \frac{\alpha_{k-1}}{1-\alpha_k}A_{k-1}\right|^{1-\alpha_k} |A_k|^{\alpha_k}.$$

Since

$$\frac{\alpha_1}{1-\alpha_k} + \cdots + \frac{\alpha_{k-1}}{1-\alpha_k} = 1,$$

it follows that

$$\left|\frac{\alpha_1}{1-\alpha_k}A_1 + \cdots + \frac{\alpha_{k-1}}{1-\alpha_k}A_{k-1}\right| \geq |A_1|^{\frac{\alpha_1}{1-\alpha_k}} \dots |A_{k-1}|^{\frac{\alpha_{k-1}}{1-\alpha_k}}. \quad \square$$

REMARK. It is possible to verify that the equality takes place if and only if $A_1 = \cdots = A_k$.

33.3.2. THEOREM. *Let λ_i be arbitrary complex numbers and $A_i \geq 0$. Then*

$$|\det(\lambda_1 A_1 + \cdots + \lambda_k A_k)| \leq \det(|\lambda_1| A_1 + \cdots + |\lambda_k| A_k).$$

PROOF ([Frank, 1965]). Let $k = 2$; we can assume that $\lambda_1 = 1$ and $\lambda_2 = \lambda$. There exists a unitary matrix U such that the matrix $UA_1U^{-1} = D$ is a diagonal one. Then $M = UA_2U^{-1} \geq 0$ and

$$\det(A_1 + \lambda A_2) = \det(D + \lambda M) = \sum_{p=0}^{n} \lambda^p \sum_{i_1 < \cdots < i_p} M\begin{pmatrix} i_1 & \dots & i_p \\ i_1 & \dots & i_p \end{pmatrix} d_{j_1} \dots d_{j_{n-p}},$$

where the set $(j_1, \dots, j_{n-p})$ complements $(i_1, \dots, i_p)$ to $(1, \dots, n)$. Since M and D are nonnegative definite, $M\begin{pmatrix} i_1 & \dots & i_p \\ i_1 & \dots & i_p \end{pmatrix} \geq 0$ and $d_j \geq 0$. Hence,

$$\begin{aligned} |\det(A_1 + \lambda A_2)| &\leq \sum_{p=0}^{n} |\lambda|^p \sum_{i_1 < \cdots < i_p} M\begin{pmatrix} i_1 & \dots & i_p \\ i_1 & \dots & i_p \end{pmatrix} \cdot d_{j_1} \dots d_{j_{n-p}} \\ &= \det(D + |\lambda| M) = \det(A_1 + |\lambda| A_2). \end{aligned}$$

Now, let us prove the inductive step. Let us again assume that $\lambda_1 = 1$. Let $A = A_1$ and $A' = \lambda_2 A_2 + \cdots + \lambda_{k+1} A_{k+1}$. There exists a unitary matrix U such that the matrix $UAU^{-1} = D$ is a diagonal one; matrices $M_j = UA_jU^{-1}$ and $M = UA'U^{-1}$ are nonnegative definite. Hence,

$$|\det(A + A')| = |\det(D + M)| \leq \sum_{p=0}^{n} \sum_{i_1 < \cdots < i_p} M\begin{pmatrix} i_1 & \dots & i_p \\ i_1 & \dots & i_p \end{pmatrix} d_{j_1} \dots d_{j_{n-p}}.$$

Since $M = \lambda_2 M_2 + \dots + \lambda_{k+1} M_{k+1}$, by the inductive hypothesis

$$M\begin{pmatrix} i_1 & \dots & i_p \\ i_1 & \dots & i_p \end{pmatrix} \le \det\left(\sum_{j=2}^{k+1} |\lambda_j| M_j \begin{pmatrix} i_1 & \dots & i_p \\ i_1 & \dots & i_p \end{pmatrix}\right).$$

It remains to notice that

$$\sum_{p=0}^{n} \sum_{i_1<\dots<i_p} d_{j_1}\dots d_{j_{n-p}} \det\left(\sum_{j=2}^{k+1} |\lambda_j| M_j \begin{pmatrix} i_1 & \dots & i_p \\ i_1 & \dots & i_p \end{pmatrix}\right)$$
$$= \det(D + |\lambda_2| M_2 + \dots + |\lambda_{k+1}| M_{k+1}) = \det\left(\sum |\lambda_i| A_i\right). \quad \square$$

33.4. THEOREM. *Let A and B be positive definite real matrices and let A_1 and B_1 be the matrices obtained from A and B, respectively, by deleting the first row and the first column. Then*

$$\frac{|A+B|}{|A_1+B_1|} \ge \frac{|A|}{|A_1|} + \frac{|B|}{|B_1|}.$$

PROOF ([Bellman, 1955]). If $A > 0$, then

$$(x, Ax)(y, A^{-1}y) \ge (x, y)^2. \tag{1}$$

Indeed, there exists a unitary matrix U such that $U^*AU = \Lambda = \operatorname{diag}(\lambda_1, \dots, \lambda_n)$, where $\lambda_i > 0$. Making the change $x = Ua$ and $y = Ub$ we get the Cauchy-Schwarz inequality

$$\left(\sum \lambda_i a_i^2\right)\left(\sum b_i^2/\lambda_i\right) \ge \left(\sum a_i b_i\right)^2. \tag{2}$$

The inequality (2) turns into equality for $a_i = b_i/\lambda_i$ and, therefore,

$$f(A) = \frac{1}{(y, A^{-1}y)} = \min_x \frac{(x, Ax)}{(x, y)^2}.$$

Now, let us prove that if $y = (1, 0, 0, \dots, 0) = e_1$, then $f(A) = |A|/|A_1|$. Indeed,

$$(e_1, A^{-1}e_1) = e_1 A^{-1} e_1^T = \frac{e_1 \operatorname{adj} A e_1^T}{|A|} = \frac{(\operatorname{adj} A)_{11}}{|A|} = \frac{|A_1|}{|A|}.$$

It remains to notice that for any functions g and h

$$\min_x g(x) + \min_x h(x) \le \min_x (g(x) + h(x))$$

and set

$$g(x) = \frac{(x, Ax)}{(x, e_1)} \text{ and } h(x) = \frac{(x, Bx)}{(x, e_1)}. \quad \square$$

Problems

33.1. Let A and B be matrices of order n $(n > 1)$, where $A > 0$ and $B \geq 0$. Prove that $|A + B| \geq |A| + |B|$ and the equality is only attained for $B = 0$.

33.2. The matrices A and B are Hermitian and $A > 0$. Prove that $\det A \leq |\det(A + iB)|$ and the equality is only attained when $B = 0$.

33.3. Let A_k and B_k be the upper left corner submatrices of order k of positive definite matrices A and B such that $A > B$. Prove that

$$|A_k| > |B_k|.$$

33.4. Let A and B be real symmetric matrices and $A \geq 0$. Prove that if $C = A + iB$ is not invertible, then $Cx = 0$ for some nonzero real vector x.

33.5. A real symmetric matrix A is positive definite. Prove that

$$\det \begin{pmatrix} 0 & x_1 & \dots & x_n \\ x_1 & & & \\ \vdots & & A & \\ x_n & & & \end{pmatrix} \leq 0.$$

33.6. Let $A > 0$ and let n be the order of A. Prove that $|A|^{1/n} = \min \frac{1}{n} \operatorname{tr}(AB)$, where the minimum is taken over all positive definite matrices B with determinant 1.

34. Inequalities for eigenvalues

34.1.1. THEOREM (Schur's inequality). *Let $\lambda_1, \dots, \lambda_n$ be eigenvalues of $A = \left\|a_{ij}\right\|_1^n$. Then $\sum_{i=1}^n |\lambda_i|^2 \leq \sum_{i,j=1}^n |a_{ij}|^2$ and the equality is attained if and only if A is a normal matrix.*

PROOF. There exists a unitary matrix U such that $T = U^*AU$ is an upper triangular matrix and T is a diagonal matrix if and only if A is a normal matrix (cf. 17.1). Since $T^* = U^*A^*U$, then $TT^* = U^*AA^*U$ and, therefore, $\operatorname{tr}(TT^*) = \operatorname{tr}(AA^*)$. It remains to notice that

$$\operatorname{tr}(AA^*) = \sum_{i,j=1}^n |a_{ij}|^2 \quad \text{and} \quad \operatorname{tr}(TT^*) = \sum_{i=1}^n |\lambda_i|^2 + \sum_{i<j} |t_{ij}|^2. \qquad \square$$

34.1.2. THEOREM. *Let $\lambda_1, \dots, \lambda_n$ be eigenvalues of $A = B + iC$, where B and C are Hermitian matrices. Then*

$$\sum_{i=1}^n |\operatorname{Re} \lambda_i|^2 \leq \sum_{i,j=1}^n |b_{ij}|^2 \quad \textit{and} \quad \sum_{i=1}^n |\operatorname{Im} \lambda_i|^2 \leq \sum_{i,j=1}^n |c_{ij}|^2.$$

PROOF. Let, as in the proof of Theorem 34.1.1, $T = U^*AU$. We have $B = \frac{1}{2}(A + A^*)$ and $iC = \frac{1}{2}(A - A^*)$; therefore, $U^*BU = (T + T^*)/2$ and $U^*(iC)U = (T - T^*)/2$. Hence,

$$\sum_{i,j=1}^n |b_{ij}|^2 = \operatorname{tr}(BB^*) = \frac{\operatorname{tr}(T + T^*)^2}{4} = \sum_{i=1}^n |\operatorname{Re} \lambda_i|^2 + \sum_{i<j} \frac{|t_{ij}|^2}{2}$$

and $\sum_{i,j=1}^n |c_{ij}|^2 = \sum_{i=1}^n |\operatorname{Im} \lambda_i|^2 + \sum_{i<j} \frac{1}{2}|t_{ij}|^2$. $\square$

34.2.1. THEOREM (H. Weyl). *Let A and B be Hermitian matrices, $C = A + B$. Let the eigenvalues of these matrices form increasing sequences: $\alpha_1 \leq \dots \leq \alpha_n$, $\beta_1 \leq \dots \leq \beta_n$, $\gamma_1 \leq \dots \leq \gamma_n$. Then*
a) $\gamma_i \geq \alpha_j + \beta_{i-j+1}$ *for* $i \geq j$;
b) $\gamma_i \leq \alpha_j + \beta_{i-j+n}$ *for* $i \leq j$.

PROOF. Select orthonormal bases $\{a_i\}$, $\{b_i\}$ and $\{c_i\}$ for each of the matrices A, B and C such that $Aa_i = \alpha_i a_i$, etc. First, suppose that $i \geq j$. Consider the subspaces $V_1 = \operatorname{Span}(a_j, \dots, a_n)$, $V_2 = \operatorname{Span}(b_{i-j+1}, \dots, b_n)$ and $V_3 = \operatorname{Span}(c_1, \dots, c_i)$. Since $\dim V_1 = n - j + 1$, $\dim V_2 = n - i + j$ and $\dim V_3 = i$, it follows that

$$\dim(V_1 \cap V_2 \cap V_3) \geq \dim V_1 + \dim V_2 + \dim V_3 - 2n = 1.$$

Therefore, the subspace $V_1 \cap V_2 \cap V_3$ contains a vector x of unit length. Clearly,

$$\alpha_j + \beta_{i-j+1} \leq (x, Ax) + (x, Bx) = (x, Cx) \leq \gamma_i.$$

Replacing matrices A, B and C by $-A$, $-B$ and $-C$ we can reduce the inequality b) to the inequality a). $\square$

34.2.2. THEOREM. *Let $A = \begin{pmatrix} B & C \\ C^* & D \end{pmatrix}$ be an Hermitian matrix. Let the eigenvalues of A and B form increasing sequences: $\alpha_1 \leq \dots \leq \alpha_n$, $\beta_1 \leq \dots \leq \beta_m$. Then*

$$\alpha_i \leq \beta_i \leq \alpha_{i+n-m}.$$

PROOF. For A and B take orthonormal eigenbases $\{a_i\}$ and $\{b_i\}$; we can assume that A and B act in the spaces V and U, where $U \subset V$. Consider the subspaces $V_1 = \operatorname{Span}(a_i, \dots, a_n)$ and $V_2 = \operatorname{Span}(b_1, \dots, b_i)$. The subspace $V_1 \cap V_2$ contains a unit vector x. Clearly,

$$\alpha_i \leq (x, Ax) = (x, Bx) \leq \beta_i.$$

Applying this inequality to the matrix $-A$ we get $-\alpha_{n-i+1} \leq -\beta_{m-i+1}$, i.e., $\beta_j \leq \alpha_{j+n-m}$. $\square$

34.3. THEOREM. *Let A and B be Hermitian projections, i.e., $A^2 = A$ and $B^2 = B$. Then the eigenvalues of AB are real and belong to the segment* $[0, 1]$.

PROOF ([Afriat, 1956]). The eigenvalues of the matrix $AB = (AAB)B$ coincide with eigenvalues of the matrix $B(AAB) = (AB)^*AB$ (see 11.6). The latter matrix is nonnegative definite and, therefore, its eigenvalues are real and nonnegative. If all eigenvalues of AB are zero, then all eigenvalues of the Hermitian matrix $(AB)^*AB$ are also zero; hence, $(AB)^*AB$ is zero itself and, therefore, $AB = 0$. Now, suppose that $ABx = \lambda x \neq 0$. Then $Ax = \lambda^{-1}AABx = \lambda^{-1}ABx = x$ and, therefore,

$$(x, Bx) = (Ax, Bx) = (x, ABx) = \lambda(x, x), \quad \text{i.e., } \lambda = \frac{(x, Bx)}{(x, x)}.$$

For B there exists an orthonormal basis such that $(x, Bx) = \beta_1|x_1|^2 + \dots + \beta_n|x_n|^2$, where either $\beta_i = 0$ or 1. Hence, $\lambda \leq 1$. $\square$

34.4. The numbers $\sigma_i = \sqrt{\mu_i}$, where μ_i are eigenvalues of A^*A, are called *singular values* of A. For an Hermitian nonnegative definite matrix the singular values and the eigenvalues coincide. If $A = SU$ is a polar decomposition of A, then the singular values of A coincide with the eigenvalues of S. For S, there exists a unitary matrix V such that $S = V\Lambda V^*$, where Λ is a diagonal matrix. Therefore, any matrix A can be represented in the form $A = V\Lambda W$, where V and W are unitary matrices and $\Lambda = \operatorname{diag}(\sigma_1, \dots, \sigma_n)$.

34.4.1. THEOREM. *Let $\sigma_1, \dots, \sigma_n$ be the singular values of A, where $\sigma_1 \geq \dots \geq \sigma_n$, and let $\lambda_1, \dots, \lambda_n$ be the eigenvalues of A, where $|\lambda_1| \geq \dots \geq |\lambda_n|$. Then $|\lambda_1 \dots \lambda_m| \leq \sigma_1 \dots \sigma_m$ for $m \leq n$.*

PROOF. Let $Ax = \lambda_1 x$. Then

$$|\lambda_1|^2(x, x) = (Ax, Ax) = (x, A^*Ax) \leq \sigma_1^2(x, x)$$

since σ_1^2 is the maximal eigenvalue of the Hermitian operator A^*A. Hence, $|\lambda_1| \leq \sigma_1$ and for $m = 1$ the inequality is proved. Let us apply the inequality obtained to the operators $\Lambda^m(A)$ and $\Lambda^m(A^*A)$ (see 28.5). Their eigenvalues are equal to $\lambda_{i_1} \dots \lambda_{i_m}$ and $\sigma_{i_1}^2 \dots \sigma_{i_m}^2$; hence, $|\lambda_1 \dots \lambda_m| \leq \sigma_1 \dots \sigma_m$.

It is also clear that $|\lambda_1 \dots \lambda_n| = |\det A| = \sqrt{\det(A^*A)} = \sigma_1 \dots \sigma_n$. □

34.4.2. THEOREM. *Let $\sigma_1 \geq \dots \geq \sigma_n$ be the singular values of A and let $\tau_1 \geq \dots \geq \tau_n$ be the singular values of B. Then $|\operatorname{tr}(AB)| \leq \sum_{i=1}^n \sigma_i \tau_i$.*

PROOF [Mirsky, 1975]). Let $A = U_1 S V_1$ and $B = U_2 T V_2$, where U_i and V_i are unitary matrices, $S = \operatorname{diag}(\sigma_1, \dots, \sigma_n)$ and $T = \operatorname{diag}(\tau_1, \dots, \tau_n)$. Then

$$\operatorname{tr}(AB) = \operatorname{tr}(U_1 S V_1 U_2 T V_2) = \operatorname{tr}(V_2 U_1 S V_1 U_2 T) = \operatorname{tr}(U^T S V T),$$

where $U = (V_2 U_1)^T$ and $V = V_1 U_2$. Hence,

$$|\operatorname{tr}(AB)| = |\sum u_{ij} v_{ij} \sigma_i \tau_j| \leq \frac{\sum |u_{ij}|^2 \sigma_i \tau_j + \sum |v_{ij}|^2 \sigma_i \tau_j}{2}.$$

The matrices whose (i, j)th elements are $|u_{ij}|^2$ and $|v_{ij}|^2$ are doubly stochastic and, therefore, $\sum |u_{ij}|^2 \sigma_i \tau_j \leq \sum \sigma_i \tau_i$ and $\sum |v_{ij}|^2 \sigma_i \tau_j \leq \sum \sigma_i \tau_j$ (see Problem 38.1). □

Problems

34.1 (Gershgorin discs). Prove that every eigenvalue of $\|a_{ij}\|_1^n$ belongs to one of the discs $|a_{kk} - z| \leq \rho_k$, where $\rho_k = \sum_{i \neq j} |a_{kj}|$.

34.2. Prove that if U is a unitary matrix and $S \geq 0$, then $|\operatorname{tr}(US)| \leq \operatorname{tr} S$.

34.3. Prove that if A and B are nonnegative definite matrices, then $|\operatorname{tr}(AB)| \leq \operatorname{tr} A \cdot \operatorname{tr} B$.

34.4. Matrices A and B are Hermitian. Prove that $\operatorname{tr}(AB)^2 \leq \operatorname{tr}(A^2 B^2)$.

34.5 ([Cullen, 1965]). Prove that $\lim_{k \to \infty} A^k = 0$ if and only if one of the following conditions holds:

a) the absolute values of all eigenvalues of A are less than 1;

b) there exists a positive definite matrix H such that $H - A^*HA > 0$.

Singular values

34.6. Prove that if all singular values of A are equal, then $A = \lambda U$, where U is a unitary matrix.

34.7. Prove that if the singular values of A are equal to $\sigma_1, \dots, \sigma_n$, then the singular values of $\operatorname{adj} A$ are equal to $\prod_{i\neq 1}\sigma_i, \dots, \prod_{i\neq n}\sigma_i$.

34.8. Let $\sigma_1, \dots, \sigma_n$ be the singular values of A. Prove that the eigenvalues of $\begin{pmatrix} 0 & A \\ A^* & 0 \end{pmatrix}$ are equal to $\sigma_1, \dots, \sigma_n, -\sigma_1, \dots, -\sigma_n$.

35. Inequalities for matrix norms

35.1. The *operator* (or *spectral*) *norm* of a matrix A is $\|A\|_s = \sup\limits_{|x|\neq 0}(|Ax|/|x|)$. The number $\rho(A) = \max|\lambda_i|$, where $\lambda_1, \dots, \lambda_n$ are the eigenvalues of A, is called the *spectral radius* of A. Since there exists a nonzero vector x such that $Ax = \lambda_i x$, it follows that $\|A\|_s \geq \rho(A)$. In the complex case this is obvious. In the real case we can express the vector x as $x_1 + ix_2$, where x_1 and x_2 are real vectors. Then

$$|Ax_1|^2 + |Ax_2|^2 = |Ax|^2 = |\lambda_i|^2(|x_1|^2 + |x_2|^2),$$

and, therefore, both inequalities $|Ax_1| < \lambda_i||x_1|$ and $|Ax_2| < \lambda_i||x_2|$ can not hold simultaneously.

It is easy to verify that if U is a unitary matrix, then $\|A\|_s = \|AU\|_s = \|UA\|_s$. To this end it suffices to observe that

$$\frac{|AUx|}{|x|} = \frac{|Ay|}{|U^{-1}y|} = \frac{|Ay|}{|y|},$$

where $y = Ux$ and $|UAx|/|x| = |Ax|/|x|$.

35.1.1. Theorem. $\|A\|_s = \sqrt{\rho(A^*A)}$.

Proof. If $\Lambda = \operatorname{diag}(\lambda_1, \dots, \lambda_n)$, then

$$\left(\frac{|\Lambda x|}{|x|}\right)^2 = \frac{\sum|\lambda_i x_i|^2}{\sum|x_i|^2} \leq \max_i|\lambda_i|.$$

Let $|\lambda_j| = \max\limits_i|\lambda_i|$ and $\Lambda x = \lambda_j x$. Then $|\Lambda x|/|x| = |\lambda_j|$. Therefore, $\|\Lambda\|_s = \rho(\Lambda)$.

Any matrix A can be represented in the form $A = U\Lambda V$, where U and V are unitary matrices and Λ is a diagonal matrix with the singular values of A standing on its diagonal (see 34.4). Hence, $\|A\|_s = \|\Lambda\|_s = \rho(\Lambda) = \sqrt{\rho(A^*A)}$. □

35.1.2. Theorem. *If A is a normal matrix, then $\|A\|_s = \rho(A)$.*

Proof. A normal matrix A can be represented in the form $A = U^*\Lambda U$, where $\Lambda = \operatorname{diag}(\lambda_1, \dots, \lambda_n)$ and U is a unitary matrix. Therefore, $A^*A = U^*\Lambda\overline{\Lambda}U$. Let $Ae_i = \lambda_i e_i$ and $x_i = U^{-1}e_i$. Then $A^*Ax_i = |\lambda_i|^2 x_i$ and, therefore, $\rho(A^*A) = \rho(A)^2$. □

35.2. The *Euclidean norm* of a matrix A is

$$\|A\|_e = \sqrt{\sum_{i,j}|a_{ij}|^2} = \sqrt{\operatorname{tr}(A^*A)} = \sqrt{\sum_i \sigma_i^2},$$

where σ_i are the singular values of A.

If U is a unitary matrix, then

$$\|AU\|_e = \sqrt{\operatorname{tr}(U^*A^*AU)} = \sqrt{\operatorname{tr}(A^*A)} = \|A\|_e$$

and $\|UA\|_e = \|A\|_e$.

THEOREM. *If A is a matrix of order n, then*

$$\|A\|_s \le \|A\|_e \le \sqrt{n}\|A\|_s.$$

PROOF. Let $\sigma_1, \dots, \sigma_n$ be the singular values of A and $\sigma_1 \ge \dots \ge \sigma_n$. Then $\|A\|_s^2 = \sigma_1^2$ and $\|A\|_e^2 = \sigma_1^2 + \dots + \sigma_n^2$. Clearly, $\sigma_1^2 \le \sigma_1^2 + \dots + \sigma_n^2 \le n\sigma_1^2$. $\square$

REMARK. The Euclidean and spectral norms are invariant with respect to the action of the group of unitary matrices. Therefore, it is not accidental that the Euclidean and spectral norms are expressed in terms of the singular values of the matrix: they are also invariant with respect to this group.

If $f(A)$ is an arbitrary matrix function and $f(A) = f(UA) = f(AU)$ for any unitary matrix U, then f only depends on the singular values of A. Indeed, $A = U\Lambda V$, where $\Lambda = \operatorname{diag}(\sigma_1, \dots, \sigma_n)$ and U and V are unitary matrices. Hence, $f(A) = f(\Lambda)$. Observe that in this case $A^* = V^*\Lambda U^*$ and, therefore, $f(A^*) = f(A)$. In particular $\|A^*\|_e = \|A\|_e$ and $\|A^*\|_s = \|A\|_s$.

35.3.1. THEOREM. *Let A be an arbitrary matrix, S an Hermitian matrix. Then $\|A - \frac{A+A^*}{2}\| \le \|A - S\|$, where $\|\cdot\|$ is either the Euclidean or the operator norm.*

PROOF.

$$\left\|A - \frac{A+A^*}{2}\right\| = \left\|\frac{A-S}{2} + \frac{S-A^*}{2}\right\| \le \frac{\|A-S\|}{2} + \frac{\|S-A^*\|}{2}.$$

Besides, $\|S - A^*\| = \|(S - A^*)^*\| = \|S - A\|$. $\square$

35.3.2. THEOREM. *Let $A = US$ be the polar decomposition of A and W a unitary matrix. Then $\|A - U\|_e \le \|A - W\|_e$ and if $|A| \ne 0$, then the equality is only attained for $W = U$.*

PROOF. It is clear that

$$\|A - W\|_e = \|SU - W\|_e = \|S - WU^*\|_e = \|S - V\|_e,$$

where $V = WU^*$ is a unitary matrix. Besides,

$$\|S - V\|_e^2 = \operatorname{tr}(S - V)(S - V^*) = \operatorname{tr} S^2 + \operatorname{tr} I - \operatorname{tr}(SV + V^*S).$$

By Problem 34.2, $|\operatorname{tr}(SV)| \le \operatorname{tr} S$ and $|\operatorname{tr}(V^*S)| \le \operatorname{tr} S$. It follows that $\|S - V\|_e^2 \le \|S - I\|_e^2$. If $S > 0$, then the equality is only attained if $V = e^{i\varphi}I$ and $\operatorname{tr} S = e^{i\varphi}\operatorname{tr} S$, i.e., $WU^* = V = I$. $\square$

35.4. THEOREM ([Franck,1961]). *Let A be an invertible matrix, X a noninvertible matrix. Then*

$$\|A - X\|_s \geq \|A^{-1}\|_s^{-1}$$

and if $\|A^{-1}\|_s = \rho(A^{-1})$, then there exists a noninvertible matrix X such that

$$\|A - X\|_s = \|A^{-1}\|_s^{-1}.$$

PROOF. Take a vector v such that $Xv = 0$ and $v \neq 0$. Then

$$\|A - X\|_s \geq \frac{|(A - X)v|}{|v|} = \frac{|Av|}{|v|} \geq \min_x \frac{|Ax|}{|x|} = \min_y \frac{|y|}{|A^{-1}y|} = \|A^{-1}\|_s^{-1}.$$

Now, suppose that $\|A^{-1}\|_s = |\lambda^{-1}|$ and $A^{-1}y = \lambda^{-1}y$, i.e., $Ay = \lambda y$. Then $\|A^{-1}\|_s^{-1} = |\lambda| = |Ay|/|y|$. The matrix $X = A - \lambda I$ is noninvertible and $\|A - X\|_s = \|\lambda I\|_s = |\lambda| = \|A^{-1}\|_s^{-1}$. $\square$

Problems

35.1. Prove that if λ is a nonzero eigenvalue of A, then $\|A^{-1}\|_s^{-1} \leq |\lambda| \leq \|A\|_s$.
35.2. Prove that $\|AB\|_s \leq \|A\|_s \|B\|_s$ and $\|AB\|_e \leq \|A\|_e \|B\|_e$.
35.3. Let A be a matrix of order n. Prove that

$$\|\operatorname{adj} A\|_e \leq n^{\frac{2-n}{2}} \|A\|_e^{n-1}.$$

36. Schur's complement and Hadamard's product. Theorems of Emily Haynsworth

36.1. Let $A = \begin{pmatrix} A_{11} & A_{12} \\ A_{21} & A_{22} \end{pmatrix}$, where $|A_{11}| \neq 0$. Recall that Schur's complement of A_{11} in A is the matrix $(A|A_{11}) = A_{22} - A_{21}A_{11}^{-1}A_{12}$ (see 3.1).

36.1.1. THEOREM. *If $A > 0$, then $(A|A_{11}) > 0$.*

PROOF. Let $T = \begin{pmatrix} I & -A_{11}^{-1}B \\ 0 & I \end{pmatrix}$, where $B = A_{12} = A_{21}^*$. Then

$$T^*AT = \begin{pmatrix} A_{11} & 0 \\ 0 & A_{22} - B^*A_{11}^{-1}B \end{pmatrix},$$

is a positive definite matrix, hence, $A_{22} - B^*A_{11}^{-1}B > 0$. $\square$

REMARK. We can similarly prove that if $A \geq 0$ and $|A_{11}| \neq 0$, then $(A|A_{11}) \geq 0$.

36.1.2. THEOREM ([Haynsworth,1970]). *If H and K are arbitrary positive definite matrices of order n and X and Y are arbitrary matrices of size $n \times m$, then*

$$X^*H^{-1}X + Y^*K^{-1}Y - (X + Y)^*(H + K)^{-1}(X + Y) \geq 0.$$

PROOF. Clearly,

$$A = T^* \begin{pmatrix} H & 0 \\ 0 & 0 \end{pmatrix} T = \begin{pmatrix} H & X \\ X^* & X^*H^{-1}X \end{pmatrix} > 0, \text{ where } T = \begin{pmatrix} I_n & H^{-1}X \\ 0 & I_m \end{pmatrix}.$$

Similarly, $B = \begin{pmatrix} K & Y \\ Y^* & Y^*K^{-1}Y \end{pmatrix} \geq 0$. It remains to apply Theorem 36.1.1 to the Schur complement of $H + K$ in $A + B$. $\square$

36.1.3. THEOREM ([Haynsworth, 1970]). *Let $A, B \geq 0$ and $A_{11}, B_{11} > 0$. Then*

$$(A+B|A_{11}+B_{11}) \geq (A|A_{11}) + (B|B_{11}).$$

PROOF. By definition,

$$(A+B|A_{11}+B_{11}) = (A_{22}+B_{22}) - (A_{21}+B_{21})(A_{11}+B_{11})^{-1}(A_{12}+B_{12}),$$

and by Theorem 36.1.2,

$$A_{21}A_{11}^{-1}A_{12} + B_{21}B_{11}^{-1}B_{12} \geq (A_{21}+B_{21})(A_{11}+B_{11})^{-1}(A_{12}+B_{12}).$$

Hence,

$$(A+B|A_{11}+B_{11}) \geq (A_{22}+B_{22}) - (A_{21}A_{11}^{-1}A_{12} + B_{21}B_{11}^{-1}B_{12}) = (A|A_{11}) + (B|B_{11}). \quad \square$$

We can apply the obtained results to the proof of the following statement.

36.1.4. THEOREM ([Haynsworth, 1970]). *Let A_k and B_k be upper left corner submatrices of order k in positive definite matrices A and B of order n, respectively. Then*

$$|A+B| \geq |A|\left(1+\sum_{k=1}^{n-1}\frac{|B_k|}{|A_k|}\right) + |B|\left(1+\sum_{k=1}^{n-1}\frac{|A_k|}{|B_k|}\right).$$

PROOF. First, observe that by Theorem 36.1.3 and Problem 33.1 we have

$$|(A+B|A_{11}+B_{11})| \geq |(A|A_{11}) + (B|B_{11})| \geq |(A|A_{11})| + |(B|B_{11})| = \frac{|A|}{|A_{11}|} + \frac{|B|}{|B_{11}|}.$$

For $n=2$ we get

$$|A+B| = |A_1+B_1|\cdot|(A+B|A_1+B_1)| \geq (|A_1|+|B_1|)\left(\frac{|A|}{|A_1|}+\frac{|B|}{|B_1|}\right) = |A|\left(1+\frac{|B_1|}{|A_1|}\right) + |B|\left(1+\frac{|A_1|}{|B_1|}\right).$$

Now let us assume that the statement is proved for matrices of order $n-1$ and let us prove it for matrices of order n. By the inductive hypothesis we have

$$|A_{n-1}+B_{n-1}| \geq |A_{n-1}|\left(1+\sum_{k=1}^{n-2}\frac{|B_k|}{|A_k|}\right) + |B_{n-1}|\left(1+\sum_{k=1}^{n-2}\frac{|A_k|}{|B_k|}\right).$$

Moreover, by the above remark

$$|(A+B|A_{n-1}+B_{n-1})| \geq \frac{|A|}{|A_{n-1}|} + \frac{|B|}{|B_{n-1}|}.$$

Therefore,

$$\begin{aligned}|A+B| &\geq \left[|A_{n-1}|\left(1+\sum_{k=1}^{n-2}\frac{|B_k|}{|A_k|}\right) + |B_{n-1}|\left(1+\sum_{k=1}^{n-2}\frac{|A_k|}{|B_k|}\right)\right]\left(\frac{|A|}{|A_{n-1}|}+\frac{|B|}{|B_{n-1}|}\right)\\ &\geq |A|\left(1+\sum_{k=1}^{n-2}\frac{|B_k|}{|A_k|}+\frac{|B_{n-1}|}{|A_{n-1}|}\right) + |B|\left(1+\sum_{k=1}^{n-2}\frac{|A_k|}{|B_k|}+\frac{|A_{n-1}|}{|B_{n-1}|}\right). \quad \square\end{aligned}$$

36.2. If $A = \|a_{ij}\|_1^n$ and $B = \|b_{ij}\|_1^n$ are square matrices, then their *Hadamard product* is the matrix $C = \|c_{ij}\|_1^n$, where $c_{ij} = a_{ij}b_{ij}$. The Hadamard product is denoted by $A \circ B$.

36.2.1. THEOREM (Schur). *If $A, B > 0$, then $A \circ B > 0$.*

PROOF. Let $U = \|u_{ij}\|_1^n$ be a unitary matrix such that $A = U^*\Lambda U$, where $\Lambda = \operatorname{diag}(\lambda_1, \dots, \lambda_n)$. Then $a_{ij} = \sum_p \overline{u}_{pi}\lambda_p u_{pj}$ and, therefore,

$$\sum_{i,j} a_{ij}b_{ij}\overline{x}_i x_j = \sum_p \lambda_p \sum_{i,j} b_{ij} y_i^p \overline{y}_j^p,$$

where $y_i^p = x_i u_{pi}$. All the numbers λ_p are positive and, therefore, it remains to prove that if not all numbers x_i are zero, then not all numbers y_i^p are zero. For this it suffices to notice that

$$\sum_{i,p} |y_i^p|^2 = \sum_{i,p} |x_i u_{pi}|^2 = \sum_i (|x_i|^2 \sum_p |u_{pi}|^2) = \sum_i |x_i|^2. \quad \square$$

36.2.2. The Oppenheim inequality.

THEOREM (Oppenheim). *If $A, B > 0$, then*

$$\det(A \circ B) \geq (\textstyle\prod a_{ii}) \det B.$$

PROOF. For matrices of order 1 the statement is obvious. Suppose that the statement is proved for matrices of order $n - 1$. Let us express the matrices A and B of order n in the form

$$A = \begin{pmatrix} a_{11} & A_{12} \\ A_{21} & A_{22} \end{pmatrix}, \quad B = \begin{pmatrix} b_{11} & B_{12} \\ B_{21} & B_{22} \end{pmatrix},$$

where a_{11} and b_{11} are numbers. Then

$$\det(A \circ B) = a_{11}b_{11} \det(A \circ B | a_{11}b_{11})$$

and

$$\begin{aligned}(A \circ B | a_{11}b_{11}) &= A_{22} \circ B_{22} - A_{21} \circ B_{21} a_{11}^{-1} b_{11}^{-1} A_{12} \circ B_{12} \\ &= A_{22} \circ (B|b_{11}) + (A|a_{11}) \circ (B_{21}B_{12}b_{11}^{-1}).\end{aligned}$$

Since $(A|a_{11})$ and $(B|b_{11})$ are positive definite matrices (see Theorem 36.1.1), by Theorem 36.2.1 the matrices $A_{22} \circ (B|b_{11})$ and $(A|a_{11}) \circ (B_{21}B_{12}b_{11}^{-1})$ are positive definite. Hence, $\det(A \circ B) \geq a_{11}b_{11} \det(A_{22} \circ (B|b_{11}))$; cf. Problem 33.1. By inductive hypothesis $\det(A_{22} \circ (B|b_{11})) \geq a_{22} \dots a_{nn} \det(B|b_{11})$; it is also clear that $\det(B|b_{11}) = \dfrac{\det B}{b_{11}}$. $\square$

REMARK. The equality is only attained if B is a diagonal matrix.

Problems

36.1. Prove that if A and B are positive definite matrices of order n and $A \geq B$, then $|A+B| \geq |A| + n|B|$.

36.2. [Djoković, 1964]. Prove that any positive definite matrix A can be represented in the form $A = B \circ C$, where B and C are positive definite matrices.

36.3. [Djoković, 1964]. Prove that if $A > 0$ and $B \geq 0$, then $\operatorname{rank}(A \circ B) \geq \operatorname{rank} B$.

37. Nonnegative matrices

37.1. A real matrix $A = \|a_{ij}\|_1^n$ is said to be *positive* (resp. *nonnegative*) if $a_{ij} > 0$ (resp. $a_{ij} \geq 0$).

In this section $A > 0$ means that the matrix A is positive and $A > B$ means that $A - B > 0$.

Observe that in all other sections the notation $A > 0$ means that A is an Hermitian (or real symmetric) positive definite matrix.

A vector $x = (x_1, \dots, x_n)$ is called *positive* and we write $x > 0$ if $x_i > 0$.

A matrix A of order n is called *reducible* if it is possible to divide the set $\{1, \dots, n\}$ into two nonempty subsets I and J such that $a_{ij} = 0$ for $i \in I$ and $j \in J$, and *irreducible* otherwise. In other words, A is reducible if by a permutation of its rows and columns it can be reduced to the form $\begin{pmatrix} A_{11} & A_{12} \\ 0 & A_{22} \end{pmatrix}$, where A_{11} and A_{22} are square matrices.

THEOREM. *If A is a nonnegative irreducible matrix of order n, then $(I+A)^{n-1} > 0$.*

PROOF. For every nonzero nonnegative vector y consider the vector $z = (I + A)y = y + Ay$. Suppose that not all coordinates of y are positive. Renumbering the vectors of the basis, if necessary, we can assume that $y = \begin{pmatrix} u \\ 0 \end{pmatrix}$, where $u > 0$. Then $Ay = \begin{pmatrix} A_{11} & A_{12} \\ A_{21} & A_{22} \end{pmatrix} \begin{pmatrix} u \\ 0 \end{pmatrix} = \begin{pmatrix} A_{11}u \\ A_{21}u \end{pmatrix}$. Since $u > 0$, $A_{21} \geq 0$, and $A_{21} \neq 0$, we have $A_{21}u \neq 0$. Therefore, z has at least one more positive coordinate than y. Hence, if $y \geq 0$ and $y \neq 0$, then $(I+A)^{n-1}y > 0$. Taking for y, first e_1, then e_2, etc., e_n we get the required solution. $\square$

37.2. Let A be a nonnegative matrix of order n and x a nonnegative vector. Further, let

$$r_x = \min_i \left\{ \sum_{j=1}^n a_{ij} \frac{x_j}{x_i} \right\} = \sup\{\rho \geq 0 | Ax \geq \rho x\}.$$

and $r = \sup\limits_{x \geq 0} r_x$. It suffices to take the supremum over the compact set $P = \{x \geq 0 | |x| = 1\}$, and not over all $x \geq 0$. Therefore, there exists a nonzero nonnegative vector z such that $Az \geq rz$ and there is no positive vector w such that $Aw > rw$.

A nonnegative vector z is called an *extremal vector* of A if $Az \geq rz$.

37.2.1. THEOREM. *If A is a nonnegative irreducible matrix, then $r > 0$ and an extremal vector of A is its eigenvector.*

PROOF. If $\xi = (1, \dots, 1)$, then $A\xi > 0$ and, therefore, $r > 0$. Let z be an extremal vector of A. Then $Az - rz = \eta \geq 0$. Suppose that $\eta \neq 0$. Multiplying both sides of the inequality $\eta \geq 0$ by $(I + A)^{n-1}$ we get $Aw - rw = (I + A)^{n-1}\eta > 0$, where $w = (I + A)^{n-1}z > 0$. Contradiction. $\square$

37.2.1.1. REMARK. A nonzero extremal vector z of A is positive. Indeed, $z \geq 0$ and $Az = rz$ and, therefore, $(1 + r)^{n-1}z = (I + A)^{n-1}z > 0$.

37.2.1.2. REMARK. An eigenvector of A corresponding to the eigenvalue r is unique up to proportionality. Indeed, let $Ax = rx$ and $Ay = ry$, where $x > 0$. If $\mu = \min(y_i/x_i)$, then $y_j \geq \mu x_j$, the vector $z = y - \mu x$ has nonnegative coordinates and at least one of them is zero. Suppose that $z \neq 0$. Then $z > 0$ since $z \geq 0$ and $Az = rz$ (see Remark 37.2.1.1). Contradiction.

37.2.2. THEOREM. *Let A be a nonnegative irreducible matrix and let a matrix B be such that $|b_{ij}| \leq a_{ij}$. If β is an eigenvalue of B, then $|\beta| \leq r$, and if $\beta = re^{i\varphi}$ then $|b_{ij}| = a_{ij}$ and $B = e^{i\varphi}DAD^{-1}$, where $D = \operatorname{diag}(d_1, \dots, d_n)$ and $|d_i| = 1$.*

PROOF. Let $By = \beta y$, where $y \neq 0$. Consider the vector $y^+ = (|y_1|, \dots, |y_n|)$. Since $\beta y_i = \sum_j b_{ij}y_j$, then $|\beta y_i| = \sum_j |b_{ij}y_j| \leq \sum_j a_{ij}|y_j|$ and, therefore, $|\beta|y^+ \leq ry^+$, i.e., $|\beta| \leq r$.

Now, suppose that $\beta = re^{i\varphi}$. Then y^+ is an extremal vector of A and, therefore, $y^+ > 0$ and $Ay^+ = ry^+$. Let $B^+ = \left\|b'_{ij}\right\|$, where $b'_{ij} = |b_{ij}|$. Then $B^+ \leq A$ and $Ay^+ = ry^+ = B^+y^+$ and since $y^+ > 0$, then $B^+ = A$. Consider the matrix $D = \operatorname{diag}(d_1, \dots, d_n)$, where $d_i = y_i/|y_i|$. Then $y = Dy^+$ and the equality $By = \beta y$ can be rewritten in the form $BDy^+ = \beta Dy^+$, i.e., $Cy^+ = ry^+$, where $C = e^{-i\varphi}D^{-1}BD$. The definition of C implies that $C^+ = B^+ = A$. Let us prove now that $C^+ = C$. Indeed, $Cy^+ = ry^+ = B^+y^+ = C^+y^+$ and since $C^+ \geq 0$ and $y^+ > 0$, then $C^+y^+ \geq Cy^+$, where equality is only possible if $C = C^+ = A$. $\square$

37.3. THEOREM. *Let A be a nonnegative irreducible matrix, k the number of its distinct eigenvalues whose absolute values are equal to the maximal eigenvalue r and $k > 1$. Then there exists a permutation matrix P such that the matrix PAP^T is of the block form*

$$\begin{pmatrix} 0 & A_{12} & 0 & \dots & 0 \\ 0 & 0 & A_{23} & \dots & 0 \\ \vdots & \vdots & \ddots & \ddots & \vdots \\ 0 & 0 & 0 & \ddots & A_{k-1,k} \\ A_{k1} & 0 & 0 & \dots & 0 \end{pmatrix}.$$

PROOF. The largest in absolute value eigenvalues of A are of the form $\alpha_j = r\exp(i\varphi_j)$. Applying Theorem 37.2.2 to $B = A$, we get $A = \exp(i\varphi_j)D_jAD_j^{-1}$. Therefore,

$$p(t) = |tI - A| = |tI - \exp(i\varphi_j)D_jAD_j^{-1}| = \lambda p(\exp(-i\varphi_j)t).$$

The numbers $\alpha_1, \dots, \alpha_k$ are roots of the polynomial p and, therefore, they are invariant with respect to rotations through angles φ_j (i.e., they constitute a group). Taking into account that the eigenvalue r is simple (see Problem 37.4), we get $\alpha_j = r\exp(\frac{2j\pi i}{k})$. Let y_1 be the eigenvector corresponding to the eigenvalue $\alpha_1 = r\exp(\frac{2\pi i}{k})$. Then $y_1^+ > 0$ and $y_1 = D_1 y_1^+$ (see the proof of Theorem 37.2.2). There exists a permutation matrix P such that

$$PD_1P^T = \operatorname{diag}(e^{i\gamma_1}I_1, \dots, e^{i\gamma_s}I_s),$$

where the numbers $e^{i\gamma_1}, \dots, e^{i\gamma_s}$ are distinct and $I_1, \dots, I_s$ are unit matrices. If instead of y_1 we take $e^{-i\gamma_1}y_1$, then we may assume that $\gamma_1 = 0$.

Let us break the matrix PAP^T into blocks A_{pq} in accordance with the blocks of the matrix PD_1P^T. Since $A = \exp(i\varphi_j)D_jAD_j^{-1}$, it follows that

$$PAP^T = \exp(i\varphi_1)(PD_1P^T)(PAP^T)(PD_1P^T)^{-1},$$

i.e.,

$$A_{pq} = \exp\left[i\left(\gamma_p - \gamma_q + \frac{2\pi}{k}\right)\right]A_{pq}.$$

Therefore, if $\frac{2\pi}{k} + \gamma_p \not\equiv \gamma_q \pmod{2\pi}$, then $A_{pq} = 0$. In particular, $s > 1$, since otherwise $A = 0$.

The numbers γ_i are distinct and, therefore, for any p there exists at most one number q such that $A_{pq} \neq 0$ (in which case $q \neq p$). The irreducibility of A implies that at least one such q exists.

Therefore, there exists a map $p \mapsto q(p)$ such that $A_{p,q(p)} \neq 0$ and $\frac{2\pi}{k} + \gamma_p \equiv \gamma_{q(p)} \pmod{2\pi}$.

For $p = 1$ we get $\gamma_{q(1)} \equiv \frac{2\pi}{k} \pmod{2\pi}$. After permutations of rows and columns of PAP^T we can assume that $\gamma_{q(1)} = \gamma_2$. By repeating similar arguments we can get

$$\gamma_{q(j-1)} = \gamma_j = \frac{2\pi(j-1)}{k} \quad \text{for } 2 \leq j \leq \min(k, s).$$

Let us prove that $s = k$. First, suppose that $1 < s < k$. Then $\frac{2\pi}{k} + \gamma_s - \gamma_r \not\equiv 0$ mod 2π for $1 \leq r \leq s-1$. Therefore, $A_{sr} = 0$ for $1 \leq r \leq s-1$, i.e., A is reducible.

Now suppose that $s > k$. Then $\gamma_i = \frac{2(i-1)\pi}{k}$ for $1 \leq i \leq k$. The numbers γ_j are distinct for $1 \leq j \leq s$ and for any i, $1 \leq i \leq k$, there exists $j (1 \leq j \leq k)$ such that $\frac{2\pi}{k} + \gamma_i \equiv \gamma_j \pmod{2\pi}$. Therefore, $\frac{2\pi}{k} + \gamma_i \not\equiv \gamma_r \pmod{2\pi}$ for $1 \leq i \leq k$ and $k < r \leq s$, i.e., $A_{ir} = 0$ for such k and r. In either case we get contradiction. Hence, $k = s$.

Now, it is clear that for the indicated choice of P the matrix PAP^T is of the required form. $\square$

COROLLARY. *If $A > 0$, then the maximal positive eigenvalue of A is strictly greater than the absolute value of any of its other eigenvalues.*

37.4. A nonnegative matrix A is called *primitive* if it is irreducible and there is only one eigenvalue whose absolute value is maximal.

37.4.1. THEOREM. *If A is primitive, then $A^m > 0$ for some m.*

PROOF ([Marcus, Minc, 1975]). Dividing, if necessary, the elements of A by the eigenvalue whose absolute value is maximal we can assume that A is an irreducible matrix whose maximal eigenvalue is equal to 1, the absolute values of the other eigenvalues being less than 1.

Let $S^{-1}AS = \begin{pmatrix} 1 & 0 \\ 0 & B \end{pmatrix}$ be the Jordan normal form of A. Since the absolute values of all eigenvalues of B are less than 1, it follows that $\lim_{n\to\infty} B^n = 0$ (see Problem 34.5 a)). The first column x^T of S is the eigenvector of A corresponding to the eigenvalue 1 (see Problem 11.6). Therefore, this vector is an extremal vector of A; hence, $x_i > 0$ for all i (see 37.2.1.2). Similarly, the first row, y, of the matrix S^{-1} consists of positive elements. Hence,

$$\lim_{n\to\infty} A^n = \lim_{n\to\infty} S \begin{pmatrix} 1 & 0 \\ 0 & B^n \end{pmatrix} S^{-1} = S \begin{pmatrix} 1 & 0 \\ 0 & 0 \end{pmatrix} S^{-1} = x^T y > 0$$

and, therefore, $A^m > 0$ for some m. □

REMARK. If $A \geq 0$ and $A^m > 0$, then A is primitive. Indeed, the irreducibility of A is obvious; besides, the maximal positive eigenvalue of A^m is strictly greater than the absolute value of any of its other eigenvalues and the eigenvalues of A^m are obtained from the eigenvalues of A by raising them to mth power.

37.4.2. THEOREM (Wielandt). *Let A be a nonnegative primitive matrix of order n. Then $A^{n^2-2n+2} > 0$.*

PROOF (Following [Sedláček, 1959]). To a nonnegative matrix A of order n we can assign a directed graph with n vertices by connecting the vertex i with the vertex j if $a_{ij} > 0$ (the case $i = j$ is not excluded). The element b_{ij} of A^s is positive if and only if on the constructed graph there exists a directed path of length s leading from vertex i to vertex j.

Indeed, $b_{ij} = \sum a_{ii_1} a_{i_1 i_2} \dots a_{i_{s-1}j}$, where $a_{ii_1} a_{i_1 i_2} \dots a_{i_{s-1}j} > 0$ if and only if the path $ii_1 i_2 \dots i_{s-1} j$ runs over the directed edges of the graph.

To a primitive matrix there corresponds a connected graph, i.e., from any vertex we can reach any other vertex along a directed path. Among all cycles, select a cycle of the least length (if $a_{ii} > 0$, then the edge ii is such a cycle). Let, for definiteness sake, this be the cycle $12\dots l1$. Then the elements $b_{11}, \dots, b_{ll}$ of A^l are positive.

From any vertex i we can reach one of the vertices $1, \dots, l$ along a directed path whose length does not exceed $n - l$. By continuing our passage along this cycle further, if necessary, we can turn this path into a path of length $n - l$.

Now, consider the matrix A^l. It is also primitive and a directed graph can also be assigned to it. Along this graph, from a vertex $j \in \{1, \dots, l\}$ (which we have reached from the vertex i) we can traverse to any given vertex k along a path whose length does not exceed $n - 1$. Since the vertex j is connected with itself, the same path can be turned into a path whose length is exactly $n - 1$. Therefore, for any vertices i and k on the graph corresponding to A there exists a directed path of length

$n - l + l(n-1) = l(n-2) + n$. If $l = n$, then the matrix A can be reduced to the form

$$\begin{pmatrix} 0 & a_{12} & 0 & \dots & 0 \\ 0 & 0 & a_{23} & \ddots & 0 \\ \vdots & \vdots & \ddots & \ddots & \vdots \\ 0 & 0 & 0 & \ddots & a_{n-1,n} \\ a_{n1} & 0 & 0 & \dots & 0 \end{pmatrix};$$

this matrix is not primitive. Therefore, $l \leq n-1$; hence, $l(n-2) + n \leq n^2 - 2n + 2$. It remains to notice that if $A \geq 0$ and $A^p > 0$, then $A^{p+1} > 0$ (Problem 37.1).

The estimate obtained in Theorem 37.4.2 is exact. It is reached, for instance, at the matrix

$$A = \begin{pmatrix} 0 & 1 & 0 & \dots & 0 \\ 0 & 0 & 1 & \dots & 0 \\ \vdots & \vdots & \ddots & \ddots & \vdots \\ 0 & 0 & 0 & \dots & 1 \\ 1 & 1 & 0 & \dots & 0 \end{pmatrix}$$

of order n, where $n \geq 3$. To this matrix we can assign the operator that acts as follows:

$$Ae_1 = e_n, \quad Ae_2 = e_1 + e_n, \quad Ae_3 = e_2, \quad \dots, \quad Ae_n = e_{n-1}.$$

Let $B = A^{n-1}$. It is easy to verify that

$$Be_1 = e_2, \quad Be_2 = e_2 + e_3, \quad Be_3 = e_3 + e_4, \quad \dots, \quad Be_n = e_n + e_1.$$

Therefore, the matrix B^{n-1} has just one zero element situated on the $(1,1)$th position and the matrix $AB^{n-1} = A^{n^2-2n+2}$ is positive. $\square$

Problems

37.1. Prove that if $A \geq 0$ and $A^k > 0$, then $A^{k+1} > 0$.

37.2. Prove that a nonnegative eigenvector of an irreducible nonnegative matrix is positive.

37.3. Let $A = \begin{pmatrix} B & C \\ D & E \end{pmatrix}$ be a nonnegative irreducible matrix and B a square matrix. Prove that if α and β are the maximal eigenvalues of A and B, then $\beta < \alpha$.

37.4. Prove that if A is a nonnegative irreducible matrix, then its maximal eigenvalue is a simple root of its characteristic polynomial.

37.5. Prove that if A is a nonnegative irreducible matrix and $a_{11} > 0$, then A is primitive.

37.6 ([Šidák, 1964]). A matrix A is primitive. Can the number of positive elements of A be greater than that of A^2?

38. Doubly stochastic matrices

38.1. A nonnegative matrix $A = \|a_{ij}\|_1^n$ is called *doubly stochastic* if $\sum_{i=1}^n a_{ik} = 1$ and $\sum_{j=1}^n a_{kj} = 1$ for all k.

38.1.1. THEOREM. *The product of doubly stochastic matrices is a doubly stochastic matrix.*

PROOF. Let A and B be doubly stochastic matrices and $C = AB$. Then

$$\sum_{i=1}^{n} c_{ij} = \sum_{i=1}^{n}\sum_{p=1}^{n} a_{ip}b_{pj} = \sum_{p=1}^{n} b_{pj}\sum_{i=1}^{n} a_{ip} = \sum_{p=1}^{n} b_{pj} = 1.$$

Similarly, $\sum_{j=1}^{n} c_{ij} = 1$. $\square$

38.1.2. THEOREM. *If $A = \left\|a_{ij}\right\|_1^n$ is a unitary matrix, then the matrix $B = \left\|b_{ij}\right\|_1^n$, where $b_{ij} = |a_{ij}|^2$, is doubly stochastic.*

PROOF. It suffices to notice that $\sum_{i=1}^{n}|a_{ij}|^2 = \sum_{j=1}^{n}|a_{ij}|^2 = 1$. $\square$

38.2.1. THEOREM (Birkhoff). *The set of all doubly stochastic matrices of order n is a convex polyhedron with permutation matrices as its vertices.*

Let $i_1, \dots, i_k$ be numbers of some of the rows of A and $j_1, \dots, j_l$ numbers of some of its columns. The matrix $\left\|a_{ij}\right\|$, where $i \in \{i_1, \dots, i_k\}$ and $j \in \{j_1, \dots, j_l\}$, is called a *submatrix of A*. By a *snake in* A we will mean the set of elements $a_{1\sigma(1)}, \dots, a_{n\sigma(n)}$, where σ is a permutation. In the proof of Birkhoff's theorem we will need the following statement.

38.2.2. THEOREM (Frobenius-König). *Each snake in a matrix A of order n contains a zero element if and only if A contains a zero submatrix of size $s \times t$, where $s+t = n+1$.*

PROOF. First, suppose that on the intersection of rows $i_1, \dots, i_s$ and columns $j_1, \dots, j_t$ there stand zeros and $s + t = n + 1$. Then at least one of s numbers $\sigma(i_1), \dots, \sigma(i_s)$ belongs to $\{j_1, \dots, j_t\}$ and, therefore, the corresponding element of the snake is equal to 0.

Now, suppose that every snake in A of order n contains 0 and prove that A contains a zero submatrix of size $s \times t$, where $s + t = n + 1$. The proof will be carried out by induction on n. For $n = 1$ the statement is obvious.

Now let us assume that the statement is true for matrices of order $n-1$ and consider a nonzero matrix of order n. In it, take a zero element and delete the row and the column which contain it. In the resulting matrix of order $n - 1$ every snake contains a zero element and, therefore, it has a zero submatrix of size $s_1 \times t_1$, where $s_1 + t_1 = n$. Hence, the initial matrix A can be reduced by permutation of rows and columns to the block form plotted on Figure 6 a).

Suppose that a matrix X has a snake without zero elements. Every snake in the matrix Z can be complemented by this snake to a snake in A. Hence, every snake in Z does contain 0. As a result, we see that either all snakes of X or all snakes of Z contain 0. Let, for definiteness sake, all snakes of X contain 0. Then X contains a zero submatrix of size $p \times q$, where $p + q = s_1 + 1$. Hence, A contains a zero submatrix of size $p \times (t_1 + q)$ (on Figure 6 b) this matrix is shaded). Clearly, $p + (t_1 + q) = s_1 + 1 + t_1 = n + 1$. $\square$

COROLLARY. *Any doubly stochastic matrix has a snake consisting of positive elements.*

PROOF. Indeed, otherwise this matrix would contain a zero submatrix of size $s \times t$, where $s + t = n + 1$. The sum of the elements of each of the rows considered and each

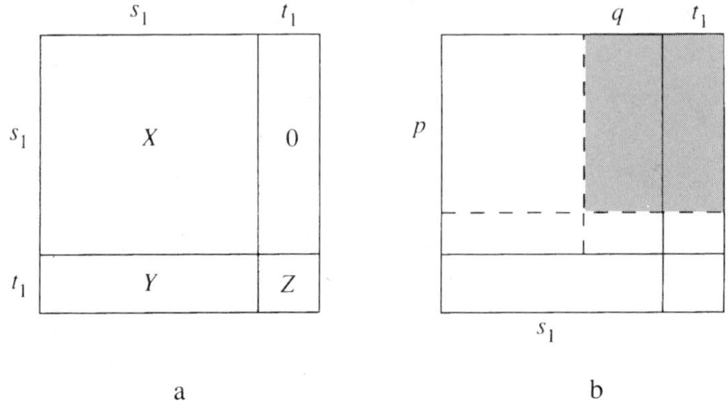

FIGURE 6

of the columns considered is equal to 1; the elements on the intersections of these rows and columns are all zeros and, therefore, the sum of the elements of these rows and columns alone is equal to $s + t = n + 1$; this exceeds the sum of all elements which is equal to n. Contradiction. □

PROOF of the Birkhoff theorem. We have to prove that any doubly stochastic matrix S can be represented in the form $S = \sum \lambda_i P_i$, where P_i is a permutation matrix, $\lambda_i \geq 0$ and $\sum \lambda_i = 1$.

We will use induction on the number k of positive elements of a matrix S of order n. For $k = n$ the statement is obvious since in this case S is a permutation matrix. Now, suppose that S is not a permutation matrix. Then this matrix has a positive snake (see Corollary 38.2.2). Let P be a permutation matrix corresponding to this snake and x the minimal element of the snake. Clearly, $x \neq 1$. The matrix $T = \frac{1}{1-x}(S - xP)$ is doubly stochastic and it has at least one positive element less than S does. By inductive hypothesis T can be represented in the desired form; and $S = xP + (1-x)T$. □

38.2.3. THEOREM. *Any doubly stochastic matrix S of order n is a convex linear hull of no more than $n^2 - 2n + 2$ permutation matrices.*

PROOF. Let us cross out from S the last row and the last column. S is uniquely recovered from the remaining $(n-1)^2$ elements and, therefore, the set of doubly stochastic matrices of order n can be considered as a convex polyhedron in the space of dimension $(n-1)^2$. It remains to make use of the result of Problem 7.2. □

As an example of an application of the Birkhoff theorem, we prove the following statement.

38.2.4. THEOREM (Hoffman-Wielandt). *Let A and B be normal matrices; let $\alpha_1, \dots, \alpha_n$ and $\beta_1, \dots, \beta_n$ be their eigenvalues. Then*

$$\|A - B\|_e^2 \geq \min_\sigma \sum_{i=1}^n (\alpha_{\sigma(i)} - \beta_i)^2,$$

where the minimum is taken over all permutations σ.

PROOF. Let $A = V\Lambda_a V^*$, $B = W\Lambda_b W^*$, where U and W are unitary matrices and $\Lambda_a = \operatorname{diag}(\alpha_1, \dots, \alpha_n)$, $\Lambda_b = \operatorname{diag}(\beta_1, \dots, \beta_n)$. Then

$$\|A - B\|_e^2 = \|W^*(V\Lambda_a V^* - W\Lambda_b W^*)W\|_e^2 = \|U\Lambda_a U^* - \Lambda_b\|_e^2,$$

where $U = W^*V$. Besides,

$$\begin{aligned}\|U\Lambda_a U^* - \Lambda_b\|_e^2 &= \operatorname{tr}(U\Lambda_a U^* - \Lambda_b)(U\Lambda_a^* U^* - \Lambda_b^*)\\ &= \operatorname{tr}(\Lambda_a\Lambda_a^* + \Lambda_b\Lambda_b^*) - 2\operatorname{Re}\operatorname{tr}(U\Lambda_a U^*\Lambda_b^*)\\ &= \sum_{i=1}^n (|\alpha_i|^2 + |\beta_i|^2) - 2\sum_{i,j=1}^n |u_{ij}|^2 \operatorname{Re}(\overline{\beta}_i\alpha_j).\end{aligned}$$

Since the matrix $\|c_{ij}\|$, where $c_{ij} = |u_{ij}|^2$, is doubly stochastic, we have

$$\|A - B\|_e^2 \geq \sum_{i=1}^n (|\alpha_i|^2 + |\beta_i|^2) - 2\min\sum_{i,j=1}^n c_{ij}\operatorname{Re}(\overline{\beta}_i\alpha_j),$$

where the minimum is taken over all doubly stochastic matrices C. For fixed sets of numbers α_i, β_j we have to find the minimum of a linear function on a convex polyhedron whose vertices are permutation matrices. This minimum is attained at one of the vertices, i.e., for a matrix $c_{ij} = \delta_{i,\sigma(i)}$. In this case

$$2\sum_{i,j=1}^n c_{ij}\operatorname{Re}(\overline{\beta}_i\alpha_j) = 2\sum_{i=1}^n \operatorname{Re}(\overline{\beta}_i\alpha_{\sigma(i)}).$$

Hence,

$$\|A - B\|_e^2 \geq \sum_{i=1}^n \left(|\alpha_{\sigma(i)}|^2 + |\beta_i|^2 - 2\operatorname{Re}(\overline{\beta}_i\alpha_{\sigma(i)})\right) = \sum_{i=1}^n |\alpha_{\sigma(i)} - \beta_i|^2. \quad \square$$

38.3.1. THEOREM. *Let $x_1 \geq \dots \geq x_n$ and $y_1 \geq \dots \geq y_n$, where $x_1 + \dots + x_k \leq y_1 + \dots + y_k$ for all $k < n$ and $x_1 + \dots + x_n = y_1 + \dots + y_n$. Then there exists a doubly stochastic matrix S such that $Sy = x$.*

PROOF. Let us assume that $x_1 \neq y_1$ and $x_n \neq y_n$, since otherwise we can ignore several first or several last coordinates. The hypothesis implies that $x_1 \leq y_1$ and $x_1 + \dots + x_{n-1} \leq y_1 + \dots + y_{n-1}$, i.e., $x_n \geq y_n$. Hence, $x_1 < y_1$ and $x_n > y_n$. Now, consider the operator which is the identity on $y_2, \dots, y_{n-1}$ and on y_1 and y_n acts by the matrix $\begin{pmatrix} \alpha & 1-\alpha \\ 1-\alpha & \alpha \end{pmatrix}$. If $0 < \alpha < 1$, then the matrix S_1 of this operator is doubly stochastic. Select a number α so that $\alpha y_1 + (1-\alpha)y_n = x_1$, i.e., $\alpha = (x_1 - y_n)(y_1 - y_n)^{-1}$. Since $y_1 > x_1 \geq x_n > y_n$, we have $0 < \alpha < 1$. As a result, with the help of S_1 we pass from the set $y_1, y_2, \dots, y_n$ to the set $x_1, y_2, \dots, y_{n-1}, y_n'$, where $y_n' = (1-\alpha)y_1 + \alpha y_n$. Since $x_1 + y_n' = y_1 + y_n$, we have $x_2 + \dots + x_{n-1} + x_n = y_2 + \dots + y_{n-1} + y_n'$ and, therefore, for the sets $x_2, \dots, x_n$ and $y_2, \dots, y_{n-1}, y_n'$ we can repeat similar arguments, etc. It remains to notice that the product of doubly stochastic matrices is a doubly stochastic matrix, see Theorem 38.1.1. $\square$

38.3.2. THEOREM (H. Weyl's inequality). *Let $\alpha_1 \geq \cdots \geq \alpha_n$ be the absolute values of the eigenvalues of an invertible matrix A, and let $\sigma_1 \geq \cdots \geq \sigma_n$ be its singular values. Then $\alpha_1^s + \cdots + \alpha_k^s \leq \sigma_1^s + \cdots + \sigma_k^s$ for all $k \leq n$ and $s > 0$.*

PROOF. By Theorem 34.4.1, $\alpha_1 \dots \alpha_n = \sigma_1 \dots \sigma_n$ and $\alpha_1 \dots \alpha_k \leq \sigma_1 \dots \sigma_k$ for $k \leq n$. Let x and y be the columns $(\ln \alpha_1, \dots, \ln \alpha_n)^T$ and $(\ln \sigma_1, \dots, \ln \sigma_n)^T$. By Theorem 38.3.1 there exists a doubly stochastic matrix S such that $x = Sy$. Fix $k \leq n$ and for $u = (u_1, \dots, u_n)$ consider the function $f(u) = f(u_1) + \cdots + f(u_k)$, where $f(t) = \exp(st)$ is a convex function; the function f is convex on a set of vectors with positive coordinates.

Now, fix a vector u with positive coordinates and consider the function $g(S) = f(Su)$ defined on the set of doubly stochastic matrices. If $0 \leq \alpha \leq 1$, then

$$\begin{aligned} g(\lambda S + (1-\lambda)T) &= f(\lambda Su + (1-\lambda)Tu) \\ &\leq \lambda f(SU) + (1-\lambda) f(Tu) = \lambda g(S) + (1-\lambda) g(T), \end{aligned}$$

i.e., g is a convex function. A convex function defined on a convex polytop takes its maximal value at one of the vertices. Therefore, $g(S) \leq g(P)$, where P is the matrix of permutation π (see Theorem 38.2.1). As the result we get

$$f(x) = f(Sy) = g(S) \leq g(P) = f(y_{\pi(1)}, \dots, y_{\pi(n)}).$$

It remains to notice that

$$f(x) = \exp(s \ln \alpha_1) + \cdots + \exp(s \ln \alpha_k) = \alpha_1^s + \cdots + \alpha_k^s$$

and

$$f(y_{\pi(1)}, \dots, y_{\pi(n)}) = \sigma_{\pi(1)}^s + \cdots + \sigma_{\pi(k)}^s \leq \sigma_1^s + \cdots + \sigma_k^s. \quad \square$$

Problems

38.1 ([Mirsky, 1975]). Let $A = \left\|a_{ij}\right\|_1^n$ be a doubly stochastic matrix; $x_1 \geq \cdots \geq x_n \geq 0$ and $y_1 \geq \cdots \geq y_n \geq 0$. Prove that $\sum_{r,s} a_{rs} x_r y_s \leq \sum_r x_r y_r$.

38.2 ([Bellman, Hoffman, 1954]). Let $\lambda_1, \dots, \lambda_n$ be eigenvalues of an Hermitian matrix H. Prove that the point with coordinates $(h_{11}, \dots, h_{nn})$ belongs to the convex hull of the points whose coordinates are obtained from $\lambda_1, \dots, \lambda_n$ under all possible permutations.

Solutions

33.1. Theorem 20.1 shows that there exists a matrix P such that $P^*AP = I$ and $P^*BP = \operatorname{diag}(\mu_1, \dots, \mu_n)$, where $\mu_i \geq 0$. Therefore, $|A+B| = d^2 \prod(1+\mu_i)$, $|A| = d^2$ and $|B| = d^2 \prod \mu_i$, where $d = |\det P|$. It is also clear that

$$\prod(1+\mu_i) = 1 + (\mu_1 + \cdots + \mu_n) + \cdots + \prod \mu_i \geq 1 + \prod \mu_i.$$

The inequality is strict if $\mu_1 + \cdots + \mu_n > 0$, i.e., at least one of the numbers $\mu_1, \dots, \mu_n$ is nonzero.

33.2. As in the preceding problem, $\det(A+iB) = d^2\prod(\alpha_k + i\beta_k)$ and $\det A = d^2\prod\alpha_k$, where $\alpha_k > 0$ and $\beta_k \in \mathbb{R}$. Since $|\alpha_k + i\beta_k|^2 = |\alpha_k|^2 + |\beta_k|^2$, we have $|\alpha_k + i\beta_k| \geq |\alpha_k|$ and the inequality is strict if $\beta_k \neq 0$.

33.3. Since $A - B = C > 0$, it follows that $A_k = B_k + C_k$, where $A_k, B_k, C_k > 0$. Therefore, $|A_k| > |B_k| + |C_k|$ (cf. Problem 33.1).

33.4. Let $x+iy$ be a nonzero eigenvector of C corresponding to the zero eigenvalue. Then

$$(A+iB)(x+iy) = (Ax - By) + i(Bx + Ay) = 0,$$

i.e., $Ax = By$ and $Ay = -Bx$. Therefore,

$$0 \leq (Ax, x) = (By, x) = (y, Bx) = -(y, Ay) \leq 0,$$

i.e., $(Ax, x) = (Ay, y) = 0$. Hence, $Ay = By = 0$. Therefore, $Ax = Bx = 0$ and $Ay = By = 0$ and at least one of the vectors x and y is nonzero.

33.5. Let $z = (z_1, \dots, z_n)$. The quadratic form Q corresponding to the matrix considered is of the form

$$2\sum_{i=1}^{n} x_i z_0 z_i + (Az, z) = 2z_0(z, x) + (Az, z).$$

The form Q is positive definite on a subspace of codimension 1 and, therefore, it remains to prove that the quadratic form Q is not positive definite. If $x \neq 0$, then $(z, x) \neq 0$ for some z. Therefore, the number z_0 can be chosen so that $2z_0(z, x) + (Az, z) < 0$.

33.6. There exists a unitary matrix U such that $U^*AU = \operatorname{diag}(\lambda_1, \dots, \lambda_n)$, where $\lambda_i \geq 0$. Besides, $\operatorname{tr}(AB) = \operatorname{tr}(U^*AUB')$, where $B' = U^*BU$. Therefore, we can assume that $A = \operatorname{diag}(\lambda_1, \dots, \lambda_n)$. In this case

$$\frac{\operatorname{tr}(AB)}{n} = \frac{(\sum \lambda_i b_{ii})}{n} \geq \left(\prod \lambda_i b_{ii}\right)^{1/n} = |A|^{1/n}\left(\prod b_{ii}\right)^{1/n}$$

and $\prod b_{ii} \geq |B| = 1$ (cf. 33.2). Thus, the minimum is attained at the matrix

$$B = |A|^{1/n}\operatorname{diag}(\lambda_1^{-1}, \dots, \lambda_n^{-1}).$$

34.1. Let λ be an eigenvalue of the given matrix. Then the system $\sum a_{ij}x_j = \lambda x_i$ $(i = 1, \dots, n)$ has a nonzero solution $(x_1, \dots, x_n)$. Among the numbers $x_1, \dots, x_n$ select the one with the greatest absolute value; let it be x_k. Since

$$a_{kk}x_k - \lambda x_k = -\sum_{j\neq k} a_{kj}x_j,$$

we have

$$|a_{kk}x_k - \lambda x_k| \leq \sum_{j\neq k} |a_{kj}x_j| \leq \rho_k|x_k|,$$

i.e., $|a_{kk} - \lambda| \leq \rho_k$.

34.2. Let $S = V^*DV$, where $D = \operatorname{diag}(\lambda_1, \dots, \lambda_n)$, and V is a unitary matrix. Then

$$\operatorname{tr}(US) = \operatorname{tr}(UV^*DV) = \operatorname{tr}(VUV^*D).$$

Let $VUV^* = W = \|w_{ij}\|_1^n$; then $\operatorname{tr}(US) = \sum w_{ii}\lambda_i$. Since W is a unitary matrix, it follows that $|w_{ii}| \le 1$ and, therefore,

$$|\sum w_{ii}\lambda_i| \le \sum |\lambda_i| = \sum \lambda_i = \operatorname{tr} S.$$

If $S > 0$, i.e., $\lambda_i \ne 0$ for all i, then $\operatorname{tr} S = \operatorname{tr}(US)$ if and only if $w_{ii} = 1$, i.e., $W = I$ and, therefore, $U = I$. The equality $\operatorname{tr} S = |\operatorname{tr}(US)|$ for a positive definite matrix S can only be satisfied if $w_{ii} = e^{i\varphi}$, i.e., $U = e^{i\varphi} I$.

34.3. Let $\alpha_1 \ge \dots \ge \alpha_n \ge 0$ and $\beta_1 \ge \dots \ge \beta_n \ge 0$ be the eigenvalues of A and B. For nonnegative definite matrices the eigenvalues coincide with the singular values and, therefore,

$$|\operatorname{tr}(AB)| \le \sum \alpha_i\beta_i \le \left(\sum \alpha_i\right)\left(\sum \beta_i\right) = \operatorname{tr} A \ \operatorname{tr} B$$

(see Theorem 34.4.2).

34.4. The matrix $C = AB - BA$ is skew-Hermitian and, therefore, its eigenvalues are purely imaginary; hence, $\operatorname{tr}(C^2) \le 0$. The inequality $\operatorname{tr}(AB - BA)^2 \le 0$ implies

$$\operatorname{tr}(AB)^2 + \operatorname{tr}(BA)^2 \le \operatorname{tr}(ABBA) + \operatorname{tr}(BAAB).$$

It is easy to verify that $\operatorname{tr}(BA)^2 = \operatorname{tr}(AB)^2$ and $\operatorname{tr}(ABBA) = \operatorname{tr}(BAAB) = \operatorname{tr}(A^2B^2)$.

34.5. a) If $A^k \longrightarrow 0$ and $Ax = \lambda x$, then $\lambda^k \longrightarrow 0$. Now, suppose that $|\lambda_i| < 1$ for all eigenvalues of A. It suffices to consider the case when $A = \lambda I + N$ is a Jordan block of order n. In this case

$$A^k = \binom{k}{0}\lambda^k I + \binom{k}{1}\lambda^{k-1}N + \dots + \binom{k}{n}\lambda^{k-n}N^n,$$

since $N^{n+1} = 0$. Each summand tends to zero since $\binom{k}{p} = k(k-1)\dots(k-p+1) \le k^p$ and $\lim\limits_{k\longrightarrow\infty} k^p\lambda^k = 0$.

b) If $Ax = \lambda x$ and $H - A^*HA > 0$ for $H > 0$, then

$$0 < (Hx - A^*HAx, x) = (Hx, x) - (H\lambda x, \lambda x) = (1 - |\lambda|^2)(Hx, x);$$

hence, $|\lambda| < 1$. Now, suppose that $A^k \longrightarrow 0$. Then $(A^*)^k \longrightarrow 0$ and $(A^*)^kA^k \longrightarrow 0$. If $Bx = \lambda x$ and $b = \max|b_{ij}|$, then $|\lambda| \le nb$, where n is the order of B. Hence, all eigenvalues of $(A^*)^kA^k$ tend to zero and, therefore, for a certain m the absolute value of every eigenvalue α_i of the nonnegative definite matrix $(A^*)^mA^m$ is less than 1, i.e., $0 \le \alpha_i < 1$. Let

$$H = I + A^*A + \dots + (A^*)^{m-1}A^{m-1}.$$

Then $H - A^*HA = I - (A^*)^mA^m$ and, therefore, the eigenvalues of the Hermitian matrix $H - A^*HA$ are equal to $1 - \alpha_i > 0$.

34.6. The eigenvalues of an Hermitian matrix A^*A are equal and, therefore, $A^*A = tI$, where $t \in \mathbb{R}$. Hence, $U = t^{-1/2}A$ is a unitary matrix.

34.7. It suffices to apply the result of Problem 11.8 to the matrix A^*A.

34.8. It suffices to notice that $\begin{vmatrix} \lambda I & -A \\ -A^* & \lambda I \end{vmatrix} = |\lambda^2 I - A^*A|$ (cf. 3.1).

35.1. Suppose that $Ax = \lambda x$, $\lambda x \neq 0$. Then $A^{-1}x = \lambda^{-1}x$; therefore, $\max_y \frac{|Ay|}{|y|} \geq \frac{|Ax|}{|x|} = \lambda$ and

$$\left(\max_y \frac{|A^{-1}y|}{|y|}\right)^{-1} = \min_y \frac{|y|}{|A^{-1}y|} \leq \frac{|x|}{|A^{-1}x|} = \lambda.$$

35.2. If $\|AB\|_s \neq 0$, then

$$\|AB\|_s = \max_x \frac{|ABx|}{|x|} = \frac{|ABx_0|}{|x_0|},$$

where $Bx_0 \neq 0$. Let $y = Bx_0$; then

$$\frac{|ABx_0|}{|x_0|} = \frac{|Ay|}{|y|} \cdot \frac{|Bx_0|}{|x_0|} \leq \|A\|_s \|B\|_s$$

To prove the inequality $\|AB\|_e \leq \|A\|_e \|B\|_e$ it suffices to use the inequality

$$\left|\sum_{k=1}^n a_{ik}b_{kj}\right|^2 \leq \left(\sum_{k=1}^n |a_{ik}|^2\right)\left(\sum_{k=1}^n |b_{kj}|^2\right).$$

35.3. Let $\sigma_1, \dots, \sigma_n$ be the singular values of the matrix A. Then the singular values of $\operatorname{adj} A$ are equal to $\prod_{i\neq 1}\sigma_i, \dots, \prod_{i\neq n}\sigma_i$ (Problem 34.7). Therefore, $\|A\|_e^2 = \sigma_1^2 + \dots + \sigma_n^2$ and

$$\|\operatorname{adj} A\|_e^2 = \prod_{i\neq 1}\sigma_i + \dots + \prod_{i\neq n}\sigma_i.$$

First, suppose that A is invertible. Then

$$\|\operatorname{adj} A\|_e^2 = (\sigma_1^2 \dots \sigma_n^2)(\sigma_1^{-2} + \dots + \sigma_n^{-2}).$$

Multiplying the inequalities

$$\sigma_1^2 \dots \sigma_n^2 \leq n^{-n}(\sigma_1^2 + \dots + \sigma_n^2)^n \text{ and } (\sigma_1^{-2} + \dots + \sigma_n^{-2})(\sigma_1^2 + \dots + \sigma_n^2) \leq n^2$$

we get

$$\|\operatorname{adj} A\|_e^2 \leq n^{2-n}\|A\|_e^{2(n-1)}.$$

Both parts of this inequality depend continuously on the elements of A and, therefore, the inequality holds for noninvertible matrices as well. The inequality becomes an equality if $\sigma_1 = \dots = \sigma_n$, i.e., if A is proportional to a unitary matrix (see Problem 34.6).

36.1. By Theorem 36.1.4

$$|A+B| \geq |A|\left(1+\sum_{k=1}^{n-1}\frac{|B_k|}{|A_k|}\right)+|B|\left(1+\sum_{k=1}^{n-1}\frac{|A_k|}{|B_k|}\right).$$

Besides, $\frac{|A_k|}{|B_k|} \geq 1$ (see Problem 33.3).

36.2. Consider a matrix $B(\lambda) = \|b_{ij}\|_1^n$, where $b_{ii} = 1$ and $b_{ij} = \lambda$ for $i \neq j$. It is possible to reduce the Hermitian form corresponding to this matrix to the form $\lambda|\sum x_i|^2 + (1-\lambda)\sum |x_i|^2$ and, therefore $B(\lambda) > 0$ for $0 < \lambda < 1$. The matrix $C(\lambda) = A \circ B(\lambda)$ is Hermitian for real λ and $\lim\limits_{\lambda \longrightarrow 1} C(\lambda) = A > 0$. Hence, $C(\lambda_0) > 0$ for a certain $\lambda_0 > 1$. Since $B(\lambda_0) \circ B(\lambda_0^{-1})$ is the matrix all of whose elements are 1, it follows that $A = C(\lambda_0) \circ B(\lambda_0^{-1}) > 0$.

36.3. If $B > 0$, then we can make use of Schur's theorem (see Theorem 36.2.1). Now, suppose that $\operatorname{rank} B = k$, where $0 < k < \operatorname{rank} A$. Then B contains a positive definite principal submatrix $M(B)$ of rank k (see Problem 19.5). Let $M(A)$ be the corresponding submatrix of A; since $A > 0$, it follows that $M(A) > 0$. By the Schur theorem the submatrix $M(A) \circ M(B)$ of $A \circ B$ is invertible.

37.1. Let $A \geq 0$ and $B > 0$. The matrix $C = AB$ has a nonzero element c_{pq} only if the pth row of A is zero. But then the pth row of A^k is also zero.

37.2. Suppose that the given eigenvector is not positive. We may assume that it is of the form $\begin{pmatrix} x \\ 0 \end{pmatrix}$, where $x > 0$. Then $\begin{pmatrix} A & B \\ C & D \end{pmatrix}\begin{pmatrix} x \\ 0 \end{pmatrix} = \begin{pmatrix} Ax \\ Cx \end{pmatrix}$, and, therefore, $Cx = 0$. Since $C \geq 0$, we have $C = 0$ and, therefore, the given matrix is decomposable. Contradiction.

37.3. Let $y \geq 0$ be a nonzero eigenvector of B corresponding to the eigenvalue β and $x = \begin{pmatrix} y \\ 0 \end{pmatrix}$. Then

$$Ax = \begin{pmatrix} B & C \\ D & E \end{pmatrix}\begin{pmatrix} y \\ 0 \end{pmatrix} = \begin{pmatrix} By \\ 0 \end{pmatrix} + \begin{pmatrix} 0 \\ Dy \end{pmatrix} = \beta x + z,$$

where $z = \begin{pmatrix} 0 \\ Dy \end{pmatrix} \geq 0$. The equality $Ax = \beta x$ cannot hold since the eigenvector of an indecomposable matrix is positive (cf. Problem 37.2). Besides,

$$\sup\{t \geq 0 \mid Ax - tx \geq 0\} \geq \beta$$

and if $\beta = \alpha$, then x is an extremal vector (cf. Theorem 37.2.1); therefore, $Ax = \beta x$. The obtained contradiction means that $\beta < \alpha$.

37.4. Let $f(\lambda) = |\lambda I - A|$. It is easy to verify that $f'(\lambda) = \sum_{i=1}^n |\lambda I - A_i|$, where A_i is a matrix obtained from A by deleting the ith row and the ith column (see Problem 11.7). If r and r_i are the greatest eigenvalues of A and A_i, respectively, then $r > r_i$ (see Problem 37.3). Therefore, all numbers $|rI - A_i|$ are positive. Hence, $f'(r) \neq 0$.

37.5. Suppose that A is not primitive. Then for a certain permutation matrix P the matrix PAP^T is of the form indicated in the assumptions of Theorem 37.3. On the other hand, the diagonal elements of PAP^T are obtained from the diagonal elements of A under a permutation. Contradiction.

37.6. Yes, it can. For instance, consider a nonnegative matrix A corresponding to the directed graph

$$\begin{gathered} 1 \longrightarrow (1,2), \ 2 \longrightarrow (3,4,5), \ 3 \longrightarrow (6,7,8), \ 4 \longrightarrow (6,7,8), \\ 5 \longrightarrow (6,7,8), \ 6 \longrightarrow (9), \ 7 \longrightarrow (9), \ 8 \longrightarrow (9), \ 9 \longrightarrow (1). \end{gathered}$$

It is easy to verify that the matrix A is indecomposable and, since $a_{11} > 0$, it is primitive (cf. Problem 37.5). The directed graph

$$\begin{gathered} 1 \longrightarrow (1,2,3,4,5), \ 2 \longrightarrow (6,7,8), \ 3 \longrightarrow (9), \ 4 \longrightarrow (9), \\ 5 \longrightarrow (9), \ 6 \longrightarrow (1), \ 7 \longrightarrow (1), \ 8 \longrightarrow (1), \ 9 \longrightarrow (1,2). \end{gathered}$$

corresponds to A^2. The first graph has 18 edges, whereas the second one has 16 edges.

38.1. There exist nonnegative numbers ξ_i and η_i such that $x_r = \xi_r + \cdots + \xi_n$ and $y_r = \eta_r + \cdots + \eta_n$. Therefore,

$$\begin{aligned} \sum_r x_r y_r - \sum_{r,s} a_{rs} x_r y_s &= \sum_{r,s} (\delta_{rs} - a_{rs}) x_r y_s \\ &= \sum_{r,s} (\delta_{rs} - a_{rs}) \sum_{i \geq r} \xi_i \sum_{j \geq s} \eta_j = \sum_{i,j} \xi_i \eta_j \sum_{r \leq i} \sum_{s \leq j} (\delta_{rs} - a_{rs}). \end{aligned}$$

It suffices to verify that $\sum_{r \leq i} \sum_{s \leq j} (\delta_{rs} - a_{rs}) \geq 0$. If $i \leq j$, then $\sum_{r \leq i} \sum_{s \leq j} \delta_{rs} = \sum_{r \leq i} \sum_{s=1}^{n} \delta_{rs}$ and, therefore,

$$\sum_{r \leq i} \sum_{s \leq j} (\delta_{rs} - a_{rs}) \geq \sum_{r \leq i} \sum_{s=1}^{n} (\delta_{rs} - a_{rs}) = 0.$$

The case $i \geq j$ is similar.

38.2. There exists a unitary matrix U such that $H = U \Lambda U^*$, where $\Lambda = \operatorname{diag}(\lambda_1, \ldots, \lambda_n)$. Since $h_{ij} = \sum_k u_{ik} \bar{u}_{jk} \lambda_k$, then $h_{ii} = \sum_k x_{ik} \lambda_k$, where $x_{ik} = |u_{ik}|^2$. Therefore, $h = X\lambda$, where h is the column $(h_{11}, \ldots, h_{nn})^T$, λ is the column $(\lambda_1, \ldots, \lambda_n)^T$, and X is a doubly stochastic matrix. By Theorem 38.2.1, $X = \sum_\sigma t_\sigma P_\sigma$, where P_σ is the matrix of the permutation σ, $t_\sigma \geq 0$, and $\sum_\sigma t_\sigma = 1$. Hence, $h = \sum_\sigma t_\sigma (P_\sigma \lambda)$.

CHAPTER VII

MATRICES IN ALGEBRA AND CALCULUS

39. Commuting matrices

39.1. Square matrices A and B of the same order are said *to be commuting* if $AB = BA$. Let us describe the set of all matrices X commuting with a given matrix A. Since the equalities $AX = XA$ and $A'X' = X'A'$, where $A' = PAP^{-1}$ and $X' = PXP^{-1}$ are equivalent, we may assume that $A = \operatorname{diag}(J_1, \dots, J_k)$, where $J_1, \dots, J_k$ are Jordan blocks. Let us represent X in the corresponding block form $X = \left\|X_{ij}\right\|_1^k$. Then the equation $AX = XA$ becomes equivalent to the system of equations

$$J_i X_{ij} = X_{ij} J_j.$$

It is not difficult to verify that if the eigenvalues of the matrices J_i and J_j are distinct, then the equation $J_i X_{ij} = X_{ij} J_j$ has only the zero solution and, if J_i and J_j are Jordan blocks of order m and n, respectively, corresponding to the same eigenvalue, then any solution of the equation $J_i X_{ij} = X_{ij} J_j$ is of the form $\begin{pmatrix} Y & 0 \end{pmatrix}$ or $\begin{pmatrix} Y \\ 0 \end{pmatrix}$, where

$$Y = \begin{pmatrix} y_1 & y_2 & \dots & y_{k-1} & y_k \\ 0 & y_1 & \dots & y_{k-2} & y_{k-1} \\ \vdots & \vdots & \ddots & \vdots & \vdots \\ 0 & 0 & \dots & y_1 & y_2 \\ 0 & 0 & \dots & 0 & y_1 \end{pmatrix}$$

and $k = \min(m, n)$. The dimension of the space of such matrices Y is equal to k.

Thus, we have obtained the following statement.

39.1.1. THEOREM. *Let Jordan blocks of size $a_1(\lambda), \dots, a_r(\lambda)$ correspond to an eigenvalue λ of a matrix A. Then the dimension of the space of solutions of the equation $AX = XA$ is equal to*

$$\sum_\lambda \sum_{i,j} \min(a_i(\lambda), a_j(\lambda)).$$

39.1.2. THEOREM. *Let m be the dimension of the space of solutions of the equation $AX = XA$, where A is a square matrix of order n. Then the following conditions are equivalent:*

a) $m = n$;

b) *the characteristic polynomial of A coincides with the minimal polynomial;*

c) *any matrix commuting with A is a polynomial in A.*

PROOF. a) $\Leftrightarrow$ b) By Theorem 39.1.1,

$$m = \sum_\lambda \sum_{i,j} \min(a_i(\lambda), a_j(\lambda)) \geq \sum_\lambda \sum_i a_i(\lambda) = n$$

with equality if and only if the Jordan blocks of A correspond to distinct eigenvalues, i.e., the characteristic polynomial coincides with the minimal polynomial.

b) $\Rightarrow$ c) If the characteristic polynomial of A coincides with the minimal polynomial, then the dimension of $\operatorname{Span}(I, A, \dots, A^{n-1})$ is equal to n and, therefore, it coincides with the space of solutions of the equation $AX = XA$, i.e., any matrix commuting with A is a polynomial in A.

c) $\Rightarrow$ a) If every matrix commuting with A is a polynomial in A, then, according to to the Cayley–Hamilton theorem, the space of solutions of the equation $AX = XA$ is contained in the space $\operatorname{Span}(I, A, \dots, A^{k-1})$ and $k \leq n$. On the other hand, $k \geq m \geq n$ and, therefore, $m = n$. $\square$

39.2.1. THEOREM. *Commuting operators A and B in a space V over $\mathbb{C}$ have a common eigenvector.*

PROOF. Let λ be an eigenvalue of A and $W \subset V$ the subspace of all eigenvectors of A corresponding to λ. Then $BW \subset W$. Indeed, if $Aw = \lambda w$, then $A(Bw) = BAw = \lambda(Bw)$. The restriction of B to W has an eigenvector w_0 and this vector is also an eigenvector of A (corresponding to the eigenvalue λ). $\square$

39.2.2. THEOREM. *Commuting diagonalizable operators A and B in a space V over $\mathbb{C}$ have a common eigenbasis.*

PROOF. For every eigenvalue λ of A consider the subspace V_λ consisting of all eigenvectors of A corresponding to the eigenvalue λ. Then $V = \oplus_\lambda V_\lambda$ and $BV_\lambda \subset V_\lambda$. The restriction of the diagonalizable operator B to V_λ is a diagonalizable operator. Indeed, the minimal polynomial of the restriction of B to V_λ is a divisor of the minimal polynomial of B and the minimal polynomial of B has no multiple roots. For every eigenvalue μ of the restriction of B to V_λ consider the subspace $V_{\lambda,\mu}$ consisting of all eigenvectors of the restriction of B to V_λ corresponding to the eigenvalue μ. Then $V_\lambda = \oplus_\mu V_{\lambda,\mu}$ and $V = \oplus_{\lambda,\mu} V_{\lambda,\mu}$. By selecting an arbitrary basis in every subspace $V_{\lambda,\mu}$, we finally obtain a common eigenbasis of A and B. $\square$

We can similarly construct a common eigenbasis for any finite family of pairwise commuting diagonalizable operators.

39.3. THEOREM. *Suppose the matrices A and B are such that any matrix commuting with A commutes also with B. Then $B = g(A)$, where g is a polynomial.*

PROOF. It is possible to consider the matrices A and B as linear operators in a certain space V. For an operator A there exists a cyclic decomposition $V = V_1 \oplus \dots \oplus V_k$ with the following property (see 14.1): $AV_i \subset V_i$ and the restriction A_i of A to V_i is a cyclic block; the characteristic polynomial of A_i is equal to p_i, where p_i is divisible by p_{i+1} and p_1 is the minimal polynomial of A.

Let the vector e_i span V_i, i.e., $V_i = \operatorname{Span}(e_i, Ae_i, A^2e_i, \dots)$ and $P_i : V \longrightarrow V_i$ be a projection. Since $AV_i \subset V_i$, we have $AP_iv = P_iAv$ and, therefore, $P_iB = BP_i$. Hence, $Be_i = BP_ie_i = P_iBe_i \in V_i$, i.e., $Be_i = g_i(A)e_i$, where g_i is a polynomial. Any vector $v_i \in V_i$ is of the form $f(A)e_i$, where f is a polynomial. Therefore, $Bv_i = g_i(A)v_i$. Let us prove that $g_i(A)v_i = g_1(A)v_i$, i.e., we can take g_1 for the required polynomial g.

Let us consider an operator $X_i : V \longrightarrow V$ that sends vector $f(A)e_i$ to $(fn_i)(A)e_1$, where $n_i = p_1p_i^{-1}$, and that sends every vector $v_j \in V_j$, where $j \neq i$, into itself.

First, we verify that the operator X_i is well defined. Let $f(A)e_i = 0$, i.e., let f be divisible by p_i. Then n_if is divisible by $n_ip_i = p_1$ and, therefore, $(fn_i)(A)e_1 = 0$. It is easy to check that $X_iA = AX_i$ and, therefore, $X_iB = BX_i$.

On the other hand, $X_iBe_i = (n_ig_i)(A)e_1$ and $BX_ie_i = (n_ig_1)(A)e_1$. Therefore, we have $n_i(A)[g_i(A) - g_1(A)]e_1 = 0$. It follows that the polynomial $n_i(g_i - g_1)$ is divisible by $p_1 = n_ip_i$, i.e., $g_i - g_1$ is divisible by p_i, hence $g_i(A)v_i = g_1(A)v_i$ for any $v_i \in V_i$. $\square$

Problems

39.1. Let $A = \operatorname{diag}(\lambda_1, \dots, \lambda_n)$, where the numbers λ_i are distinct, and let a matrix X commute with A.

a) Prove that X is a diagonal matrix.

b) Moreover, let the numbers λ_i be nonzero and let X commute with NA, where $N = |\delta_{i+1,j}|_1^n$. Prove that $X = \lambda I$.

39.2. Prove that if X commutes with all matrices, then $X = \lambda I$.

39.3. Find all matrices commuting with E, where E is the matrix all elements of which are equal to 1.

39.4. Let P_σ be the matrix corresponding to a permutation σ. Prove that if $AP_\sigma = P_\sigma A$ for all σ, then $A = \lambda I + \mu E$, where E is the matrix all elements of which are equal to 1.

39.5. Prove that for any complex matrix A there exists a matrix B such that $AB = BA$ and the characteristic polynomial of B coincides with the minimal polynomial.

39.6. a) Let A and B be commuting nilpotent matrices. Prove that $A + B$ is a nilpotent matrix.

b) Let A and B be commuting diagonalizable matrices. Prove that $A + B$ is diagonalizable.

39.7. In a space of dimension n, there are given (distinct) pairwise commuting involutions $A_1, \dots, A_m$. Prove that $m \leq 2^n$.

39.8. Diagonalizable operators $A_1, \dots, A_n$ commute with each other. Prove that all these operators can be polynomially expressed in terms of a diagonalizable operator.

39.9. In the space of matrices of order $2m$, indicate a subspace of dimension $m^2 + 1$ consisting of matrices commuting with each other.

40. Commutators

40.1. Let A and B be square matrices of the same order. The matrix

$$[A, B] = AB - BA$$

is called the *commutator* of the matrices A and B. The equality $[A, B] = 0$ means that A and B *commute*.

It is easy to verify that $\operatorname{tr}[A, B] = 0$ for any A and B; cf. 11.1.

It is subject to an easy direct verification that the following *Jacobi identity* holds:

$$[A,[B,C]]+[B,[C,A]]+[C,[A,B]]=0.$$

An algebra (not necessarily matrix) is called a *Lie algebra* if the multiplication (usually called *bracket* and denoted by $[\cdot,\cdot]$) in this algebra is a skew-commutative, i.e., $[A, B] = -[B, A]$, and satisfies Jacobi identity. The map $\operatorname{ad}_A : M_{n,n} \longrightarrow M_{n,n}$ defined by the formula $\operatorname{ad}_A(X) = [A, X]$ is a linear operator in the space of matrices. The map which to every matrix A assigns the operator ad_A is called the *adjoint representation* of $M_{n,n}$. The adjoint representation has important applications in the theory of Lie algebras.

The following properties of ad_A are easy to verify:

1) $\operatorname{ad}_{[A,B]} = \operatorname{ad}_A \operatorname{ad}_B - \operatorname{ad}_B \operatorname{ad}_A$ (this equality is equivalent to the Jacobi identity);

2) the operator $D = \operatorname{ad}_A$ is a *derivation* of the matrix algebra, i.e.,

$$D(XY) = XD(Y) + (DX)Y;$$

3) $D^n(XY) = \sum_{k=0}^{n} \binom{n}{k}(D^k X)(D^{n-k} Y)$;

4) $D(X^n) = \sum_{k=0}^{n-1} X^k(DX)X^{n-1-k}$.

40.2. If $A = [X, Y]$, then $\operatorname{tr} A = 0$. It turns out that the converse is also true: if $\operatorname{tr} A = 0$, then there exist matrices X and Y such that $A = [X, Y]$. Moreover, we can impose various restrictions on the matrices X and Y.

40.2.1. THEOREM ([Fregus, 1966]). *Let* $\operatorname{tr} A = 0$. *Then there exist matrices X and Y such that X is an Hermitian matrix,* $\operatorname{tr} Y = 0$, *and* $A = [X, Y]$.

PROOF. There exists a unitary matrix U such that all the diagonal elements of $UAU^* = B = \|b_{ij}\|_1^n$ are zeros (see 15.2). Consider a matrix $D = \operatorname{diag}(d_1, \dots, d_n)$, where $d_1, \dots, d_n$ are arbitrary distinct real numbers. Let $Y_1 = \|y_{ij}\|_1^n$, where $y_{ii} = 0$ and $y_{ij} = \dfrac{b_{ij}}{d_i - d_j}$ for $i \neq j$. Then

$$DY_1 - Y_1 D = \|(d_i - d_j)y_{ij}\|_1^n = \|b_{ij}\|_1^n = UAU^*.$$

Therefore,

$$A = U^*DY_1U - U^*Y_1DU = XY - YX,$$

where $X = U^*DU$ and $Y = U^*Y_1U$. Clearly, X is an Hermitian matrix and $\operatorname{tr} Y = 0$. □

REMARK. If A is a real matrix, then the matrices X and Y can be selected to be real ones.

40.2.2. THEOREM ([Gibson, 1975]). *Let* $\operatorname{tr} A = 0$ *and* $\lambda_1, \dots, \lambda_n, \mu_1, \dots, \mu_n$ *be given complex numbers such that* $\lambda_i \neq \lambda_j$ *for* $i \neq j$. *Then there exist complex matrices X and Y with eigenvalues* $\lambda_1, \dots, \lambda_n$ *and* $\mu_1, \dots, \mu_n$, *respectively, such that* $A = [X, Y]$.

PROOF. There exists a matrix P such that all diagonal elements of the matrix $PAP^{-1} = B = \|b_{ij}\|_1^n$ are zero (see 15.1). Let $D = \operatorname{diag}(\lambda_1, \dots, \lambda_n)$ and $c_{ij} =$

$\dfrac{b_{ij}}{(\lambda_i - \lambda_j)}$ for $i \neq j$. The diagonal elements c_{ii} of C can be selected so that the eigenvalues of C are $\mu_1, \dots, \mu_n$ (see 48.2). Then

$$DC - CD = \left\|(\lambda_i - \lambda_j)c_{ij}\right\|_1^n = B.$$

It remains to set $X = P^{-1}DP$ and $Y = P^{-1}CP$. □

REMARK. This proof is valid over any algebraically closed field.

40.3. THEOREM ([Smiley, 1961]). *Suppose the matrices A and B are such that for a certain integer $s > 0$ the identity $\operatorname{ad}_A^s X = 0$ implies $\operatorname{ad}_X^s B = 0$. Then B can be expressed as a polynomial of A.*

PROOF. The case $s = 1$ was considered in 39.3; therefore, in what follows we will assume that $s \geq 2$. Observe that for $s \geq 2$ the identity $\operatorname{ad}_A^s X = 0$ does not necessarily imply $\operatorname{ad}_X^s A = 0$.

We may assume that $A = \operatorname{diag}(J_1, \dots, J_t)$, where J_i is a Jordan block. Let $X = \operatorname{diag}(1, \dots, n)$. It is easy to verify that $\operatorname{ad}_A^2 X = 0$ (see Problem 40.1); therefore, $\operatorname{ad}_A^s X = 0$ and $\operatorname{ad}_X^s B = 0$. The matrix X is diagonalizable and, therefore, $\operatorname{ad}_X B = 0$ (see Problem 40.6). Hence, B is a diagonal matrix (see Problem 39.1 a)). In accordance with the block notation $A = \operatorname{diag}(J_1, \dots, J_t)$ let us express the matrices B and X in the form $B = \operatorname{diag}(B_1, \dots, B_t)$ and $X = \operatorname{diag}(X_1, \dots, X_t)$. Let

$$Y = \operatorname{diag}((J_1 - \lambda_1 I)X_1, \dots, (J_t - \lambda_t I)X_t),$$

where λ_i is the eigenvalue of the Jordan block J_i. Then $\operatorname{ad}_A^2 Y = 0$ (see Problem 40.1). Hence, $\operatorname{ad}_A^2(X + Y) = 0$ and, therefore, $\operatorname{ad}_{X+Y}^s B = 0$. The matrix $X + Y$ is diagonalizable, since its eigenvalues are equal to $1, \dots, n$. Hence, $\operatorname{ad}_{X+Y} B = 0$ and, therefore, $\operatorname{ad}_Y B = 0$.

The equations $[X, B] = 0$ and $[Y, B] = 0$ imply that $B_i = b_i I$ (see Problem 39.1). Let us prove that if the eigenvalues of J_i and J_{i+1} are equal, then $b_i = b_{i+1}$. Consider the matrix

$$U = \begin{pmatrix} 0 & \dots & 0 & 1 \\ 0 & \dots & 0 & 0 \\ \vdots & \dots & \vdots & \vdots \\ 0 & \dots & 0 & 0 \end{pmatrix}$$

of order equal to the sum of the orders of J_i and J_{i+1}. In accordance with the block expression $A = \operatorname{diag}(J_1, \dots, J_t)$ introduce the matrix $Z = \operatorname{diag}(0, U, 0)$. It is easy to verify that $ZA = AZ = \lambda Z$, where λ is the common eigenvalue of J_i and J_{i+1}. Hence,

$$\operatorname{ad}_A(X + Z) = \operatorname{ad}_A Z = 0, \quad \operatorname{ad}_A^s(X + Y) = 0,$$

and $\operatorname{ad}_{X+Z}^s B = 0$. Since the eigenvalues of $X + Z$ are equal to $1, \dots, n$, it follows that $X + Z$ is diagonalizable and, therefore, $\operatorname{ad}_{X+Z} B = 0$. Since $[X, B] = 0$, we have $[Z, B] = [X + Z, B] = 0$, i.e., $b_i = b_{i+1}$.

We can assume that $A = \operatorname{diag}(M_1, \dots, M_q)$, where M_i is the union of Jordan blocks with equal eigenvalues. Then $B = \operatorname{diag}(B_1', \dots, B_q')$, where $B_i' = b_i' I$. The identity $[W, A] = 0$ implies that $W = \operatorname{diag}(W_1, \dots, W_q)$ (see 39.1) and, therefore, $[W, B] = 0$. Thus, the case $s \geq 2$ reduces to the case $s = 1$. □

40.4. Matrices $A_1, \dots, A_m$ are said to be *simultaneously triangularizable* if there exists a matrix P such that all matrices $P^{-1}A_iP$ are upper triangular.

THEOREM ([Drazin, Dungey, Greunberg, 1951]). *Matrices $A_1, \dots, A_m$ are simultaneously triangularizable if and only if the matrix $p(A_1, \dots, A_m)[A_i, A_j]$ is nilpotent for every polynomial $p(x_1, \dots, x_m)$ in noncommuting indeterminates.*

PROOF. If the matrices $A_1, \dots, A_m$ are simultaneously triangularizable, then the matrices $P^{-1}[A_i, A_j]P$ and $P^{-1}p(A_1, \dots, A_m)P$ are upper triangular and all diagonal elements of the first matrix are zeros. Hence, the product of these matrices is a nilpotent matrix, i.e., the matrix $p(A_1, \dots, A_m)[A_i, A_j]$ is nilpotent.

Now we suppose that every matrix of the form $p(A_1, \dots, A_m)[A_i, A_j]$ is nilpotent and prove that in this case the matrices $A_1, \dots, A_m$ are simultaneously triangularizable.

First, let us prove that for every nonzero vector u there exists a polynomial $h(x_1, \dots, x_m)$ such that $h(A_1, \dots, A_m)u$ is a nonzero common eigenvector of the matrices $A_1, \dots, A_m$.

We use induction on m. For $m = 1$ there exists a number k such that the vectors u, $A_1u, \dots, A_1^{k-1}u$ are linearly independent and $A_1^k u = a_{k-1}A_1^{k-1}u + \dots + a_0u$. Let $g(x) = x^k - a_{k-1}x^{k-1} - \dots - a_0$ and $g_0(x) = \dfrac{g(x)}{(x - x_0)}$, where x_0 is a root of the polynomial g. Then $g_0(A_1)u \neq 0$ and $(A_1 - x_0I)g_0(A_1)u = g(A_1)u = 0$, i.e., $g_0(A_1)u$ is an eigenvector of A_1.

Suppose that our statement holds for any $m - 1$ matrices $A_1, \dots, A_{m-1}$.

For a given nonzero vector u a certain nonzero vector $v_1 = h(A_1, \dots, A_{m-1})u$ is a common eigenvector of the matrices $A_1, \dots, A_{m-1}$. The following two cases are possible.

1) $[A_i, A_m]f(A_m)v_1 = 0$ for all i and any polynomial f. For $f = 1$ we get $A_iA_mv_1 = A_mA_iv_1$; hence, $A_iA_m^kv_1 = A_m^kA_iv_1$, i.e., $A_ig(A_m)v_1 = g(A_m)A_iv_1$ for any g. For a matrix A_m there exists a polynomial g_1 such that $g_1(A_m)v_1$ is an eigenvector of this matrix. Since $A_ig_1(A_m)v_1 = g_1(A_m)A_iv_1$ and v_1 is an eigenvector of $A_1, \dots, A_m$, we have $g_1(A_m)v_1 = g_1(A_m)h(A_1, \dots, A_{m-1})u$ is an eigenvector of $A_1, \dots, A_m$.

2) $[A_i, A_m]f_1(A_m)v_1 \neq 0$ for a certain f_1 and certain i. The vector $C_1f_1(A_m)v_1$, where $C_1 = [A_i, A_m]$, is nonzero and, therefore, the matrices $A_1, \dots, A_{m-1}$ have a common eigenvector $v_2 = g_1(A_1, \dots, A_{m-1})C_1f_1(A_m)v_1$. We can apply the same argument to the vector v_2, etc. As a result we get a sequence $v_1, v_2, v_3, \dots$, where v_k is an eigenvector of the matrices $A_1, \dots, A_{m-1}$ and where

$$v_{k+1} = g_k(A_1, \dots, A_{m-1})C_kf_k(A_m)v_k, \quad C_k = [A_s, A_m] \quad \text{for a certain } s.$$

This sequence terminates with a vector v_p if $[A_i, A_m]f(A_m)v_p = 0$ for all i and all polynomials f.

For A_m there exists a polynomial $g_p(x)$ such that $g_p(A_m)v_p$ is an eigenvector of A_m. As in case 1), we see that this vector is an eigenvector of $A_1, \dots, A_m$ and

$$g_p(A_m)v_p = g_p(A_m)g(A_1, \dots, A_m)h(A_1, \dots, A_{m-1})u.$$

It remains to show that the sequence $v_1, v_2, \dots$ terminates. Suppose that this is not so. Then there exist numbers $\lambda_1, \dots, \lambda_{n+1}$ not all equal to zero for which

$\lambda_1 v_1 + \dots + \lambda_{n+1} v_{n+1} = 0$ and, therefore, there exists a number j such that $\lambda_j \neq 0$ and

$$-\lambda_j v_j = \lambda_{j+1} v_{j+1} + \dots + \lambda_{n+1} v_{n+1}.$$

Clearly,

$$v_{j+1} = g_j(A_1, \dots, A_{m-1}) C_j f_j(A_m) v_j, v_{j+2} = u_{j+1}(A_1, \dots, A_m) C_j f_j(A_m) v_j,$$

etc. Hence,

$$-\lambda_j v_j = u(A_1, \dots, A_m) C_j f_j(A_m) v_j$$

and, therefore,

$$f_j(A_m) u(A_1, \dots, A_m) C_j f_j(A_m) v_j = -\lambda_j f_j(A_m) v_j.$$

It follows that the nonzero vector $f_j(A_m)v_j$ is an eigenvector of the operator $f_j(A_m)u(A_1, \dots, A_m)C_j$ coresponding to the nonzero eigenvalue $-\lambda_j$. But by hypothesis this operator is nilpotent and, therefore, it has no nonzero eigenvalues. Contradiction.

Now we turn to the proof of the theorem by induction on n. For $n = 1$ the statement is obvious. As we have already demonstrated the operators $A_1, \dots, A_m$ have a common eigenvector y corresponding to certain eigenvalues $\alpha_1, \dots, \alpha_m$. We can assume that $|y| = 1$, i.e., $y^* y = 1$. There exists a unitary matrix Q whose first column is y. Clearly,

$$Q^* A_i Q = Q^*(\alpha_i y \dots) = \begin{pmatrix} \alpha_i & * \\ 0 & A'_i \end{pmatrix}$$

and the matrices $A'_1, \dots, A'_m$ of order $n - 1$ satisfy the condition of the theorem. By inductive hypothesis there exists a unitary matrix P_1 of order $n - 1$ such that the matrices $P_1^* A'_i P_1$ are upper triangular. Then $P = Q \begin{pmatrix} 1 & 0 \\ 0 & P_1 \end{pmatrix}$ is the desired matrix. (It even turned out to be unitary.) $\square$

40.5. THEOREM. *Let A and B be operators in a vector space V over $\mathbb{C}$ and let* $\operatorname{rank}[A, B] \leq 1$. *Then A and B are simultaneously triangularizable.*

PROOF. It suffices to prove that the operators A and B have a common eigenvector $v \in V$. Indeed, then the operators A and B induce operators A_1 and B_1 in the space $V_1 = V/\operatorname{Span}(v)$ and $\operatorname{rank}[A_1, B_1] \leq 1$. It follows that A_1 and B_1 have a common eigenvector in V_1, etc. Besides, we can assume that $\operatorname{Ker} A \neq 0$ (otherwise we can replace A by $A - \lambda I$).

The proof will be carried out by induction on $n = \dim V$. If $n = 1$, then the statement is obvious. Let $C = [A, B]$. In the proof of the inductive step we will consider two cases.

1) $\operatorname{Ker} A \subset \operatorname{Ker} C$. In this case $B(\operatorname{Ker} A) \subset \operatorname{Ker} A$, since if $Ax = 0$, then $Cx = 0$ and $ABx = BAx + Cx = 0$. Therefore, we can consider the restriction of B to $\operatorname{Ker} A \neq 0$ and select in $\operatorname{Ker} A$ an eigenvector v of B; then the vector v is also an eigenvector of A.

2) $\operatorname{Ker} A \not\subset \operatorname{Ker} C$, i.e., $Ax = 0$ and $Cx \neq 0$ for a vector x. Since $\operatorname{rank} C = 1$, we have $\operatorname{Im} C = \operatorname{Span}(y)$, where $y = Cx$. Besides,

$$y = Cx = ABx - BAx = ABx \in \operatorname{Im} A.$$

It follows that $B(\operatorname{Im} A) \subset \operatorname{Im} A$. Indeed, $BAz = ABz - Cz$, where $ABz \in \operatorname{Im} A$ and $Cz \in \operatorname{Im} C \subset \operatorname{Im} A$. We have $\operatorname{Ker} A \neq 0$; hence, $\dim \operatorname{Im} A < n$. Let A' and B' be the restrictions of A and B to $\operatorname{Im} A$. Then $\operatorname{rank}[A', B'] \leq 1$ and, therefore, by the inductive hypothesis the operators A' and B' have a common eigenvector. □

Problems

40.1. Let $J = N + \lambda I$ be a Jordan block of order n, $A = \operatorname{diag}(1, 2, \dots, n)$ and $B = NA$. Prove that $\operatorname{ad}_J^2 A = \operatorname{ad}_J^2 B = 0$.

40.2. Prove that if $C = [A_1, B_1] + \dots + [A_n, B_n]$ and C commutes with the matrices $A_1, \dots, A_n$, then C is nilpotent.

40.3. Prove that $\operatorname{ad}_A^n(B) = \sum_{i=0}^n (-1)^{n-i}\binom{n}{i} A^i B A^{n-i}$.

40.4. ([Kleinecke, 1957].) Prove that if $\operatorname{ad}_A^2(B) = 0$, then

$$\operatorname{ad}_A^n(B^n) = n!(\operatorname{ad}_A(B))^n.$$

40.5. Prove that if $[A, [A, B]] = 0$ and m and n are natural numbers such that $m > n$, then $n[A^m, B] = m[A^n, B]A^{m-n}$.

40.6. Prove that if A is a diagonalizable matrix and $\operatorname{ad}_A^n X = 0$, then $\operatorname{ad}_A X = 0$.

40.7. a) Prove that if $\operatorname{tr}(AXY) = \operatorname{tr}(AYX)$ for any X and Y, then $A = \lambda I$.

b) Let f be a linear function on the space of matrices of order n. Prove that if $f(XY) = f(YX)$ for any matrices X and Y, then $f(X) = \lambda \operatorname{tr} X$.

41. Quaternions and Cayley numbers. Clifford algebras

41.1. Let A be an algebra with unit over $\mathbb{R}$ endowed with a conjugation operation $a \mapsto \overline{a}$ satisfying $\overline{\overline{a}} = a$ and $\overline{ab} = \overline{b}\overline{a}$.

Let us consider the space $A \oplus A = \{(a, b) \mid a, b \in A\}$ and define a multiplication in it setting

$$(a, b)(u, v) = (au - \overline{v}b, b\overline{u} + va).$$

The obtained algebra is called the *double* of A. This construction is of interest because, as we will see, the algebra of complex numbers $\mathbb{C}$ is the double of $\mathbb{R}$, the algebra of *quaternions* $\mathbb{H}$ is the double of $\mathbb{C}$, and the *Cayley algebra* $\mathbb{O}$ is the double of $\mathbb{H}$.

It is easy to verify that the element $(1, 0)$ is a two-sided unit. Let $e = (0, 1)$. Then $(b, 0)e = (0, b)$ and, therefore, by identifying an element x of A with the element $(x, 0)$ of the double of A we have a representation of every element of the double in the form

$$(a, b) = a + be.$$

In the double of A we can define a *conjugation* by the formula

$$\overline{(a, b)} = (\overline{a}, -b),$$

i.e., by setting $\overline{a+be} = \overline{a} - be$. If $x = a + be$ and $y = u + ve$, then

$$\begin{aligned}\overline{xy} &= \overline{au} + \overline{(be)u} + \overline{a(ve)} + \overline{(be)(ve)} = \\ &= \overline{u}\cdot\overline{a} - \overline{u}(be) - (ve)\overline{a} + (ve)(be) = \overline{y}\cdot\overline{x}.\end{aligned}$$

It is easy to verify that $ea = \overline{a}e$ and $a(be) = (ba)e$. Therefore, the double of A is noncommutative, and if the conjugation in A is nonidentical and A is noncommutative, then the double is nonassociative. If A is both commutative and associative, then its double is associative.

41.2. Since $(0,1)(0,1) = (-1,0)$, we have $e^2 = -1$ and, therefore, the double of the algebra $\mathbb{R}$ with the identity conjugation is $\mathbb{C}$. Let us consider the double of $\mathbb{C}$ with the standard conjugation. Any element of this double can be expressed in the form

$$q = a + be, \text{ where } a = a_0 + a_1 i,\ b = a_2 + a_3 i \text{ and } a_0, \dots, a_3 \in \mathbb{R}.$$

Setting $j = e$ and $k = ie$ we get the conventional expression of a quaternion

$$q = a_0 + a_1 i + a_2 j + a_3 k.$$

The number a_0 is called the *real part* of the quaternion q and the quaternion $a_1 i + a_2 j + a_3 k$ is called its *imaginary part*. A quaternion is *real* if $a_1 = a_2 = a_3 = 0$ and *purely imaginary* if $a_0 = 0$.

The multiplication in the quaternion algebra $\mathbb{H}$ is given by the formulae

$$i^2 = j^2 = k^2 = -1, \quad ij = -ji = k, \quad jk = -kj = i, \quad ki = -ik = j.$$

The quaternion algebra is the double of an associative and commutative algebra and, therefore, it is associative.

The quaternion $\overline{q}$ conjugate to $q = a + be$ is equal to $\overline{a} - be = a_0 - a_1 i - a_2 j - a_3 k$. In 41.1 it was shown that $\overline{q_1 q_2} = \overline{q_2}\,\overline{q_1}$.

41.2.1. THEOREM. *The inner product (q, r) of quaternions q and r is equal to $\frac{1}{2}(\overline{q}r + \overline{r}q)$; in particular $|q|^2 = (q,q) = \overline{q}q$.*

PROOF. The function $B(q,r) = \frac{1}{2}(\overline{q}r + \overline{r}q)$ is symmetric and bilinear. Therefore, it suffices to verify that $B(q,r) = (q,r)$ for basis elements. It is easy to see that $B(1,i) = 0, B(i,i) = 1$ and $B(i,j) = 0$ and the remaining equalities are similarly checked. □

COROLLARY. *The element $\dfrac{\overline{q}}{|q|^2}$ is a two-sided inverse for q.*

Indeed, $\overline{q}q = |q|^2 = q\overline{q}$. □

41.2.2. THEOREM. $|qr| = |q| \cdot |r|$.

PROOF. Clearly,

$$|qr|^2 = qr\overline{qr} = qr\overline{r}\,\overline{q} = q|r|^2\overline{q} = |q|^2|r|^2. \quad \square$$

COROLLARY. *If $q \neq 0$ and $r \neq 0$, then $qr \neq 0$.*

41.3. To any quaternion $q = \alpha + xi + yj + zk$ we can assign the matrix $C(q) = \begin{pmatrix} u & v \\ -\overline{v} & \overline{u} \end{pmatrix}$, where $u = \alpha + ix$ and $v = y + iz$. For these matrices we have $C(qr) = C(q)C(r)$ (see Problem 41.4).

To a purely imaginary quaternion $q = xi + yj + zk$ we can assign the matrix $R(q) = \begin{pmatrix} 0 & -z & y \\ z & 0 & -x \\ -y & x & 0 \end{pmatrix}$. Since the product of imaginary quaternions can have a nonzero real part, the matrix $R(qr)$ is not determined for all q and r. However, since, as is easy to verify,

$$R(qr - rq) = R(q)R(r) - R(r)R(q),$$

the vector product $[q, r] = \frac{1}{2}(qr - rq)$ corresponds to the commutator of skew-symmetric 3×3 matrices. A linear subspace in the space of matrices is called a *matrix Lie algebra* if together with any two matrices A and B the commutator $[A, B]$ also belongs to it. It is easy to verify that the set of real skew-symmetric matrices and the set of complex skew-Hermitian matrices are matrix Lie algebras denoted by $\mathfrak{so}(n, \mathbb{R})$ and $\mathfrak{su}(n)$, respectively.

41.3.1. THEOREM. *The algebras $\mathfrak{so}(3, \mathbb{R})$ and $\mathfrak{su}(2)$ are isomorphic.*

PROOF. As is shown above these algebras are both isomorphic to the algebra of purely imaginary quaternions with the bracket $[q, r] = (qr - rq)/2$. $\square$

41.3.2. THEOREM. *The Lie algebras $\mathfrak{so}(4, \mathbb{R})$ and $\mathfrak{so}(3, \mathbb{R}) \oplus \mathfrak{so}(3, \mathbb{R})$ are isomorphic.*

PROOF. The Lie algebra $\mathfrak{so}(3, \mathbb{R})$ can be identified with the Lie algebra of purely imaginary quaternions. Let us assign to a quaternion $q \in \mathfrak{so}(3, \mathbb{R})$ the transformation $P(q) : u \mapsto qu$ of the space $\mathbb{R}^4 = \mathbb{H}$. As is easy to verify,

$$P(xi + yj + zk) = \begin{pmatrix} 0 & -x & -y & -z \\ x & 0 & -z & y \\ y & z & 0 & -x \\ z & -y & x & 0 \end{pmatrix} \in \mathfrak{so}(4, \mathbb{R}).$$

Similarly, the map $Q(q) : u \mapsto u\overline{q}$ belongs to $\mathfrak{so}(4, \mathbb{R})$. It is easy to verify that the maps $q \mapsto P(q)$ and $q \mapsto Q(q)$ are Lie algebra homomorphisms, i.e.,

$$P(qr - rq) = P(q)P(r) - P(r)P(q) \text{ and } Q(qr - rq) = Q(q)Q(r) - Q(r)Q(q).$$

Therefore, the map

$$\begin{aligned} \mathfrak{so}(3, \mathbb{R}) \oplus \mathfrak{so}(3, \mathbb{R}) &\longrightarrow \mathfrak{so}(4, \mathbb{R}) \\ (q, r) &\mapsto P(q) + Q(r) \end{aligned}$$

is a Lie algebra homomorphism. Since the dimensions of these algebras coincide, it suffices to verify that this map is a monomorphism. The identity $P(q) + Q(r) = 0$ means that $qx + x\overline{r} = 0$ for all x. For $x = 1$ we get $q = -\overline{r}$ and, therefore, $qx - xq = 0$ for all x. Hence, q is a real quaternion; on the other hand, by definition, q is a purely imaginary quaternion and, therefore, $q = r = 0$. $\square$

41.4. Let us consider the algebra of quaternions $\mathbb{H}$ as a space over $\mathbb{R}$. In $\mathbb{H} \otimes \mathbb{H}$, we can introduce an algebra structure by setting

$$(x_1 \otimes x_2)(y_1 \otimes y_2) = x_1 y_1 \otimes x_2 y_2.$$

Let us identify $\mathbb{R}^4$ with $\mathbb{H}$. It is easy to check that the map $w : \mathbb{H} \otimes \mathbb{H} \longrightarrow M_4(\mathbb{R})$ given by the formula $[w(x_1 \otimes x_2)]x = x_1 x \overline{x}_2$ is an algebra homomorphism, i.e., $w(uv) = w(u)w(v)$.

THEOREM. *The map $w : \mathbb{H} \otimes \mathbb{H} \longrightarrow M_4(\mathbb{R})$ is an algebra isomorphism.*

PROOF. The dimensions of $\mathbb{H} \otimes \mathbb{H}$ and $M_4(\mathbb{R})$ are equal. Still, unlike the case considered in 41.3, the calculation of the kernel of w is not as easy as the calculation of the kernel of the map $(q, r) \mapsto P(q) + Q(r)$, since the space $\mathbb{H} \otimes \mathbb{H}$ contains not only elements of the form $x \otimes y$. Instead, we should better prove that the image of w coincides with $M_4(\mathbb{R})$. The matrices

$$e = \begin{pmatrix} 1 & 0 \\ 0 & 1 \end{pmatrix}, \quad \varepsilon = \begin{pmatrix} 1 & 0 \\ 0 & -1 \end{pmatrix}, \quad a = \begin{pmatrix} 0 & 1 \\ 1 & 0 \end{pmatrix}, \quad b = \begin{pmatrix} 0 & 1 \\ -1 & 0 \end{pmatrix}$$

form a basis in the space of matrices of order 2. The images of $x \otimes y$, where $x, y \in \{1, i, j, k\}$, under the map w are given in Table 1.

TABLE 1. Values of $x \otimes y$

$x \backslash y$	1	i	j	k
1	$\begin{pmatrix} e & 0 \\ 0 & e \end{pmatrix}$	$\begin{pmatrix} b & 0 \\ 0 & -b \end{pmatrix}$	$\begin{pmatrix} 0 & e \\ -e & 0 \end{pmatrix}$	$\begin{pmatrix} 0 & b \\ b & 0 \end{pmatrix}$
i	$\begin{pmatrix} -b & 0 \\ 0 & -b \end{pmatrix}$	$\begin{pmatrix} e & 0 \\ 0 & -e \end{pmatrix}$	$\begin{pmatrix} 0 & -b \\ b & 0 \end{pmatrix}$	$\begin{pmatrix} 0 & e \\ e & 0 \end{pmatrix}$
j	$\begin{pmatrix} 0 & -\varepsilon \\ \varepsilon & 0 \end{pmatrix}$	$\begin{pmatrix} 0 & a \\ a & 0 \end{pmatrix}$	$\begin{pmatrix} \varepsilon & 0 \\ 0 & \varepsilon \end{pmatrix}$	$\begin{pmatrix} -a & 0 \\ 0 & a \end{pmatrix}$
k	$\begin{pmatrix} 0 & -a \\ a & 0 \end{pmatrix}$	$\begin{pmatrix} 0 & -\varepsilon \\ -\varepsilon & 0 \end{pmatrix}$	$\begin{pmatrix} a & 0 \\ 0 & a \end{pmatrix}$	$\begin{pmatrix} \varepsilon & 0 \\ 0 & -\varepsilon \end{pmatrix}$

From this table it is clear that among the linear combinations of the pairs of images of these elements we encounter all matrices with three zero blocks, the fourth block being one of the matrices e, ε, a, or b. Among the linear combinations of these matrices we encounter all matrices containing precisely one nonzero element and this element is equal to 1. Such matrices obviously form a basis of $M_4(\mathbb{R})$. □

41.5. The double of the quaternion algebra with the natural conjugation operation is the *Cayley* or *octonion algebra*. A basis of this algebra as a space over $\mathbb{R}$ is formed by the elements

$$1, \ i, \ j, \ k, \ e, \ f = ie, \ g = je, \ \text{and } h = ke.$$

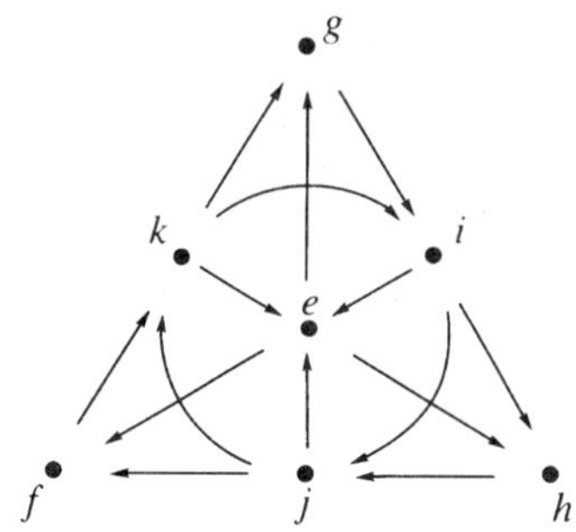

FIGURE 7

The multiplication table of these basis elements can be conveniently given with the help of Figure 7.

The product of two elements belonging to one line or one circle is the third element that belongs to the same line or circle and the sign is determined by the orientation; for example $ie = f$, $if = -e$.

Let $\xi = a + be$, where a and b are quaternions. The conjugation in $\mathbb{O}$ is given by the formula $\overline{(a, b)} = (\overline{a}, -b)$, i.e., $\overline{a + be} = \overline{a} - be$. Clearly,

$$\xi\overline{\xi} = (a, b)\overline{(a, b)} = (a, b)(\overline{a}, -b) = (a\overline{a} + b\overline{b}, ba - ba) = a\overline{a} + b\overline{b},$$

i.e., $\xi\overline{\xi}$ is the sum of squares of coordinates of ξ. Therefore, $|\xi| = \sqrt{\xi\overline{\xi}} = \sqrt{\overline{\xi}\xi}$ is the length of ξ.

THEOREM. $|\xi\eta| = |\xi| \cdot |\eta|$.

PROOF. For quaternions a similar theorem is proved quite simply, cf. 41.2. In our case the lack of associativity is a handicap. Let $\xi = a + be$ and $\eta = u + ve$, where a, b, u, v are quaternions. Then

$$|\xi\eta|^2 = (au - \overline{v}b)(\overline{u}\,\overline{a} - \overline{b}v) + (b\overline{u} + va)(u\overline{b} + \overline{a}\,\overline{v}).$$

Let us express a quaternion v in the form $v = \lambda + v_1$, where λ is a real number and $\overline{v}_1 = -v_1$. Then

$$\begin{aligned} |\xi\eta|^2 =& (au - \lambda b + v_1 b)(\overline{u}\,\overline{a} - \lambda\overline{b} - \overline{b}v_1) \\ &+ (b\overline{u} + \lambda a + v_1 a)(u\overline{b} + \lambda\overline{a} - \overline{a}v_1). \end{aligned}$$

Besides,

$$|\xi|^2|\eta|^2 = (a\overline{a} + b\overline{b})(u\overline{u} + \lambda^2 - v_1 v_1).$$

Since $u\overline{u}$ and $b\overline{b}$ are real numbers, $au\overline{u}\,\overline{a} = a\overline{a}u\overline{u}$ and $b\overline{b}v_1 = v_1 b\overline{b}$. Making use of similar equalities we get

$$\begin{aligned} &|\xi\eta|^2 - |\xi|^2|\eta|^2 \\ &\quad = \lambda(-b\overline{u}\,\overline{a} - au\overline{b} + b\overline{u}\,\overline{a} + au\overline{b}) + v_1(b\overline{u}\,\overline{a} + au\overline{b}) - (au\overline{b} + b\overline{u}\,\overline{a})v_1 = 0 \end{aligned}$$

because $b\overline{u}\overline{a} + au\overline{b}$ is a real number. $\square$

41.5.1. COROLLARY. *If $\xi \neq 0$, then $\overline{\xi}/|\xi|^2$ is a two-sided inverse for ξ.*

41.5.2. COROLLARY. *If $\xi \neq 0$ and $\eta \neq 0$, then $\xi\eta \neq 0$.*

The quaternion algebra is noncommutative and, therefore, $\mathbb{O}$ is a nonassociative algebra. Instead, the elements of $\mathbb{O}$ satisfy the conditions

$$x(yy) = (xy)y, \quad x(xy) = (xx)y \text{ and } (yx)y = y(xy)$$

(see Problem 41.8). It is possible to show that any subalgebra of $\mathbb{O}$ generated by two elements is associative.

41.6. By analogy with the vector product in the space of purely imaginary quaternions, we can define the vector product in the 7-dimensional space of purely imaginary octanions. Let x and y be purely imaginary octanions. Their *vector product* is the imaginary part of xy; it is denoted by $x \times y$. Clearly,

$$x \times y = \frac{1}{2}(xy - \overline{xy}) = \frac{1}{2}(xy - yx).$$

It is possible to verify that the *inner product* (x, y) *of octanions* x and y is equal to $\frac{1}{2}(\overline{x}y + \overline{y}x)$ and for purely imaginary octanions we get $(x, y) = -\frac{1}{2}(xy + yx)$.

THEOREM. *The vector product of purely imaginary octanions possesses the following properties:*

a) $x \times y \perp x$, $x \times y \perp y$;

b) $|x \times y|^2 = |x|^2|y|^2 - |(x, y)|^2$.

PROOF. a) We have to prove that

$$x(xy - yx) + (xy - yx)x = 0. \tag{1}$$

Since $x(yx) = (xy)x$ (see Problem 41.8 b)), we see that (1) is equivalent to $x(xy) = (yx)x$. By Problem 41.8, a) we have $x(xy) = (xx)y$ and $(yx)x = y(xx)$. It remains to notice that $xx = -x\overline{x} = -(x, x)$ is a real number.

b) We have to prove that

$$-(xy - yx)(xy - yx) = 4|x|^2|y|^2 - (xy + yx)(xy + yx),$$

i.e.,

$$2|x|^2|y|^2 = (xy)(yx) + (yx)(xy).$$

Let $a = xy$. Then $\overline{a} = yx$ and

$$2|x|^2|y|^2 = 2(a, a) = a\overline{a} + \overline{a}a = (xy)(yx) + (yx)(xy). \quad \square$$

41.7. The remaining part of this section will be devoted to the solution of the following

PROBLEM (Hurwitz-Radon). *What is the maximal number of orthogonal operators $A_1, \dots, A_m$ in $\mathbb{R}^n$ satisfying the relations $A_i^2 = -I$ and $A_iA_j + A_jA_i = 0$ for $i \neq j$?*

This problem might look quite artificial. There are, however, many important problems in one way or another related to quaternions or octonions that reduce to this problem. (Observe that the operators of multiplication by $i, j, \dots, h$ satisfy the required relations.)

We will first formulate the answer and then tell which problems reduce to our problem.

THEOREM (Hurwitz-Radon). *Let us express an integer n in the form $n = (2a+1)2^b$, where $b = c + 4d$ and $0 \le c \le 3$. Let $\rho(n) = 2^c + 8d$; then the maximal number of required operators in $\mathbb{R}^n$ is equal to $\rho(n) - 1$.*

41.7.1. The product of quadratic forms. Let $a = x_1 + ix_2$ and $b = y_1 + iy_2$. Then the identity $|a|^2|b|^2 = |ab|^2$ can be rewritten in the form

$$(x_1^2 + x_2^2)(y_1^2 + y_2^2) = z_1^2 + z_2^2,$$

where $z_1 = x_1y_1 - x_2y_2$ and $z_2 = x_1y_2 + x_2y_1$. Similar identities can be written for quaternions and octonions.

THEOREM. *Let m and n be fixed natural numbers; let $z_1(x, y), \dots, z_n(x, y)$ be real bilinear functions of $x = (x_1, \dots, x_m)$ and $y = (y_1, \dots, y_n)$. Then the identity of the form*

$$(x_1^2 + \dots + x_m^2)(y_1^2 + \dots + y_n^2) = z_1^2 + \dots + z_n^2$$

exists if and only if $m \le \rho(n)$.

PROOF. Let $z_i = \sum_j b_{ij}(x)y_j$, where $b_{ij}(x)$ are linear functions. Then

$$z_i^2 = \sum_j b_{ij}^2(x)y_j^2 + 2\sum_{j<k} b_{ij}(x)b_{ik}(x)y_jy_k.$$

Therefore, $\sum_i b_{ij}^2 = x_1^2 + \dots + x_m^2$ and $\sum_{j<k} b_{ij}(x)b_{ik}(x) = 0$. Let $B(x) = \left\| b_{ij}(x) \right\|_1^n$. Then $B^T(x)B(x) = (x_1^2 + \dots + x_m^2)I$. The matrix $B(x)$ can be expressed in the form $B(x) = x_1B_1 + \dots + x_mB_m$. Hence,

$$B^T(x)B(x) = x_1^2B_1^TB_1 + \dots + x_m^2B_m^TB_m + \sum_{i<j}(B_i^TB_j + B_j^TB_i)x_ix_j;$$

therefore, $B_i^TB_i = I$ and $B_i^TB_j + B_j^TB_i = 0$. The operators B_i are orthogonal and $B_i^{-1}B_j = -B_j^{-1}B_i$ for $i \neq j$.

Let us consider the orthogonal operators $A_1, \dots, A_{m-1}$, where $A_i = B_m^{-1}B_i$. Then $B_m^{-1}B_i = -B_i^{-1}B_m$ and, therefore, $A_i = -A_i^{-1}$, i.e., $A_i^2 = -I$. Besides, $B_i^{-1}B_j = -B_j^{-1}B_i$ for $i \neq j$; hence,

$$A_iA_j = B_m^{-1}B_iB_m^{-1}B_j = -B_i^{-1}B_mB_m^{-1}B_j = B_j^{-1}B_i = -A_jA_i.$$

It is also easy to verify that if the orthogonal operators $A_1, \dots, A_{m-1}$ are such that $A_i^2 = -I$ and $A_iA_j + A_jA_i = 0$, then the operators $B_1 = A_1, \dots, B_{m-1} = A_{m-1}$, $B_m = I$ possess the required properties. To complete the proof of Theorem 41.7.1 it remains to make use of Theorem 41.7. □

41.7.2. Normed algebras.

THEOREM. *Let a real algebra A be endowed with the Euclidean space structure so that $|xy| = |x| \cdot |y|$ for any $x, y \in A$. Then the dimension of A is equal to* 1, 2, 4 *or* 8.

PROOF. Let $e_1, \dots, e_n$ be an orthonormal basis of A. Then

$$(x_1 e_1 + \cdots + x_n e_n)(y_1 e_1 + \cdots + y_n e_n) = z_1 e_1 + \cdots + z_n e_n,$$

where $z_1, \dots, z_n$ are bilinear functions in x and y. The equality $|z|^2 = |x|^2 |y|^2$ implies that

$$(x_1^2 + \cdots + x_n^2)(y_1^2 + \cdots + y_n^2) = z_1^2 + \cdots + z_n^2.$$

It remains to make use of Theorem 41.7.1 and notice that $\rho(n) = n$ if and only if $n = 1$, 2, 4 or 8. □

41.7.3. The vector product.

THEOREM ([Massey, 1983]). *Let a bilinear operation $f(v, w) = v \times w \in \mathbb{R}^n$ be defined in $\mathbb{R}^n$, $n \geq 3$, satisfying the following properties. $v \times w$ is perpendicular to v and w and $|v \times w|^2 = |v|^2 |w|^2 - (v, w)^2$. Then $n = 3$ or 7.*

The product $\times$ determined by the above operator f is called the *vector product* of vectors.

PROOF. Consider the space $\mathbb{R}^{n+1} = \mathbb{R} \oplus \mathbb{R}^n$ and define a product in it by the formula

$$(a, v)(b, w) = (ab - (v, w), \quad aw + bv + v \times w),$$

where (v, w) is the inner product in $\mathbb{R}^n$. It is easy to verify that in the resulting algebra of dimension $n + 1$ the identity $|xy|^2 = |x|^2 |y|^2$ holds. It remains to make use of Theorem 41.7.2. □

REMARK. In spaces of dimension 3 or 7 a bilinear operation with the above properties does exist; cf. 41.6.

41.7.4. Vector fields on spheres. A vector field on a sphere S^n (say, unit sphere $S^n = \{v \in \mathbb{R}^{n+1} \mid |v| = 1\}$) is a map that to every point $v \in S^n$ assigns a vector $F(v)$ in the tangent space to S^n at v. The tangent space to S^n at v consists of vectors perpendicular to v; hence, $F(v) \perp v$. A vector field is said *linear* if $F(v) = Av$ for a linear operator A. It is easy to verify that $Av \perp v$ for all v if and only if A is a skew-symmetric operator (see Theorem 21.1.2). Therefore, any linear vector field on S^{2n} vanishes at some point.

To exclude vector fields that vanish at a point we consider orthogonal operators only; in this case $|Av| = 1$. It is easy to verify that an orthogonal operator A is skew-symmetric if and only if $A^2 = -I$. Recall that an operator whose square is equal to $-I$ is called a *complex structure* (see 10.4).

Vector fields $F_1, \dots, F_m$ are said to be *linearly independent* if the vectors $F_1(v)$, $\dots$, $F_m(v)$ are linearly independent at every point v. In particular, the vector fields corresponding to orthogonal operators $A_1, \dots, A_m$ such that $A_i v \perp A_j v$ for all $i \neq j$ are linearly independent. The equality $(A_i v, A_j v) = 0$ means that $(v, A_i^T A_j v) = 0$. Hence, $A_i^T A_j + (A_i^T A_j)^T = 0$, i.e., $A_i A_j + A_j A_i = 0$.

Thus, to construct m linearly independent vector fields on S^n it suffices to indicate orthogonal operators $A_1, \dots, A_m$ in $(n+1)$-dimensional space satisfying the relations $A_i^2 = -I$ and $A_i A_j + A_j A_i = 0$ for $i \neq j$. Thus, we have proved the following statement.

THEOREM. *On S^{n-1}, there exists $\rho(n)-1$ linearly independent vector fields.*

REMARK. It is far more difficult to prove that there do not exist $\rho(n)$ linearly independent continuous vector fields on S^{n-1}; see [Adams, 1962].

41.7.5. Linear subspaces in the space of matrices.

THEOREM. *In the space of real matrices of order n there is a subspace of dimension $m \leq \rho(n)$ all nonzero matrices of which are invertible.*

PROOF. If the matrices $A_1, \dots, A_{m-1}$ are such that $A_i^2 = -I$ and $A_iA_j + A_jA_i = 0$ for $i \neq j$, then

$$\left(\sum x_iA_i + x_mI\right)\left(-\sum x_iA_i + x_mI\right) = (x_1^2 + \dots + x_m^2)I.$$

Therefore, the matrix $\sum x_iA_i + x_mI$, where not all numbers $x_1, \dots, x_m$ are zero, is invertible. In particular, the matrices $A_1, \dots, A_{m-1}, I$ are linearly independent. □

41.8. Now we turn to the proof of Theorem 41.7. Consider the algebra C_m over $\mathbb{R}$ with generators $e_1, \dots, e_m$ and relations $e_i^2 = -1$ and $e_ie_j + e_je_i = 0$ for $i \neq j$. To every set of orthogonal matrices $A_1, \dots, A_m$ satisfying $A_i^2 = -I$ and $A_iA_j + A_jA_i = 0$ for $i \neq j$ there corresponds a *representation* (see 42.1) of C_m that maps the elements $e_1, \dots, e_m$ to orthogonal matrices $A_1, \dots, A_m$. In order to study the structure of C_m, we introduce an auxiliary algebra C'_m with generators $\varepsilon_1, \dots, \varepsilon_m$ and relations $\varepsilon_i^2 = 1$ and $\varepsilon_i\varepsilon_j + \varepsilon_j\varepsilon_i = 0$ for $i \neq j$.

The algebras C_m and C'_m are called *Clifford algebras.*

41.8.1. LEMMA. *$C_1 \cong \mathbb{C}$, $C_2 \cong \mathbb{H}$, $C'_1 \cong \mathbb{R} \oplus \mathbb{R}$ and $C'_2 \cong M_2(\mathbb{R})$.*

PROOF. The isomorphisms are explicitely given as follows:

$$\begin{array}{ll}
C_1 \longrightarrow \mathbb{C} & 1 \mapsto 1, e_1 \mapsto i; \\
C_2 \longrightarrow \mathbb{H} & 1 \mapsto 1, e_1 \mapsto i, e_2 \mapsto j; \\
C'_1 \longrightarrow \mathbb{R} \oplus \mathbb{R} & 1 \mapsto (1,1), \varepsilon_1 \mapsto (1,-1); \\
C'_2 \longrightarrow M_2(\mathbb{R}) & 1 \mapsto \begin{pmatrix} 1 & 0 \\ 0 & 1 \end{pmatrix}, \varepsilon_1 \mapsto \begin{pmatrix} 1 & 0 \\ 0 & -1 \end{pmatrix}, \varepsilon_2 \mapsto \begin{pmatrix} 0 & 1 \\ 1 & 0 \end{pmatrix}.
\end{array} \quad \square$$

COROLLARY. $\mathbb{C} \otimes \mathbb{H} \cong M_2(\mathbb{C})$.

Indeed, the complexifications of C_2 and C'_2 are isomorphic. □

41.8.2. LEMMA. *$C_{k+2} \cong C'_k \otimes C_2$ and $C'_{k+2} \cong C_k \otimes C'_2$.*

PROOF. The first isomorphism is given by the formulas

$$f(e_i) = 1 \otimes e_i \text{ for } i = 1, 2 \text{ and } f(e_i) = \varepsilon_{i-2} \otimes e_1e_2 \text{ for } i \geq 3.$$

The second isomorphism is given by the formulas

$$g(\varepsilon_i) = 1 \otimes \varepsilon_i \text{ for } i = 1, 2 \text{ and } g(\varepsilon_i) = e_{i-2} \otimes \varepsilon_1\varepsilon_2 \text{ for } i \geq 3. \quad \square$$

41.8.3. LEMMA. *$C_{k+4} \cong C_k \otimes M_2(\mathbb{H})$ and $C'_{k+4} \cong C'_k \otimes M_2(\mathbb{H})$.*

PROOF. By Lemma 41.8.2 we have

$$C_{k+4} \cong C'_{k+2} \otimes C_2 \cong C_k \otimes C'_2 \otimes C_2.$$

Since

$$C'_2 \otimes C_2 \cong \mathbb{H} \otimes M_2(\mathbb{R}) \cong M_2(\mathbb{H}),$$

we have $C_{k+4} \cong C_k \otimes M_2(\mathbb{H})$. Similarly, $C'_{k+4} \cong C'_k \otimes M_2(\mathbb{H})$. □

41.8.4. LEMMA. *$C_{k+8} \cong C_k \otimes M_{16}(\mathbb{R})$.*

PROOF. By Lemma 41.8.3

$$C_{k+8} \cong C_{k+4} \otimes M_2(\mathbb{H}) \cong C_k \otimes M_2(\mathbb{H}) \otimes M_2(\mathbb{H}).$$

Since $\mathbb{H} \otimes \mathbb{H} \cong M_4(\mathbb{R})$ (see 41.4), it follows that

$$M_2(\mathbb{H}) \otimes M_2(\mathbb{H}) \cong M_2(M_4(\mathbb{R})) \cong M_{16}(\mathbb{R}). \quad \square$$

Lemmas 41.8.1–41.8.3 make it possible to calculate C_k for $1 \leq k \leq 8$. For example,

$$C_5 \cong C_1 \otimes M_2(\mathbb{H}) \cong \mathbb{C} \otimes M_2(\mathbb{H}) \cong M_2(\mathbb{C} \otimes \mathbb{H}) \cong M_2(M_2(\mathbb{C})) \cong M_4(\mathbb{C});$$
$$C_6 \cong C_2 \otimes M_2(\mathbb{H}) \cong M_2(\mathbb{H} \otimes \mathbb{H}) \cong M_8(\mathbb{R}),$$

etc. The results of calculations are given in Table 2.

TABLE 2

k	1	2	3	4
C_k	$\mathbb{C}$	$\mathbb{H}$	$\mathbb{H} \oplus \mathbb{H}$	$M_2(\mathbb{H})$
C'_k	$\mathbb{R} \oplus \mathbb{R}$	$M_2(\mathbb{R})$	$M_2(\mathbb{C})$	$M_2(\mathbb{H})$
k	5	6	7	8
C_k	$M_4(\mathbb{C})$	$M_8(\mathbb{R})$	$M_8(\mathbb{R}) \oplus M_8(\mathbb{R})$	$M_{16}(\mathbb{R})$
C'_k	$M_2(\mathbb{H}) \oplus M_2(\mathbb{H})$	$M_4(\mathbb{H})$	$M_8(\mathbb{C})$	$M_{16}(\mathbb{R})$

Lemma 41.8.4 makes it possible now to calculate C_k for any k. The algebras $C_1, \dots, C_8$ have natural representations in the spaces $\mathbb{C}$, $\mathbb{H}$, $\mathbb{H}$, $\mathbb{H}^2$, $\mathbb{C}^4$, $\mathbb{R}^8$, $\mathbb{R}^8$ and $\mathbb{R}^{16}$ whose dimensions over $\mathbb{R}$ are equal to 2, 4, 4, 8, 8, 8, 8 and 16. Besides, under the passage from C_k to C_{k+8} the dimension of the space of the natural representation is multiplied by 16. The simplest case-by-case check indicates that for $n = 2^k$ the largest m for which C_m has the natural representation in $\mathbb{R}^n$ is equal to $\rho(n) - 1$.

Now, let us show that under these natural representations of C_m in $\mathbb{R}^n$ the elements $e_1, \dots, e_m$ turn into orthogonal matrices if we choose an appropriate basis in $\mathbb{R}^n$. First, let us consider the algebra $\mathbb{H} = \mathbb{R}^4$. Let us assign to an element $a \in \mathbb{H}$ the map $x \mapsto ax$ of the space $\mathbb{H}$ into itself. If we select the basis $1, i, j, k$ in the space $\mathbb{H} = \mathbb{R}^4$, then to elements $1, i, j, k$ the correspondence indicated assigns orthogonal matrices. We may proceed similarly in case of the algebra $\mathbb{C} = \mathbb{R}^2$.

We have shown how to select bases in $\mathbb{C} = \mathbb{R}^2$ and $\mathbb{H} = \mathbb{R}^4$ in order that the elements e_i and ε_j of the algebras C_1, C_2, C_1', and C_2' be represented by orthogonal matrices. Lemmas 41.8.2–4 show that the elements e_i and ε_j of the algebras C_m and C_m' are represented by matrices obtained consecutively with the help of the Kronecker product, and the initial matrices are orthogonal. It is clear that the Kronecker product of two orthogonal matrices is an orthogonal matrix (cf. 27.4).

Let $f : C_m \longrightarrow M_n(\mathbb{R})$ be a representation of C_m under which the elements $e_1, \dots, e_m$ turn into orthogonal matrices. Then $f(1 \cdot e_i) = f(1)f(e_i)$ and the matrix $f(e_i)$ is invertible. Hence, $f(1) = f(1 \cdot e_i)f(e_i)^{-1} = I$ is the unit matrix. The algebra C_m is either of the form $M_p(F)$ or of the form $M_p(F) \oplus M_p(F)$, where $F = \mathbb{R}, \mathbb{C}$ or $\mathbb{H}$. Therefore, if f is a representation of C_m such that $f(1) = I$, then f is completely reducible and its irreducible components are isomorphic to F^p (see 42.1); so its dimension is divisible by p. Therefore, for any n the largest m for which C_m has a representation in $\mathbb{R}^n$ such that $f(1) = I$ is equal to $\rho(n) - 1$.

Problems

41.1. Prove that the real part of the product of quaternions $x_1 i + y_1 j + z_1 k$ and $x_2 i + y_2 j + z_2 k$ is equal to the inner product of the vectors (x_1, y_1, z_1) and (x_2, y_2, z_2) taken with the minus sign, and that the imaginary part is equal to their vector product.

41.2. a) Prove that a quaternion q is purely imaginary if and only if $q^2 \leq 0$.

b) Prove that a quaternion q is real if and only if $q^2 \geq 0$.

41.3. Find all solutions of the equation $q^2 = -1$ in quaternions.

41.4. Prove that a quaternion that commutes with all purely imaginary quaternions is real.

41.5. A matrix A with quaternion elements can be represented in the form $A = Z_1 + Z_2 j$, where Z_1 and Z_2 are complex matrices. Assign to a matrix A the matrix $A_c = \begin{pmatrix} Z_1 & Z_2 \\ -\overline{Z}_2 & \overline{Z}_1 \end{pmatrix}$. Prove that $(AB)_c = A_c B_c$.

41.6. Consider the map in the space $\mathbb{R}^4 = \mathbb{H}$ which sends a quaternion x to qx, where q is a fixed quaternion.

a) Prove that this map sends orthogonal vectors to orthogonal vectors.

b) Prove that the determinant of this map is equal to $|q|^4$.

41.7. Given a tetrahedron $ABCD$ prove with the help of quaternions that

$$AB \cdot CD + BC \cdot AD \geq AC \cdot BD.$$

41.8. Let x and y be octonions. Prove that a) $x(yy) = (xy)y$ and $x(xy) = (xx)y$; b) $(yx)y = y(xy)$.

42. Representations of matrix algebras

42.1. Let A be an associative algebra and $\mathrm{Mat}(V)$ the associative algebra of linear transformations of a vector space V. A homomorphism $f : A \longrightarrow \mathrm{Mat}(V)$ of associative algebras is called a *representation* of A. Given a homomorphism f we define the action of A in V by the formula $av = f(a)v$. We have $(ab)v = a(bv)$. Thus, the space V is an A-module.

A subspace $W \subset V$ is an *invariant subspace of the representation* f if $AW \subset W$, i.e., if W is a submodule of the A-module V. A representation is said to be *irreducible* if any nonzero invariant subspace of it coincides with the whole space V. A representation $f : A \longrightarrow \mathrm{Mat}(V)$ is called *completely reducible* if the space V is the direct sum of invariant subspaces such that the restriction of f to each of them is irreducible.

42.1.1. THEOREM. *Let $A = \mathrm{Mat}(V^n)$ and $f : A \longrightarrow \mathrm{Mat}(W^m)$ a representation such that $f(I_n) = I_m$. Then $W^m = W_1 \oplus \cdots \oplus W_k$, where the W_i are invariant subspaces isomorphic to V^n.*

PROOF. Let $e_1, \dots, e_m$ be a basis of W. Since $f(I_n)e_i = e_i$, it follows that $W \subset \mathrm{Span}(Ae_1, \dots, Ae_m)$. It is possible to represent the space of A in the form of the direct sum of subspaces F_i consisting of matrices whose columns are all zero, except the ith one. Clearly, $AF_i = F_i$ and if a is a nonzero element of F_i, then $Aa = F_i$. The space W is the sum of spaces F_ie_j. These spaces are invariant, since $AF_i = F_i$. If $x = ae_j$, where $a \in F_i$ and $x \neq 0$, then $Ax = Aae_j = F_ie_j$.

Therefore, any two spaces of the form F_ie_j either do not have common nonzero elements or coincide. It is possible to represent W in the form of the direct sum of certain nonzero subspaces F_ie_j. For this we have to add at each stage subspaces not contained in the direct sum of the previously chosen subspaces. It remains to demonstrate that every nonzero space F_ie_j is isomorphic to V. Consider the map $h : F_i \longrightarrow F_ie_j$ for which $h(a) = ae_j$. Clearly, $A \,\mathrm{Ker}\, h \subset \mathrm{Ker}\, h$. Suppose that $\mathrm{Ker}\, h \neq 0$. In $\mathrm{Ker}\, h$, select a nonzero element a. Then $Aa = F_i$. On the other hand, $Aa \subset \mathrm{Ker}\, h$. Therefore, $\mathrm{Ker}\, h = F_i$, i.e., h is the zero map. Hence, either h is an isomorphism or the zero map.

This proof remains valid for the algebra of matrices over $\mathbb{H}$, i.e., when V and W are spaces over $\mathbb{H}$. Note that if $A = \mathrm{Mat}(V^n)$, where V^n is a space over $\mathbb{H}$ and $f : A \longrightarrow \mathrm{Mat}(W^m)$ a representation such that $f(I_n) = I_m$, then W^m necessarily has the structure of a vector space over $\mathbb{H}$. Indeed, the multiplication of elements of W^m by i, j, k is determined by operators $f(iI_n), f(jI_n), f(kI_n)$. $\square$

In §41 we used not only Theorem 42.1.1 but also of the following statement.

42.1.2. THEOREM. *Let $A = \mathrm{Mat}(V^n) \oplus \mathrm{Mat}(V^n)$ and $f : A \longrightarrow \mathrm{Mat}(W^m)$ a representation such that $f(I_n) = I_m$. Then $W^m = W_1 \oplus \cdots \oplus W_k$, where the W_i are invariant subspaces isomorphic to V^n.*

PROOF. Let F_i be the set of matrices defines in the proof of Theorem 42.1.1. The space A can be represented as the direct sum of its subspaces $F_i^1 = F_i \oplus 0$ and $F_i^2 = 0 \oplus F_i$. Similarly to the proof of Theorem 42.1.1 we see that the space W can be represented as the direct sum of certain nonzero subspaces $F_i^k e_j$ each of which is invariant and isomorphic to V^n. $\square$

43. The resultant

43.1. Consider polynomials $f(x) = \sum_{i=0}^n a_ix^{n-i}$ and $g(x) = \sum_{i=0}^m b_ix^{m-i}$, where $a_0 \neq 0$ and $b_0 \neq 0$. Over an algebraically closed field, f and g have a common divisor if and only if they have a common root. If the field is not algebraically closed, then the common divisor can happen to be a polynomial without roots.

The presence of a common divisor for f and g is equivalent to the fact that there exist polynomials p and q such that $fq = gp$, where $\deg p \leq n - 1$ and $\deg q \leq m - 1$.

Let $q = u_0x^{m-1} + \dots + u_{m-1}$ and $p = v_0x^{n-1} + \dots + v_{n-1}$. The equality $fq = gp$ can be expressed in the form of a system of equations

$$\begin{aligned} a_0u_0 &= b_0v_0 \\ a_1u_0 + a_0u_1 &= b_1v_0 + b_0v_1 \\ a_2u_0 + a_1u_1 + a_0u_2 &= b_2v_0 + b_1v_1 + b_0v_2 \\ &\dots\dots \end{aligned}$$

The polynomials f and g have a common root if and only if this system of equations has a nonzero solution $(u_0, u_1, \dots, v_0, v_1, \dots)$. If, for example, $m = 3$ and $n = 2$, then the determinant of this system is of the form

$$\begin{vmatrix} a_0 & 0 & 0 & -b_0 & 0 \\ a_1 & a_0 & 0 & -b_1 & -b_0 \\ a_2 & a_1 & a_0 & -b_2 & -b_1 \\ 0 & a_2 & a_1 & -b_3 & -b_2 \\ 0 & 0 & a_2 & 0 & -b_3 \end{vmatrix} = \pm \begin{vmatrix} a_0 & a_1 & a_2 & 0 & 0 \\ 0 & a_0 & a_1 & a_2 & 0 \\ 0 & 0 & a_0 & a_1 & a_2 \\ b_0 & b_1 & b_2 & b_3 & 0 \\ 0 & b_0 & b_1 & b_2 & b_3 \end{vmatrix} = \pm|S(f,g)|.$$

The matrix $S(f,g)$ is called the *Sylvester matrix* of polynomials f and g. The determinant of $S(f,g)$ is called the *resultant* of f and g and is denoted by $R(f,g)$. It is clear that $R(f,g)$ is a homogeneous polynomial of degree m with respect to the variables a_i and of degree n with respect to the variables b_j. The polynomials f and g have a common divisor if and only if the determinant of the above system is zero, i.e., if $R(f,g) = 0$.

The resultant has a number of applications. For example, given polynomial relations $P(x,z) = 0$ and $Q(y,z) = 0$ we can use the resultant to obtain a polynomial relation $R(x,y) = 0$. Indeed, consider given polynomials $P(x,z)$ and $Q(y,z)$ as polynomials in z considering x and y as constant parameters. Then the resultant $R(x,y)$ of these polynomials gives the required relation $R(x,y) = 0$.

The resultant also allows one to reduce the problem of solution of a system of algebraic equations to the search of roots of polynomials. In fact, let $P(x_0, y_0) = 0$ and $Q(x_0, y_0) = 0$. Consider $P(x,y)$ and $Q(x,y)$ as polynomials in y. At $x = x_0$ they have a common root y_0. Therefore, their resultant $R(x)$ vanishes at $x = x_0$.

43.2. THEOREM. *Let x_i be the roots of a polynomial f and let y_j be the roots of a polynomial g. Then*

$$R(f,g) = a_0^m b_0^n \prod (x_i - y_j) = a_0^m \prod g(x_i) = b_0^n \prod f(y_j).$$

PROOF. Since $f = a_0(x - x_1)\dots(x - x_n)$, we have $a_k = \pm a_0\sigma_k(x_1, \dots, x_n)$, where σ_k is an elementary symmetric function. Similarly, $b_k = \pm b_0\sigma_k(y_1, \dots, y_m)$. The resultant is a homogeneous polynomial of degree m with respect to variables a_i and of degree n with respect to the b_j; hence,

$$R(f,g) = a_0^m b_0^n P(x_1, \dots, x_n, y_1, \dots, y_m),$$

where P is a symmetric polynomial in the totality of variables $x_1, \dots, x_n$ and $y_1, \dots, y_m$ which vanishes for $x_i = y_j$. The formula

$$x_i^k = (x_i - y_j)x_i^{k-1} + y_j x_i^{k-1}$$

shows that

$$P(x_1, \dots, y_m) = (x_i - y_j)Q(x_1, \dots, y_m) + R(x_1, \dots, \widehat{x_i}, \dots, y_m).$$

Substituting $x_i = y_j$ in this equation we see that $R(x_1, \dots, \widehat{x_i}, \dots, y_m)$ is identically equal to zero, i.e., R is the zero polynomial. Similar arguments demonstrate that P is divisible by $S = a_0^m b_0^n \prod(x_i - y_j)$.

Since $g(x) = b_0 \prod_{j=1}^m (x - y_j)$, it follows that $\prod_{i=1}^n g(x_i) = b_0^n \prod_{i,j}(x_i - y_j)$; hence,

$$S = a_0^m \prod_{i=1}^n g(x_i) = a_0^m \prod_{i=1}^n (b_0 x_i^m + b_1 x_i^{m-1} + \dots + b_m)$$

is a homogeneous polynomial of degree n with respect to $b_0, \dots, b_m$.

For the variables $a_0, \dots, a_n$ our considerations are similar. It is also clear that the symmetric polynomial $a_0^m \prod(b_0 x_i^m + b_1 x_i^{m-1} + \dots + b_m)$ is a polynomial in $a_0, \dots, a_n, b_0, \dots, b_m$. Hence, $R(a_0, \dots, b_m) = \lambda S(a_0, \dots, b_m)$, where λ is a number. On the other hand, the coefficient of $\prod x_i^m$ in the polynomials $a_0^m b_0^n P(x_1, \dots, y_m)$ and $S(x_1, \dots, y_m)$ is equal to $a_0^m b_0^n$; hence, $\lambda = 1$. $\square$

43.3. The Bezout matrix. The size of the Sylvester matrix is too large and, therefore, to compute the resultant with its help is inconvenient. There are many various ways to diminish the order of the matrix used to compute the resultant. For example, we can replace the polynomial g by the remainder of its division by f (see Problem 43.1).

There are other ways to decrease the order of the matrix used for the computations. Suppose that $m = n$.

Let us express the Sylvester matrix in the form $\begin{pmatrix} A_1 & A_2 \\ B_1 & B_2 \end{pmatrix}$, where the A_i, B_i are square matrices. It is easy to verify that

$$A_1 B_1 = \begin{pmatrix} c_0 & c_1 & \dots & c_{n-1} & c_n \\ 0 & c_0 & \dots & c_{n-2} & c_{n-1} \\ \vdots & \vdots & \ddots & \vdots & \vdots \\ 0 & 0 & \dots & c_0 & c_1 \\ 0 & 0 & \dots & 0 & c_0 \end{pmatrix} = B_1 A_1, \quad \text{where } c_k = \sum_{i=0}^k a_i b_{k-i};$$

hence,

$$\begin{pmatrix} I & 0 \\ -B_1 & A_1 \end{pmatrix} \begin{pmatrix} A_1 & A_2 \\ B_1 & B_2 \end{pmatrix} = \begin{pmatrix} A_1 & A_2 \\ 0 & A_1 B_2 - B_1 A_2 \end{pmatrix}$$

and since $|A_1| = a_0^n$, we have $R(f, g) = |A_1 B_2 - B_1 A_2|$.

Let $c_{pq} = a_p b_q - a_q b_p$. It is easy to see that $A_1 B_2 - B_1 A_2 = \left\| w_{ij} \right\|_1^n$, where $w_{ij} = \sum c_{pq}$ and the summation runs over the pairs (p, q) such that $p + q = n + j - i$,

$p \le n-1$ and $q \ge j$. Since $c_{\alpha\beta} + c_{\alpha+1,\beta-1} + \dots + c_{\beta\alpha} = 0$ for $\alpha \le \beta$, we can confine ourselves to the pairs for which $p \le \min(n-1, j-1)$. For example, for $n = 4$ we get the matrix

$$\begin{pmatrix} c_{04} & c_{14} & c_{24} & c_{34} \\ c_{03} & c_{04}+c_{13} & c_{14}+c_{23} & c_{24} \\ c_{02} & c_{03}+c_{12} & c_{04}+c_{13} & c_{14} \\ c_{01} & c_{02} & c_{03} & c_{04} \end{pmatrix}.$$

Let $J = \operatorname{antidiag}(1, \dots, 1)$, i.e.,

$$J = \left\|a_{ij}\right\|_1^n, \quad \text{where } a_{ij} = \begin{cases} 1 & \text{for } i+j = n+1, \\ 0 & \text{otherwise.} \end{cases}$$

Then the matrix $Z = |w_{ij}|_1^n J$ is symmetric. It is called the *Bezoutian* or the *Bezout matrix* of f and g.

43.4. The Barnett matrix. Let us describe one more way to decrease the order of the matrix to compute the resultant ([Barnett, 1971]). For simplicity, let us assume that $a_0 = 1$, i.e., $f(x) = x^n + a_1x^{n-1} + \dots + a_n$ and $g(x) = b_0x^m + b_1x^{m-1} + \dots + b_m$. To f and g assign the *Barnett matrix* $R = g(A)$, where

$$A = \begin{pmatrix} 0 & 1 & 0 & \dots & 0 \\ 0 & 0 & 1 & \dots & 0 \\ \vdots & \vdots & \ddots & \ddots & \vdots \\ \vdots & \vdots & \vdots & \ddots & 0 \\ 0 & 0 & 0 & \ddots & 1 \\ -a_n & -a_{n-2} & -a_{n-3} & \dots & -a_1 \end{pmatrix}.$$

43.4.1. THEOREM. $\det R = R(f, g)$.

PROOF. Let $\beta_1, \dots, \beta_m$ be the roots of g. Then $g(x) = b_0(x-\beta_1)\dots(x-\beta_m)$. Hence, $g(A) = b_0(A - \beta_1 I)\dots(A-\beta_m I)$. Since $\det(A - \lambda I) = \prod_i(\alpha_i - \lambda)$ (see 1.5), we have $\det g(A) = b_0^m \prod(\alpha_i - \beta_i) = R(f, g)$. □

43.4.2. THEOREM. *For $m \le n$ it is possible to calculate the matrix R in the following recursive way. Let $r_1, \dots, r_n$ be the rows of R. Then*

$$r_1 = \begin{cases} (b_m, b_{m-1}, \dots, b_1, b_0, 0, \dots, 0) & \textit{for } m < n \\ (d_n, \dots, d_1) & \textit{for } m = n, \end{cases}$$

where $d_i = b_i - b_0 a_i$. Moreover, $r_i = r_{i-1}A$ for $i = 2, \dots, n$.

PROOF. Let $e_i = (0, \dots, 1, \dots, 0)$, where 1 occupies the ith position. For $k < n$ the first row of A^k is equal to e_{k+1}. Therefore, the structure of r_1 for $m < n$ is obvious. For $m = n$ we have to make use of the identity $A^n + \sum_{i=1}^n a_i A^{n-i} = 0$.

Since $e_i = e_{i-1}A$ for $i = 2, \dots, n$, it follows that

$$r_i = e_i R = e_{i-1}AR = e_{i-1}RA = r_{i-1}A. \quad \square$$

43.4.3. THEOREM. *The degree of the greatest common divisor of f and g is equal to $n - \operatorname{rank} R$.*

PROOF. Let $\beta_1, \dots, \beta_s$ be the roots of g with multiplicities $k_1, \dots, k_s$, respectively. Then $g(x) = b_0 \prod (x - \beta_i)^{k_i}$ and $R = g(A) = b_0 \prod_i (A - \beta_i I)^{k_i}$. Under the passage to the basis in which the matrix A is of the Jordan normal form J, the matrix R is replaced by $b_0 \prod (J - \beta_i I)^{k_i}$. The characteristic polynomial of A coincides with the minimal polynomial and, therefore, if β_i is a root of multiplicity l_i of f, then the Jordan block J_i of J corresponding to the eigenvalue β_i is of order l_i. It is also clear that

$$\operatorname{rank}(J_i - \beta_i I)^{k_i} = l_i - \min(k_i, l_i).$$

Now, considering the Jordan blocks of J separately, we easily see that $n - \operatorname{rank} R = \sum_i \min(k_i, l_i)$ and the latter sum is equal to the degree of the greatest common divisor of f and g. $\square$

43.5. Discriminant. Let $x_1, \dots, x_n$ be the roots of $f(x) = a_0 x^n + \dots + a_n$ and let $a_0 \neq 0$. The number $D(f) = a_0^{2n-2} \prod_{i<j} (x_i - x_j)$ is called the *discriminant* of f. It is also clear that $D(f) = 0$ if and only if f has multiple roots, i.e., $R(f, f') = 0$.

43.5.1. THEOREM. $R(f, f') = \pm a_0 D(f)$.

PROOF. By Theorem 43.2, $R(f, f') = a_0^{n-1} \prod_i f'(x_i)$.

It is easy to verify that if x_i is a root of f, then $f'(x_i) = a_0 \prod_{j \neq i} (x_j - x_i)$. Therefore,

$$R(f, f') = a_0^{2n-1} \prod_{j \neq i} (x_i - x_j) = \pm a_0^{2n-1} \prod_{i<j} (x_i - x_j)^2. \quad \square$$

COROLLARY. *The discriminant is a polynomial in the coefficients of f.*

43.5.2. THEOREM. *Any matrix is the limit of matrices with simple* (i.e., nonmultiple) *eigenvalues.*

PROOF. Let $f(x)$ be the characteristic polynomial of a matrix A. The polynomial f has multiple roots if and only if $D(f) = 0$. Therefore, we get an algebraic equation for elements of A. The restriction of the equation $D(f) = 0$ to the line $\{\lambda A + (1 - \lambda) B\}$, where B is a matrix with simple eigenvalues, has finitely many roots. Therefore, A is the limit of matrices with simple eigenvalues. $\square$

Problems

43.1. Let $r(x)$ be the remainder of the division of $g(x)$ by $f(x)$ and let $\deg r(x) = k$. Prove that $R(f, g) = a_0^{m-k} R(f, r)$.

43.2. Let $f(x) = a_0 x^n + \dots + a_n$, $g(x) = b_0 x^m + \dots + b_m$ and let $r_k(x) = a_{k0} x^{n-1} + a_{k1} x^{n-2} + \dots + a_{k,n-1}$ be the remainder of the division of $x^k g(x)$ by $f(x)$. Prove that

$$R(f, g) = a_0^m \begin{vmatrix} a_{n-1,0} & \dots & a_{n-1,n-1} \\ \vdots & \vdots & \vdots \\ a_{00} & \dots & a_{0,n-1} \end{vmatrix}.$$

43.3. The characteristic polynomials of matrices A and B of size $n \times n$ and $m \times m$ are equal to f and g, respectively. Prove that the resultant of the polynomials f and g is equal to the determinant of the operator $X \mapsto AX - XB$ in the space of matrices of size $n \times m$.

43.4. Let $\alpha_1, \dots, \alpha_n$ be the roots of a polynomial $f(x) = \sum_{i=0}^{n} a_i x^{n-i}$ and $s_k = \alpha_1^k + \dots + \alpha_n^k$. Prove that $D(f) = a_0^{2n-2} \det S$, where

$$S = \begin{pmatrix} s_0 & s_1 & \dots & s_{n-1} \\ s_1 & s_2 & \dots & s_n \\ \vdots & \vdots & \vdots & \vdots \\ s_{n-1} & s_n & \dots & s_{2n-2} \end{pmatrix}.$$

44. The generalized inverse matrix. Matrix equations

44.1. A matrix X is called a *generalized inverse* for a (not necessarily square) matrix A, if $XAX = X$, $AXA = A$ and the matrices AX and XA are Hermitian. It is easy to verify that for an invertible A its generalized inverse matrix coincides with the inverse matrix.

44.1.1. THEOREM. *A matrix X is a generalized inverse for A if and only if the matrices $P = AX$ and $Q = XA$ are Hermitian projections onto* $\operatorname{Im} A$ *and* $\operatorname{Im} A^*$, *respectively.*

PROOF. First, suppose that P and Q are Hermitian projections to $\operatorname{Im} A$ and $\operatorname{Im} A^*$, respectively. If v is an arbitrary vector, then $Av \in \operatorname{Im} A$ and, therefore, $PAv = Av$, i.e., $AXAv = Av$. Besides, $Xv \in \operatorname{Im} XA = \operatorname{Im} A^*$; hence, $QXv = Xv$, i.e., $XAXv = Xv$.

Now, suppose that X is a generalized inverse for A. Then $P^2 = (AXA)X = AX = P$ and $Q^2 = (XAX)A = XA = Q$, where P and Q are Hermitian matrices. It remains to show that $\operatorname{Im} P = \operatorname{Im} A$ and $\operatorname{Im} Q = \operatorname{Im} A^*$. Since $P = AX$ and $Q = Q^* = A^*X^*$, we have $\operatorname{Im} P \subset \operatorname{Im} A$ and $\operatorname{Im} Q \subset \operatorname{Im} A^*$. On the other hand, $A = AXA = PA$ and $A^* = A^*X^*A^* = Q^*A^* = QA^*$; hence, $\operatorname{Im} A \subset \operatorname{Im} P$ and $\operatorname{Im} A^* \subset \operatorname{Im} Q$. $\square$

44.1.2. THEOREM (Moore-Penrose). *For any matrix A there exists a unique generalized inverse matrix X.*

PROOF. If $\operatorname{rank} A = r$, then A can be represented in the form of the product of matrices C and D of size $m \times r$ and $r \times n$, respectively, where $\operatorname{Im} A = \operatorname{Im} C$ and $\operatorname{Im} A^* = \operatorname{Im} D^*$. It is also clear that C^*C and DD^* are invertible. Set

$$X = D^*(DD^*)^{-1}(C^*C)^{-1}C^*.$$

Then $AX = C(C^*C)^{-1}C^*$ and $XA = D^*(DD^*)^{-1}D$, i.e., the matrices AX and XA are Hermitian projections onto $\operatorname{Im} C = \operatorname{Im} A$ and $\operatorname{Im} D^* = \operatorname{Im} A^*$, respectively, (see 25.3) and, therefore, X is a generalized inverse for A.

Now, suppose that X_1 and X_2 are generalized inverses for A. Then AX_1 and AX_2 are Hermitian projections onto $\operatorname{Im} A$, implying $AX_1 = AX_2$. Similarly, $X_1A = X_2A$. Therefore,

$$X_1 = X_1(AX_1) = (X_1A)X_2 = X_2AX_2 = X_2. \quad \square$$

The *generalized inverse of A* will be denoted[1] by $A^{\text{“}-1\text{”}}$.

[1] There is no standard notation for the generalized inverse of a matrix A. Many authors took after R. Penrose who denoted it by A^+ which is confusing: might be mistaken for the Hermitian conjugate. In the original manuscript of this book Penrose's notation was used. I suggest a more dynamic and noncontroversal notation approved by the author. *Translator's remark.*

44.2. The generalized inverse matrix $A^{\text{"}-1\text{"}}$ is applied to solve systems of linear equations, both inconsistent and consistent. The most interesting are its applications to solving inconsistent systems.

44.2.1. THEOREM. *Consider a system of linear equations $Ax = b$. The value $|Ax - b|$ is minimal for x such that $Ax = AA^{\text{"}-1\text{"}}b$ and among all such x the least value of $|x|$ is attained at the vector $x_0 = A^{\text{"}-1\text{"}}b$.*

PROOF. The operator $P = AA^{\text{"}-1\text{"}}$ is a projection and therefore, $I - P$ is also a projection and $\operatorname{Im}(I - P) = \operatorname{Ker} P$ (see Theorem 25.1.2). Since P is an Hermitian operator, $\operatorname{Ker} P = (\operatorname{Im} P)^{\perp}$. Hence,

$$\operatorname{Im}(I - P) = \operatorname{Ker} P = (\operatorname{Im} P)^{\perp} = (\operatorname{Im} A)^{\perp},$$

i.e., for any vectors x and y the vectors Ax and $(I - AA^{\text{"}-1\text{"}})y$ are perpendicular and

$$|Ax + (I - AA^{\text{"}-1\text{"}})y|^2 = |Ax|^2 + |y - AA^{\text{"}-1\text{"}}y|^2.$$

Similarly,

$$|A^{\text{"}-1\text{"}}x + (I - A^{\text{"}-1\text{"}}A)y|^2 = |A^{\text{"}-1\text{"}}x|^2 + |y - A^{\text{"}-1\text{"}}Ay|^2.$$

Since

$$Ax - b = A(x - A^{\text{"}-1\text{"}}b) - (I - AA^{\text{"}-1\text{"}})b,$$

it follows that

$$|Ax - b|^2 = |Ax - AA^{\text{"}-1\text{"}}b|^2 + |b - AA^{\text{"}-1\text{"}}b|^2 \geq |b - AA^{\text{"}-1\text{"}}b|^2$$

and equality is attained if and only if $Ax = AA^{\text{"}-1\text{"}}b$. If $Ax = AA^{\text{"}-1\text{"}}b$, then

$$|x|^2 = |A^{\text{"}-1\text{"}}b + (I - A^{\text{"}-1\text{"}}A)x|^2 = |A^{\text{"}-1\text{"}}b|^2 + |x - A^{\text{"}-1\text{"}}Ax|^2 \geq |A^{\text{"}-1\text{"}}b|^2$$

and equality is attained if and only if

$$x = A^{\text{"}-1\text{"}}Ax = A^{\text{"}-1\text{"}}AA^{\text{"}-1\text{"}}b = A^{\text{"}-1\text{"}}b. \quad \square$$

REMARK. The equality $Ax = AA^{\text{"}-1\text{"}}b$ is equivalent to the equality $A^*Ax = A^*x$. Indeed, if $Ax = AA^{\text{"}-1\text{"}}b$, then $A^*b = A^*(A^{\text{"}-1\text{"}})^*A^*b = A^*AA^{\text{"}-1\text{"}}b = A^*Ax$ and if $A^*Ax = A^*b$, then

$$Ax = AA^{\text{"}-1\text{"}}Ax = (A^{\text{"}-1\text{"}})^*A^*Ax = (A^{\text{"}-1\text{"}})^*A^*b = AA^{\text{"}-1\text{"}}b.$$

With the help of the generalized inverse matrix we can write a criterion for consistency of a system of linear equations and find all its solutions.

44.2.2. THEOREM. *The matrix equation*

$$AXB = C \tag{1}$$

has a solution if and only if $AA^{“-1”}CB^{“-1”}B = C$. *The solutions of* (1) *are of the form*

$$X = A^{“-1”}CB^{“-1”} + Y - A^{“-1”}AYBB^{“-1”}, \text{ where } Y \text{ is an arbitrary matrix.}$$

PROOF. If $AXB = C$, then

$$C = AXB = AA^{“-1”}(AXB)B^{“-1”}B = AA^{“-1”}CB^{“-1”}B.$$

Conversely, if $C = AA^{“-1”}CB^{“-1”}B$, then $X_0 = A^{“-1”}CB^{“-1”}$ is a particular solution of the equation

$$AXB = C.$$

It remains to demonstrate that the general solution of the equation $AXB = 0$ is of the form $X = Y - A^{“-1”}AYBB^{“-1”}$. Clearly, $A(Y - A^{“-1”}AYBB^{“-1”})B = 0$. On the other hand, if $AXB = 0$, then $X = Y - A^{“-1”}AYBB^{“-1”}$, where $Y = X$. □

REMARK. The notion of generalized inverse matrix appeared independently in the papers of [Moore, 1935] and [Penrose, 1955]. The equivalence of Moore's and Penrose's definitions was demonstrated in the paper [Rado, 1956].

44.3. THEOREM ([Roth, 1952]). *Let* $A \in M_{m,m}$, $B \in M_{n,n}$ *and* $C \in M_{m,n}$.

a) *The equation* $AX - XB = C$ *has a solution* $X \in M_{m,n}$ *if and only if the matrices* $\begin{pmatrix} A & 0 \\ 0 & B \end{pmatrix}$ *and* $\begin{pmatrix} A & C \\ 0 & B \end{pmatrix}$ *are similar.*

b) *The equation* $AX - YB = C$ *has a solution* $X, Y \in M_{m,n}$ *if and only if matrices* $\begin{pmatrix} A & 0 \\ 0 & B \end{pmatrix}$ *and* $\begin{pmatrix} A & C \\ 0 & B \end{pmatrix}$ *are of the same rank.*

PROOF (Following [Flanders, Wimmer, 1977]). a) Let $K = \begin{pmatrix} P & Q \\ R & S \end{pmatrix}$, where $P \in M_{m,m}$ and $S \in M_{n,n}$. First, suppose that the matrices from the theorem are similar. Consider the maps $\varphi_i : M_{m,n} \longrightarrow M_{m,n}$, $i = 0, 1$, given by the formulas

$$\varphi_0(K) = \begin{pmatrix} A & 0 \\ 0 & B \end{pmatrix} K - K \begin{pmatrix} A & 0 \\ 0 & B \end{pmatrix} = \begin{pmatrix} AP - PA & AQ - QB \\ BR - RA & BS - SB \end{pmatrix},$$

$$\varphi_1(K) = \begin{pmatrix} A & C \\ 0 & B \end{pmatrix} K - K \begin{pmatrix} A & 0 \\ 0 & B \end{pmatrix} = \begin{pmatrix} AP + CR - PA & AQ + CS - QB \\ BR - RA & BS - SB \end{pmatrix}.$$

The equations $FK = KF$ and $GFG^{-1}K' = K'F$ have isomorphic spaces of solutions; this isomorphism is given by the formula $K = G^{-1}K'$. Hence, $\dim \operatorname{Ker} \varphi_0 = \dim \operatorname{Ker} \varphi_1$. If $K \in \operatorname{Ker} \varphi_i$, then $BR = RA$ and $BS = SB$. Therefore, we can consider the space

$$V = \{(R, S) \in M_{n,m+n} \mid BR = RA,\ BS = SB\}$$

and determine the projection $\mu_i : \operatorname{Ker}\varphi_i \longrightarrow V$, where $\mu_i(X) = (R, S)$. It is easy to verify that

$$\operatorname{Ker}\mu_i = \left\{ \begin{pmatrix} P & Q \\ 0 & 0 \end{pmatrix} \mid AP = PA,\ AQ = QB \right\}.$$

For μ_0 this is obvious and for μ_1 it follows from the fact that $CR = 0$ and $CS = 0$ since $R = 0$ and $S = 0$.

Let us prove that $\operatorname{Im}\mu_0 = \operatorname{Im}\mu_1$. If $(R, S) \in V$, then $\begin{pmatrix} 0 & 0 \\ R & S \end{pmatrix} \in \operatorname{Ker}\varphi_0$. Hence, $\operatorname{Im}\mu_0 = V$ and, therefore, $\operatorname{Im}\mu_1 \subset \operatorname{Im}\mu_0$. On the other hand,

$$\dim \operatorname{Im}\mu_0 + \dim \operatorname{Ker}\mu_0 = \dim \operatorname{Ker}\varphi_0 = \dim \operatorname{Ker}\varphi_1 = \dim \operatorname{Im}\mu_1 + \dim \operatorname{Ker}\mu_1.$$

The matrix $\begin{pmatrix} I & 0 \\ 0 & -I \end{pmatrix}$ belongs to $\operatorname{Ker}\varphi_0$ and, therefore, $(0, -I) \in \operatorname{Im}\mu_0 = \operatorname{Im}\mu_1$. Hence, there is a matrix of the form $\begin{pmatrix} P & Q \\ 0 & -I \end{pmatrix}$ in $\operatorname{Ker}\varphi_1$. Thus, $AQ + CS - QB = 0$, where $S = -I$. Therefore, $X = Q$ is a solution of the equation $AX - XB = C$.

Conversely, if X is a solution of this equation, then

$$\begin{pmatrix} A & 0 \\ 0 & B \end{pmatrix}\begin{pmatrix} I & X \\ 0 & I \end{pmatrix} = \begin{pmatrix} A & AX \\ 0 & B \end{pmatrix} = \begin{pmatrix} A & C + XB \\ 0 & B \end{pmatrix} = \begin{pmatrix} I & X \\ 0 & I \end{pmatrix}\begin{pmatrix} A & C \\ 0 & B \end{pmatrix}$$

and, therefore

$$\begin{pmatrix} I & X \\ 0 & I \end{pmatrix}^{-1}\begin{pmatrix} A & 0 \\ 0 & B \end{pmatrix}\begin{pmatrix} I & X \\ 0 & I \end{pmatrix} = \begin{pmatrix} A & C \\ 0 & B \end{pmatrix}.$$

b) First, suppose that the indicated matrices are of the same rank. Consider the maps $\psi_i : M_{m+n,2(m+n)} \longrightarrow M_{m+n,m+n}$, $i = 0, 1$, given by formulas

$$\psi_0(U, W) = \begin{pmatrix} A & 0 \\ 0 & B \end{pmatrix} U - W \begin{pmatrix} A & 0 \\ 0 & B \end{pmatrix} = \begin{pmatrix} AU_{11} - W_{11}A & AU_{12} - W_{12}B \\ BU_{21} - W_{21}A & BU_{22} - W_{22}B \end{pmatrix},$$

$$\begin{aligned}\psi_1(U, W) &= \begin{pmatrix} A & C \\ 0 & B \end{pmatrix} U - W \begin{pmatrix} A & 0 \\ 0 & B \end{pmatrix} \\ &= \begin{pmatrix} AU_{11} + CU_{21} - W_{11}A & AU_{12} + CU_{22} - W_{12}B \\ BU_{21} - W_{21}A & BU_{22} - W_{22}B \end{pmatrix},\end{aligned}$$

where

$$U = \begin{pmatrix} U_{11} & U_{12} \\ U_{21} & U_{22} \end{pmatrix} \text{ and } W = \begin{pmatrix} W_{11} & W_{12} \\ W_{21} & W_{22} \end{pmatrix}.$$

The spaces of solutions of equations $FU = WF$ and $GFG^{-1}U' = W'F$ are isomorphic and this isomorphism is given by the formulas $U = G^{-1}U'$ and $W = G^{-1}W'$. Hence, $\dim \operatorname{Ker}\psi_0 = \dim \operatorname{Ker}\psi_1$.

Consider the space

$$Z = \{(U_{21}, U_{22} W_{21}, W_{22}) \mid BU_{21} = W_{21}A,\ BU_{22} = W_{22}B\}$$

and define a map $\nu_i : \operatorname{Ker}\varphi_i \longrightarrow Z$, where $\nu_i(U, W) = (U_{21}, U_{22}, W_{21}, W_{22})$. Then $\operatorname{Im}\nu_1 \subset \operatorname{Im}\nu_0 = Z$ and $\operatorname{Ker}\nu_1 = \operatorname{Ker}\nu_0$. Therefore, $\operatorname{Im}\nu_1 = \operatorname{Im}\nu_0$. The matrix (U, W), where $U = W = \begin{pmatrix} I & 0 \\ 0 & -I \end{pmatrix}$, belongs to $\operatorname{Ker}\psi_0$. Hence, $\operatorname{Ker}\psi_1$ also contains an element for which $U_{22} = -I$. For this element the equality $AU_{12} + CU_{22} = W_{12}B$ is equivalent to the equality $AU_{12} - W_{12}B = C$.

Conversely, if a solution X, Y of the given equation exists, then

$$\begin{pmatrix} I & -Y \\ 0 & I \end{pmatrix}\begin{pmatrix} A & 0 \\ 0 & B \end{pmatrix}\begin{pmatrix} I & X \\ 0 & I \end{pmatrix} = \begin{pmatrix} A & AX - YB \\ 0 & B \end{pmatrix} = \begin{pmatrix} A & C \\ 0 & B \end{pmatrix}. \quad \square$$

Problems

44.1. Prove that if $C = AX = YB$, then there exists a matrix Z such that $C = AZB$.

44.2. Prove that any solution of a system of matrix equations $AX = 0$, $BX = 0$ is of the form $X = (I - A^{\text{“}-1\text{”}}A)Y(I - BB^{\text{“}-1\text{”}})$, where Y is an arbitrary matrix.

44.3. Prove that the system of equations $AX = C$, $XB = D$ has a solution if and only if each of the equations $AX = C$ and $XB = D$ has a solution and $AD = CB$.

45. Hankel matrices and rational functions

Consider a proper rational function

$$R(z) = \frac{a_1 z^{m-1} + \dots + a_m}{b_0 z^m + b_1 z^{m-1} + \dots + b_m},$$

where $b_0 \neq 0$. It is possible to expand this function in a series

$$R(z) = s_0 z^{-1} + s_1 z^{-2} + s_2 z^{-3} + \dots,$$

where

$$(1) \qquad \begin{aligned} & b_0 s_0 = a_1, \\ & b_0 s_1 + b_1 s_0 = a_2, \\ & b_0 s_2 + b_1 s_1 + b_2 s_0 = a_3, \\ & \dots\dots\dots\dots\dots\dots \\ & b_0 s_{m-1} + \dots + b_{m-1} s_0 = a_m. \end{aligned}$$

Moreover, $b_0 s_q + \dots + b_m s_{q-m} = 0$ for $q \geq m$. Thus, for all $q \geq m$ we have

$$(2) \qquad s_q = \alpha_1 s_{q-1} + \dots + \alpha_m s_{q-m},$$

where $\alpha_i = -b_i/b_0$. Consider the infinite matrix

$$S = \left\| \begin{matrix} s_0 & s_1 & s_2 & \dots \\ s_1 & s_2 & s_3 & \dots \\ s_2 & s_3 & s_4 & \dots \\ \vdots & \vdots & \vdots & \dots \end{matrix} \right\|.$$

A matrix of such a form is called a *Hankel matrix*. Relation (2) means that the $(m+1)$th row of S is a linear combination of the first m rows (with coefficients $\alpha_1, \dots, \alpha_m$). If we delete the first element of each of these rows, we see that the $(m+2)$th row of S is a linear combination of the m rows preceding it and therefore, the linear combination of the first m rows. Continuing these arguments, we deduce that any row of the matrix S is expressed in terms of its first m rows, i.e., $\operatorname{rank} S \leq m$.

Thus, if the series

$$R(z) = s_0 z^{-1} + s_1 z^{-2} + s_2 z^{-3} + \dots \tag{3}$$

corresponds to a rational function $R(z)$, then the Hankel matrix S constructed from $s_0, s_1, \dots$ is of finite rank.

Now, let us assume that the Hankel matrix S is of finite rank m. Let us construct from S a series (3). Let us prove that this series corresponds to a rational function. The first $m+1$ rows of S are linearly dependent and, therefore, there exists a number $h \leq m$ such that the $(m+1)$th row can be expressed linearly in terms of the first m rows. As has been demonstrated, in this case all rows of S are expressed in terms of the first h rows. Hence, $h = m$. Thus, the numbers s_i are connected by relation (2) for all $q \geq m$. The coefficients α_i in this relation enable us to determine the numbers $b_0 = 1, b_1 = \alpha_1, \dots, b_m = \alpha_m$. Next, with the help of relation (1) we can determine the numbers $a_1, \dots, a_m$. For the numbers a_i and b_j determined in this way we have

$$\frac{s_0}{z} + \frac{s_1}{z^2} + \dots = \frac{a_1 z^{m-1} + \dots + a_m}{b_0 z^m + \dots + b_m},$$

i.e., $R(z)$ is a rational function.

REMARK. Matrices of finite size of the form

$$\begin{pmatrix} s_0 & s_1 & \dots & s_n \\ s_1 & s_2 & \dots & s_{n+1} \\ \vdots & \vdots & \vdots & \vdots \\ s_n & s_{n+1} & \dots & s_{2n} \end{pmatrix}.$$

are also sometimes referred to as *Hankel* matrices. Let $J = \operatorname{antidiag}(1, \dots, 1)$, i.e.,

$$J = \left\| a_{ij} \right\|_0^n, \quad \text{where } a_{ij} = \begin{cases} 1 & \text{for } i + j = n, \\ 0 & \text{otherwise.} \end{cases}$$

If H is a Hankel matrix, then the matrix JH is called a Toeplitz matrix; it is of the form

$$\begin{pmatrix} a_0 & a_1 & a_2 & \dots & a_n \\ a_{-1} & a_0 & a_1 & \dots & a_{n-1} \\ a_{-2} & a_{-1} & a_0 & \dots & a_{n-2} \\ \vdots & \vdots & \vdots & \vdots & \vdots \\ a_{-n} & a_{-n+1} & a_{-n+2} & \dots & a_0 \end{pmatrix}.$$

46. Functions of matrices. Differentiation of matrices

46.1. By analogy with the exponent of a number, we can define the *expontent of a matrix* A to be the sum of the series

$$\sum_{k=0}^{\infty} \frac{A^k}{k!}.$$

Let us prove that this series converges. If A and B are square matrices of order n and $|a_{ij}| \leq a$, $|b_{ij}| \leq b$, then the absolute value of each element of AB does not exceed nab. Hence, the absolute value of the elements of A^k does not exceed $n^{k-1}a^k = (na)^k/n$ and, since

$$\frac{1}{n}\sum_{k=0}^{\infty} \frac{(na)^k}{k!} = \frac{1}{n}e^{na},$$

the series $\sum_{k=0}^{\infty} \frac{A^k}{k!}$ converges to a matrix denoted by $e^A = \exp A$; this matrix is called *the exponent* of A.

If $A_1 = P^{-1}AP$, then $A_1^k = P^{-1}A^kP$. Therefore, $\exp(P^{-1}AP) = P^{-1}(\exp A)P$. Hence, the computation of the exponent of an arbitrary matrix reduces to the computation of the exponent of its Jordan blocks.

Let $J = \lambda I + N$ be a Jordan block of order n. Then

$$(\lambda I + N)^k = \sum_{m=0}^{k} \binom{k}{m} \lambda^{k-m} N^m.$$

Hence,

$$\begin{aligned}\exp(tJ) &= \sum_{k=0}^{\infty} \frac{t^k J^k}{k!} = \sum_{k,m=0}^{\infty} \frac{t^k \binom{k}{m} \lambda^{k-m} N^m}{k!} \\ &= \sum_{m=0}^{\infty}\sum_{k=m}^{\infty} \frac{(\lambda t)^{k-m}}{(k-m)!}\frac{t^m N^m}{m!} = \sum_{m=0}^{\infty} \frac{t^m}{m!} e^{\lambda t} N^m = \sum_{m=0}^{n-1} \frac{t^m}{m!} e^{\lambda t} N^m,\end{aligned}$$

since $N^m = 0$ for $m \geq n$.

By reducing a matrix A to the Jordan normal form we get the following statement.

46.1.1. THEOREM. *If the minimal polynomial of A is equal to*

$$(x-\lambda_1)^{n_1} \dots (x-\lambda_k)^{n_k},$$

then the elements of e^{At} are of the form $p_1(t)e^{\lambda_1 t} + \dots + p_k(t)e^{\lambda_k t}$, where $p_i(t)$ is a polynomial of degree not greater than $n_i - 1$.

46.1.2. THEOREM. $\det(e^A) = e^{\operatorname{tr} A}$.

PROOF. We may assume that A is an upper triangular matrix with elements $\lambda_1, \dots, \lambda_n$ on the diagonal. Then A^k is an upper triangular matrix with elements $\lambda_1^{k_1}, \dots, \lambda_n^{k_n}$ of the diagonal. Hence, e^A is an upper triangular matrix with elements $\exp \lambda_1, \dots, \exp \lambda_n$ on the diagonal. $\square$

46.2. Consider a family of matrices $X(t) = \|x_{ij}(t)\|_1^n$ whose elements are differentiable functions of t. Let $\dot{X}(t) = \frac{dX(t)}{dt}$ be the element-wise derivative of the matrix-valued function $X(t)$.

46.2.1. THEOREM. $(XY)^{\cdot} = \dot{X}Y + X\dot{Y}$.

PROOF. If $Z = XY$, then $z_{ij} = \sum_k x_{ik}y_{kj}$; hence, $\dot{z}_{ij} = \sum_k \dot{x}_{ik}y_{kj} + \sum_k x_{ik}\dot{y}_{kj}$. Therefore, $\dot{Z} = \dot{X}Y + X\dot{Y}$. □

46.2.2. THEOREM. a) $(X^{-1})^{\cdot} = -X^{-1}\dot{X}X^{-1}$.
b) $\operatorname{tr}(X^{-1}\dot{X}) = -\operatorname{tr}((X^{-1})^{\cdot}X)$.

PROOF. a) On one hand, $(X^{-1}X)^{\cdot} = \dot{I} = 0$. On the other hand, $(X^{-1}X)^{\cdot} = (X^{-1})^{\cdot}X + X^{-1}\dot{X}$. Therefore, $(X^{-1})^{\cdot}X = -X^{-1}\dot{X}$ and $(X^{-1})^{\cdot} = -X^{-1}\dot{X}X^{-1}$.
b) Since $\operatorname{tr}(X^{-1}X) = n$, it follows that

$$0 = [\operatorname{tr}(X^{-1}X)]^{\cdot} = \operatorname{tr}((X^{-1})^{\cdot}X) + \operatorname{tr}(X^{-1}\dot{X}). \quad \square$$

46.2.3. THEOREM. $(e^{At})^{\cdot} = Ae^{At}$.

PROOF. Since the series $\sum_{k=0}^{\infty} \frac{(tA)^k}{k!}$ converges absolutely,

$$\frac{d}{dt}(e^{At}) = \sum_{k=0}^{\infty} \frac{d}{dt}\left(\frac{(tA)^k}{k!}\right) = \sum_{k=0}^{\infty} \frac{kt^{k-1}A^k}{k!} = A\sum_{k=1}^{\infty} \frac{(tA)^{k-1}}{(k-1)!} = Ae^{At}. \quad \square$$

46.3. A system of n first order linear differential equations in n variables can be expressed in the form $\dot{X} = AX$, where X is a column of length n and A is a matrix of order n. If A is a constant matrix, then $X(t) = e^{At}C$ is the solution of this equation with the initial condition $X(0) = C$ (see Theorem 46.2.3); the solution of this equation with a given initial value is unique.

The general form of the elements of the matrix e^{At} is given by Theorem 46.1.1; using the same theorem, we get the following statement.

46.3.1. THEOREM. *Consider the equation $\dot{X} = AX$. If the minimal polynomial of A is equal to $(\lambda-\lambda_1)^{n_1}\dots(\lambda-\lambda_k)^{n_k}$, then the solution $x_1(t), \dots, x_n(t)$* (i.e., the coordinates of the vector X) *is of the form*

$$x_i(t) = p_{i1}(t)e^{\lambda_1 t} + \dots + p_{ik}(t)e^{\lambda_k t},$$

where $p_{ij}(t)$ is a polynomial whose degree does not exceed $n_j - 1$.

It is easy to verify by a direct substitution that $X(t) = e^{At}Ce^{Bt}$ is a solution of $\dot{X} = AX + XB$ with the initial condition $X(0) = C$.

46.3.2. THEOREM. *Let $X(t)$ be a solution of $\dot{X} = A(t)X$. Then*

$$\det X = \exp\left(\int_0^t \operatorname{tr} A(s)\, ds\right) \det X(0).$$

PROOF. By Problem 46.6 a) $(\det X)^{\cdot} = (\det X)(\operatorname{tr}\dot{X}X^{-1})$. In our case $\dot{X}X^{-1} = A(t)$. Therefore, the function $y(t) = \det X(t)$ satisfies the condition $(\ln y)^{\cdot} = \dot{y}/y = \operatorname{tr} A(t)$. Therefore, $y(t) = c\exp(\int_0^t \operatorname{tr} A(s)ds)$, where $c = y(0) = \det X(0)$. □

Problems

46.1. Let $A = \begin{pmatrix} 0 & -t \\ t & 0 \end{pmatrix}$. Compute e^A.

46.2. a) Prove that if $[A, B] = 0$, then $e^{A+B} = e^A e^B$.

b) Prove that if $e^{(A+B)t} = e^{At}e^{Bt}$ for all t, then $[A, B] = 0$.

46.3. Prove that for any unitary matrix U there exists an Hermitian matrix H such that $U = e^{iH}$.

46.4. a) Prove that if a real matrix X is skew-symmetric, then e^X is orthogonal.

b) Prove that any orthogonal matrix U with determinant 1 can be represented in the form e^X, where X is a real skew-symmetric matrix.

46.5. a) Let A be a real matrix. Prove that $\det e^A = 1$ if and only if $\operatorname{tr} A = 0$.

b) Let B be a real matrix and $\det B = 1$. Is there a real matrix A such that $B = e^A$?

46.6. a) Prove that

$$(\det A)^{\cdot} = \operatorname{tr}(\dot{A} \operatorname{adj} A^T) = (\det A)\operatorname{tr}(\dot{A}A^{-1}).$$

b) Let A be an $n \times n$-matrix. Prove that $\operatorname{tr}(A(\operatorname{adj} A^T)^{\cdot}) = (n-1)\operatorname{tr}(\dot{A} \operatorname{adj} A^T)$.

46.7. [Aitken, 1953]. Consider a map $F : M_{n,n} \longrightarrow M_{n,n}$. Let $\Omega F(X) = \|\omega_{ij}(X)\|_1^n$, where $\omega_{ij}(X) = \frac{\partial}{\partial x_{ji}} \operatorname{tr} F(X)$. Prove that if $F(X) = X^m$, where m is an integer, then $\Omega F(X) = mX^{m-1}$.

47. Lax pairs and integrable systems

47.1. Consider a system of differential equations

$$\dot{x}(t) = f(x, t), \text{ where } x = (x_1, \dots, x_n), \quad f = (f_1, \dots, f_n).$$

A nonconstant function $F(x_1, \dots, x_n)$ is called a *first integral* of this system if

$$\frac{d}{dt}F(x_1(t), \dots, x_n(t)) = 0$$

for any solution $(x_1(t), \dots, x_n(t))$ of the system. The existence of a first integral enables one to reduce the order of the system by 1.

Let A and L be square matrices whose elements depend on $x_1, \dots, x_n$. The differential equation

$$\dot{L} = AL - LA$$

is called *the Lax differential equation* and the pair of operators L, A in it a *Lax pair*.

THEOREM. *If the functions $f_k(x_1, \dots, x_n) = \operatorname{tr}(L^k)$ are nonconstant, then they are first integrals of the Lax equation.*

PROOF. Let $B(t)$ be a solution of the equation $\dot{B} = -BA$ with the initial condition $B(0) = I$. Then

$$\det B(t) = \exp\left(\int_0^t A(s)ds\right) \neq 0$$

(see Theorem 46.3.2) and

$$\begin{aligned}(BLB^{-1})^{\cdot} &= \dot{B}LB^{-1} + B\dot{L}B^{-1} + BL(B^{-1})^{\cdot}\\ &= -BALB^{-1} + B(AL - LA)B^{-1} + BLB^{-1}(BA)B^{-1} = 0.\end{aligned}$$

Therefore, the Jordan normal form of L does not depend on t; hence, its eigenvalues are constants. □

Representation of systems of differential equations in the Lax form is an important method for finding first integrals of Hamiltonian systems of differential equations.

For example, the *Euler equations* $\dot{M} = M \times \omega$, which describe the motion of a solid body around a fixed point, are easy to express in the Lax form. For this we should take

$$L = \begin{pmatrix} 0 & -M_3 & M_2 \\ M_3 & 0 & -M_1 \\ -M_2 & M_1 & 0 \end{pmatrix} \text{ and } A = \begin{pmatrix} 0 & \omega_3 & -\omega_2 \\ -\omega_3 & 0 & \omega_1 \\ \omega_2 & -\omega_1 & 0 \end{pmatrix}.$$

The first integral of this equation is $\operatorname{tr} L^2 = -2(M_1^2 + M_2^2 + M_3^2)$.

47.2. A more instructive example is that of the *Toda lattice*:

$$\ddot{x}_i = -\frac{\partial}{\partial x_i} U, \text{ where } U = \exp(x_1 - x_2) + \dots + \exp(x_{n-1} - x_n).$$

This system of equations can be expressed in the Lax form with the following L and A:

$$L = \begin{pmatrix} b_1 & a_1 & 0 & & 0 \\ a_1 & b_2 & a_2 & & \\ 0 & a_2 & b_3 & \ddots & 0 \\ & & \ddots & \ddots & a_{n-1} \\ 0 & & 0 & a_{n-1} & b_n \end{pmatrix}, \quad A = \begin{pmatrix} 0 & a_1 & 0 & & 0 \\ -a_1 & 0 & a_2 & & \\ 0 & -a_2 & 0 & \ddots & 0 \\ & & \ddots & \ddots & a_{n-1} \\ 0 & & 0 & -a_{n-1} & 0 \end{pmatrix},$$

where $2a_k = \exp \frac{1}{2}(x_k - x_{k+1})$ and $2b_k = -\dot{x}_k$. Indeed, the equation $\dot{L} = [A, L]$ is equivalent to the system of equations

$$\dot{b}_1 = 2a_1^2, \quad \dot{b}_2 = 2(a_2^2 - a_1^2), \quad \dots, \quad \dot{b}_n = -2a_{n-1}^2,$$

$$\dot{a}_1 = a_1(b_2 - b_1), \quad \dots, \quad \dot{a}_{n-1} = a_{n-1}(b_n - b_{n-1}).$$

The equation

$$\dot{a}_k = a_k(b_{k+1} - b_k) = a_k \frac{\dot{x}_k - \dot{x}_{k+1}}{2}$$

implies that $\ln a_k = \frac{1}{2}(x_k - x_{k+1}) + c_k$, i.e., $a_k = d_k \exp \frac{1}{2}(x_k - x_{k+1})$. Therefore, the equation $\dot{b}_k = 2(a_k^2 - a_{k-1}^2)$ is equivalent to the equation

$$-\frac{\ddot{x}_k}{2} = 2(d_k^2 \exp(x_k - x_{k+1}) - d_{k-1}^2 \exp(x_{k-1} - x_k)).$$

If $d_1 = \dots = d_{n-1} = \frac{1}{2}$ we get the required equations.

47.3. The motion of a multidimensional solid body with the inertia matrix J is described by the equation

$$(1) \qquad \dot{M} = [M, \omega],$$

where ω is a skew-symmetric matrix and $M = J\omega + \omega J$; here we can assume that J is a diagonal matrix. The equation (1) is already in the Lax form; therefore, $I_k = \operatorname{tr} M^{2k}$ for $k = 1, \dots, [n/2]$ are first integrals of this equation (if $p > [n/2]$, then the functions $\operatorname{tr} M^{2p}$ can be expressed in terms of the functions I_k indicated; if p is odd, then $\operatorname{tr} M^p = 0$). But we can get many more first integrals by expressing (1) in the form

$$(2) \qquad (M + \lambda J^2)^{\cdot} = [M + \lambda J^2, \omega + \lambda J],$$

where λ is an arbitrary constant, as it was first done in [Manakov, 1976]. To prove that (1) and (2) are equivalent, it suffices to notice that

$$[M, J] = -J^2\omega + \omega J^2 = -[J^2, \omega].$$

The first integrals of (2) are all nonzero coefficients of the polynomials

$$P_k(\lambda) = \operatorname{tr}(M + \lambda J^2)^k = \sum b_s \lambda^s.$$

Since $M^T = -M$ and $J^T = J$, it follows that

$$P_k(\lambda) = \operatorname{tr}(-M + \lambda J^2)^k = \sum (-1)^{k-s} b_s \lambda^s.$$

Therefore, if $k - s$ is odd, then $b_s = 0$.

47.4. The system of *Volterra equations*

$$(1) \qquad \dot{a}_i = a_i \left(\sum_{k=1}^{p-1} a_{i+k} - \sum_{k=1}^{p-1} a_{i-k} \right),$$

where $p \geq 2$ and $a_{i+n} = a_i$, can also be expressed in the form of a family of Lax equations depending on a parameter λ. Such a representation is given in the book [Bogoyavlenskiĭ, 1991]. Let $M = \|m_{ij}\|_1^n$ and $A = \|a_{ij}\|_1^n$, where in every matrix only n elements are nonzero, namely $m_{i,i+1} = 1$ and $a_{i,i+1-p} = a_i$. Consider the equation

$$(2) \qquad (A + \lambda M)^{\cdot} = [A + \lambda M, -B - \lambda M^p].$$

If $B = \sum_{j=0}^{p-1} M^{p-1-j} A M^j$, then $[M, B] + [A, M^p] = 0$ and, therefore, equation (2) is equivalent to the equation $\dot{A} = -[A, B]$. It is easy to verify that $b_{ij} = a_{i+p-1,j} + \dots + a_{i,j+p-1}$. Therefore, $b_{ij} = 0$ for $i \neq j$ and $b_i = b_{ii} = \sum_{k=0}^{p-1} a_{i+k}$. The equation $\dot{A} = -[A, B]$ is equivalent to the system of equations

$$\dot{a}_{ij} = a_{ij}(b_i - b_j), \text{ where } a_{ij} \neq 0 \text{ only for } j = i + 1 - p.$$

As a result, we get a system of equations (here $j = i + 1 - p$):

$$\dot{a}_i = a_i \left(\sum_{k=0}^{p-1} a_{i+k} - \sum_{k=0}^{p-1} a_{j+k} \right) = a_i \left(\sum_{k=1}^{p-1} a_{i+k} - \sum_{k=1}^{p-1} a_{i-k} \right).$$

Thus, $I_k = \operatorname{tr}(A + \lambda M)^{kp}$ are first integrals of (1).

It is also possible to verify that the system of equations

$$\dot{a}_i = a_i \left(\prod_{k=1}^{p-1} a_{i+k} - \prod_{k=1}^{p-1} a_{i-k} \right)$$

is equivalent to the Lax equation

$$(A + \lambda M)^{\cdot} = [A + \lambda M, \lambda^{-1} A^p],$$

where $a_{i,i+1} = a_i$ and $m_{i,i+1-p} = -1$.

48. Matrices with prescribed eigenvalues

48.1.1. Theorem ([Farahat, Lederman, 1958]). *For any polynomial $f(x) = x^n + c_1 x^{n-1} + \dots + c_n$ and any numbers $a_1, \dots, a_{n-1}$ there exists a matrix of order n with characteristic polynomial f and elements $a_1, \dots, a_n$ on the diagonal (the last diagonal element a_n is defined by the relation $a_1 + \dots + a_n = -c_1$).*

Proof. The polynomials

$$u_0 = 1,\ u_1 = x - a_1, \dots,\ u_n = (x - a_1) \dots (x - a_n)$$

constitute a basis in the space of polynomials of degree not exceeding n and, therefore, $f = u_n + \lambda_1 u_{n-1} + \dots + \lambda_n u_0$. Equating the coefficients of x^{n-1} in the left-hand side and the right-hand side we get $c_1 = -(a_1 + \dots + a_n) + \lambda_1$, i.e., $\lambda_1 = c_1 + (a_1 + \dots + a_n) = 0$. Let

$$A = \begin{pmatrix} a_1 & 1 & 0 & & & 0 \\ 0 & a_2 & 1 & & & 0 \\ & & \ddots & \ddots & \ddots & 0 \\ & & & \ddots & a_{n-1} & 1 \\ -\lambda_n & -\lambda_{n-1} & \dots & \dots & -\lambda_2 & a_n \end{pmatrix}.$$

Expanding the determinant of $xI - A$ with respect to the last row we get

$$|xI - A| = \lambda_n + \lambda_{n-1} u_1 + \dots + \lambda_2 u_{n-2} + u_n = f,$$

i.e., A is the desired matrix. $\square$

48.1.2. THEOREM ([Farahat, Lederman, 1958]). *For any polynomial* $f(x) = x^n + c_1x^{n-1} + \dots + c_n$ *and any matrix* B *of order* $n-1$ *whose characteristic and minimal polynomials coincide there exists a matrix* A *such that* B *is a submatrix of* A *and the characteristic polynomial of* A *is equal to* f.

PROOF. Let us seek A in the form $A = \begin{pmatrix} B & P \\ Q^T & b \end{pmatrix}$, where P and Q are arbitrary columns of length $n-1$ and b is an arbitrary number. Clearly,

$$\det(xI_n - A) = (x-b)\det(xI_{n-1} - B) - Q^T \operatorname{adj}(xI_{n-1} - B)P$$

(see Theorem 3.1.3). Let us prove that $\operatorname{adj}(xI_{n-1} - B) = \sum_{r=0}^{n-2} u_r(x)B^r$, where the polynomials $u_0, \dots, u_{n-2}$ form a basis in the space of polynomials of degree not exceeding $n-2$. Let

$$g(x) = \det(xI_{n-1} - B) = x^{n-1} + t_1x^{n-2} + \dots$$

and $\varphi(x, \lambda) = (g(x) - g(\lambda))/(x - \lambda)$. Then

$$(xI_{n-1} - B)\varphi(x, B) = g(x)I_{n-1} - g(B) = g(x)I_{n-1},$$

since $g(B) = 0$ by the Cayley-Hamilton theorem. Therefore,

$$\varphi(x, B) = g(x)(xI_{n-1} - B)^{-1} = \operatorname{adj}(xI_{n-1} - B).$$

Besides, since $(x^k - \lambda^k)/(x - \lambda) = \sum_{s=0}^{k-1} x^{k-1-s}\lambda^s$, it follows that

$$\varphi(x, \lambda) = \sum_{r=0}^{n-2} t_{n-r-2} \sum_{s=0}^{r} x^{r-s}\lambda^s = \sum_{s=0}^{n-2} \lambda^s \sum_{r=s}^{n-2} t_{n-r-2}x^{r-s}$$

and, therefore, $\varphi(x, \lambda) = \sum_{s=0}^{n-2} \lambda^s u_s(x)$, where

$$u_s = x^{n-s-2} + t_1x^{n-s-3} + \dots + t_{n-s-2}.$$

Thus,

$$\begin{aligned} \det(xI_n - A) &= (x-b)(x^{n-1} + t_1x^{n-2} + \dots) - \sum_{s=0}^{n-2} u_s Q^T B^s P \\ &= x^n + (t_1 - b)x^{n-1} + h(x) - \sum_{s=0}^{n-2} u_s Q^T B^s P, \end{aligned}$$

where h is a polynomial of degree less than $n-1$ and the polynomials $u_0, \dots, u_{n-2}$ form a basis in the space of polynomials of degree less than $n-1$. Since the characteristic polynomial of B coincides with the minimal polynomial, the columns Q and P can be selected so that $(Q^T P, \dots, Q^T B^{n-2}P)$ is an arbitrary given set of numbers; cf. 13.3. □

48.2. THEOREM ([Friedland, 1972]). *Given all off-diagonal elements in a complex matrix A, it is possible to select diagonal elements $x_1, \dots, x_n$ so that the eigenvalues of A are given complex numbers; there are finitely many sets $\{x_1, \dots, x_n\}$ satisfying this condition.*

PROOF. Clearly,

$$\begin{aligned}\det(A + \lambda I) &= (x_1 + \lambda)\dots(x_n + \lambda) + \sum_{k \le n-2} \alpha(x_{i_1} + \lambda)\dots(x_{i_k} + \lambda) \\ &= \sum \lambda^{n-k} \sigma_k(x_1, \dots, x_n) + \sum_{k \le n-2} \lambda^{n-k} p_k(x_1, \dots, x_n),\end{aligned}$$

where p_k is a polynomial of degree $\le k - 2$. The equation $\det(A + \lambda I) = 0$ has the numbers $\lambda_1, \dots, \lambda_n$ as its roots if and only if

$$\sigma_k(\lambda_1, \dots, \lambda_n) = \sigma_k(x_1, \dots, x_n) + p_k(x_1, \dots, x_n).$$

Thus, our problem reduces to the system of equations

$$\sigma_k(x_1, \dots, x_n) = q_k(x_1, \dots, x_n), \text{ where } k = 1, \dots, n \text{ and } \deg q_k \le k - 1.$$

Let $\sigma_k = \sigma_k(x_1, \dots, x_n)$. Then the equality

$$x^n - \sigma_1 x^{n-1} + \sigma_2 x^{n-2} + \dots + (-1)^n \sigma_n = 0$$

holds for $x = x_1, \dots, x_n$. Let

$$f_i(x_1, \dots, x_n) = x_i^n + q_1 x_i^{n-1} - q_2 x_i^{n-2} + \dots + (-1)^n q_n = x_i^n + r_i(x_1, \dots, x_n),$$

where $\deg r_i < n$. Then

$$f_i = f_i - (x_i^n - \sigma_1 x_i^{n-1} + \sigma_2 x_i^{n-2} - \dots + (-1)^n \sigma_n) = x_i^{n-1} g_1 + x_i^{n-2} g_2 + \dots + g_n,$$

where $g_i = (-1)^{i-1}(\sigma_i + q_i)$. Therefore, $F = VG$, where F and G are columns $(f_1, \dots, f_n)^T$ and $(g_1, \dots, g_n)^T$, $V = \left\| x_i^{j-1} \right\|_1^n$. Therefore, $G = V^{-1}F$ and since $V^{-1} = W^{-1}V_1$, where $W = \det V = \prod_{i>j}(x_i - x_j)$ and V_1 is the matrix whose elements are polynomials in $x_1, \dots, x_n$, then $Wg_1, \dots, Wg_n \in I[f_1, \dots, f_n]$, where $I[f_1, \dots, f_n]$ is the ideal of the polynomial ring over $\mathbb{C}$ generated by $f_1, \dots, f_n$.

Suppose that the polynomials $g_1, \dots, g_n$ have no common roots. Then Hilbert's Nullstellensatz (see Appendix 4) shows that there exist polynomials $v_1, \dots, v_n$ such that $1 = \sum v_i g_i$; hence, $W = \sum v_i(Wg_i) \in I[f_1, \dots, f_n]$.

On the other hand, $W = \sum a_{i_1 \dots i_n} x_1^{i_1} \dots x_n^{i_n}$, where $i_k < n$. Therefore, $W \notin I[f_1, \dots, f_n]$ (see Appendix 5). It follows that the polynomials $g_1, \dots, g_n$ have a common root.

Let us show that the polynomials $g_1, \dots, g_n$ have finitely many common roots. Let $\xi = (x_1, \dots, x_n)$ be a root of polynomials $g_1, \dots, g_n$. Then ξ is a root of polynomials $f_1, \dots, f_n$ because $f_i = x_i^{n-1} g_1 + \dots + g_n$. Therefore, $x_i^n + r_i(x_1, \dots, x_n) = f_i = 0$ and $\deg r_i < n$. But such a system of equations has only finitely many solutions (see Appendix 5). Therefore, the number of distinct sets $x_1, \dots, x_n$ is finite. $\square$

48.3. THEOREM. *Let*

$$\lambda_1 \le \dots \le \lambda_n, \quad d_1 \le \dots \le d_n, \quad d_1 + \dots + d_k \ge \lambda_1 + \dots + \lambda_k$$

for $k = 1, \dots, n-1$ *and* $d_1 + \dots + d_n = \lambda_1 + \dots + \lambda_n$. *Then there exists an orthogonal matrix* P *such that the diagonal of the matrix* $P^T \Lambda P$, *where* $\Lambda = \operatorname{diag}(\lambda_1, \dots, \lambda_n)$, *is occupied by the numbers* $d_1, \dots, d_n$.

PROOF ([Chan, Kim-Hung Li, 1983]). First, let $n = 2$. Then $\lambda_1 \le d_1 \le d_2 \le \lambda_2$ and $d_2 = \lambda_1 + \lambda_2 - d_1$. If $\lambda_1 = \lambda_2$, then we can set $P = I$. If $\lambda_1 < \lambda_2$ then the matrix

$$P = (\lambda_2 - \lambda_1)^{-1/2} \begin{pmatrix} \sqrt{\lambda_2 - d_1} & -\sqrt{d_1 - \lambda_1} \\ \sqrt{d_1 - \lambda_1} & \sqrt{\lambda_2 - d_1} \end{pmatrix}$$

is the desired one.

Now, let us assume that the statement holds for some $n \ge 2$ and consider sets of $n+1$ numbers. Since $\lambda_1 \le d_1 \le d_{n+1} \le \lambda_{n+1}$, there exists a number $j > 1$ such that $\lambda_{j-1} \le d_1 \le \lambda_j$. Let P_1 be a permutation matrix such that

$$P_1^T \Lambda P_1 = \operatorname{diag}(\lambda_1, \lambda_j, \lambda_2, \dots, \widehat{\lambda_j}, \dots, \lambda_{n+1}).$$

It is easy to verify that

$$\lambda_1 \le \min(d_1, \lambda_1 + \lambda_j - d_1) \le \max(d_1, \lambda_1 + \lambda_j - d_1) \le \lambda_j.$$

Therefore, there exists an orthogonal 2×2 matrix Q such that on the diagonal of the matrix $Q^T \operatorname{diag}(\lambda_1, \lambda_j) Q$ there stand the numbers d_1 and $\lambda_1 + \lambda_j - d_1$. Consider the matrix $P_2 = \begin{pmatrix} Q & 0 \\ 0 & I_{n-1} \end{pmatrix}$. Clearly, $P_2^T (P_1^T \Lambda P_1) P_2 = \begin{pmatrix} d_1 & b^T \\ b & \Lambda_1 \end{pmatrix}$, where $\Lambda_1 = \operatorname{diag}(\lambda_1 + \lambda_j - d_1, \lambda_2, \dots, \widehat{\lambda_j}, \dots, \lambda_{n+1})$.

The diagonal elements of Λ_1 arranged in increasing order and the numbers $d_2, \dots, d_{n+1}$ satisfy the conditions of the theorem. Indeed,

$$d_2 + \dots + d_k \ge (k-1)d_1 \ge \lambda_2 + \dots + \lambda_k \tag{1}$$

for $k = 2, \dots, j-1$ and

$$\begin{aligned} d_2 + \dots + d_k &= d_1 + \dots + d_k - d_1 \ge \lambda_1 + \dots + \lambda_k - d_1 \\ &= (\lambda_1 + \lambda_j - d_1) + \lambda_2 + \dots + \lambda_{j-1} + \lambda_{j+1} + \dots + \lambda_k \end{aligned} \tag{2}$$

for $k = j, \dots, n+1$. In both cases (1), (2) the right-hand sides of the inequalities, i.e., $\lambda_2 + \dots + \lambda_k$ and $(\lambda_1 + \lambda_j - d_1) + \lambda_2 + \dots + \lambda_{j-1} + \lambda_{j+1} + \dots + \lambda_k$, are not less than the sum of $k-1$ minimal diagonal elements of Λ_1. Therefore, there exists an orthogonal matrix Q_1 such that the diagonal of $Q_1^T \Lambda_1 Q_1$ is occupied by the numbers $d_2, \dots, d_{n+1}$. Let $P_3 = \begin{pmatrix} 1 & 0 \\ 0 & Q_1 \end{pmatrix}$; then $P = P_1 P_2 P_3$ is the desired matrix. □

Solutions

39.1. a) Clearly, $AX = \left\|\lambda_i x_{ij}\right\|_1^n$ and $XA = \left\|\lambda_j x_{ij}\right\|_1^n$; therefore, $\lambda_i x_{ij} = \lambda_j x_{ij}$. Hence, $x_{ij} = 0$ for $i \neq j$.

b) By a) $X = \operatorname{diag}(x_1, \dots, x_n)$. As is easy to verify, $(NAX)_{i,i+1} = \lambda_{i+1}x_{i+1}$ and $(XNA)_{i,i+1} = \lambda_{i+1}x_i$. Hence, $x_i = x_{i+1}$ for $i = 1, 2, \dots, n-1$.

39.2. It suffices to use the result of Problem 39.1.

39.3. Let $p_1, \dots, p_n$ be the sums of the elements of the rows of the matrix X and $q_1, \dots, q_n$ the sums of the elements of its columns. Then

$$EX = \begin{pmatrix} q_1 & \dots & q_n \\ \vdots & \dots & \vdots \\ q_1 & \dots & q_n \end{pmatrix} \text{ and } XE = \begin{pmatrix} p_1 & \dots & p_1 \\ \vdots & \dots & \vdots \\ p_n & \dots & p_n \end{pmatrix}.$$

Therefore, $AX = XA$ if and only if

$$q_1 = \dots = q_n = p_1 = \dots = p_n.$$

39.4. The equality $AP_\sigma = P_\sigma A$ can be rewritten in the form $A = P_\sigma^{-1}AP_\sigma$. If $P_\sigma^{-1}AP_\sigma = \left\|b_{ij}\right\|_1^n$, then $b_{ij} = a_{\sigma(i)\sigma(j)}$. For any numbers p and q there exists a permutation σ such that $p = \sigma(q)$. Therefore, $a_{qq} = b_{qq} = a_{\sigma(q)\sigma(q)} = a_{pp}$, i.e., all diagonal elements of A are equal. If $i \neq j$ and $p \neq q$, then there exists a permutation σ such that $i = \sigma(p)$ and $j = \sigma(q)$. Hence, $a_{pq} = b_{pq} = a_{\sigma(p)}a_{\sigma(q)} = a_{ij}$, i.e., all off-diagonal elements of A are equal. It follows that

$$A = \alpha I + \beta(E - I) = (\alpha - \beta)I + \beta E.$$

39.5. We may assume that $A = \operatorname{diag}(A_1, \dots, A_k)$, where A_i is a Jordan block. Let $\mu_1, \dots, \mu_k$ be distinct numbers and B_i the Jordan block corresponding to the eigenvalue μ_i and of the same size as A_i. Then for B we can take the matrix $\operatorname{diag}(B_1, \dots, B_k)$.

39.6. a) For commuting matrices A and B we have

$$(A + B)^n = \sum \binom{n}{k} A^k B^{n-k}.$$

Let $A^m = B^m = 0$. If $n = 2m - 1$, then either $k \geq m$ or $n - k \geq m$; hence, $(A + B)^n = 0$.

b) By Theorem 39.2.2, the operators A and B have a common eigenbasis; this basis is the eigenbasis for the operator $A + B$.

39.7. Involutions are diagonalizable operators whose diagonal form has ± 1 on the diagonal (see 26.1). Therefore, there exists a basis in which all matrices A_i are of the form $\operatorname{diag}(\pm 1, \dots, \pm 1)$. There are 2^n such matrices.

39.8. Let us decompose the space V into the direct sum of invariant subspaces V_i such that every operator A_j has on every subspace V_i only one eigenvalue λ_{ij}. Consider the diagonal operator D whose restriction to V_i is of the form $\mu_i I$ and all numbers μ_i are distinct. For every j there exists an interpolation polynomial f_j such that $f_j(\mu_i) = \lambda_{ij}$ for all i (see Appendix 3). Clearly, $f_j(D) = A_j$.

39.9. It is easy to verify that all matrices of the form $\begin{pmatrix} \lambda I & A \\ 0 & \lambda I \end{pmatrix}$, where A is an arbitrary matrix of order m, commute.

40.1. It is easy to verify that $[N, A] = N$. Therefore,

$$\operatorname{ad}_J A = [J, A] = [N, A] = N = J - \lambda I.$$

It is also clear that $\operatorname{ad}_J(J - \lambda I) = 0$.

For any matrices X and Y we have

$$\operatorname{ad}_Y((Y - \lambda I)X) = (Y - \lambda I)\operatorname{ad}_Y X.$$

Hence,

$$\operatorname{ad}_Y^2((Y - \lambda I)X) = (Y - \lambda I)\operatorname{ad}_Y^2 X.$$

Setting $Y = J$ and $X = A$ we get $\operatorname{ad}_J^2(NA) = (NA)\operatorname{ad}_J^2 A = 0$.

40.2. Since

$$\begin{aligned} C^n = C^{n-1}\sum [A_i, B_i] &= \sum C^{n-1}A_iB_i - \sum C^{n-1}B_iA_i \\ &= \sum A_i(C^{n-1}B_i) - \sum (C^{n-1}B_i)A_i = \sum [A_i, C^{n-1}B_i], \end{aligned}$$

it follows that $\operatorname{tr} C^n = 0$ for $n \geq 1$. It follows that C is nilpotent; cf. Theorem 24.2.1.

40.3. For $n = 1$ the statement is obvious. It is also clear that if the statement holds for n, then

$$\begin{aligned} \operatorname{ad}_A^{n+1}(B) &= \sum_{i=0}^{n}(-1)^{n-i}\binom{n}{i}A^{i+1}BA^{n-i} - \sum_{i=0}^{n}(-1)^{n-i}\binom{n}{i}A^iBA^{n-i+1} \\ &= \sum_{i=1}^{n+1}(-1)^{n-i+1}\binom{n}{i-1}A^iBA^{n-i+1} + \sum_{i=0}^{n}(-1)^{n-i+1}\binom{n}{i}A^iBA^{n-i+1} \\ &= \sum_{i=0}^{n+1}(-1)^{n+1-i}\binom{n+1}{i}A^iBA^{n+1-i}. \end{aligned}$$

40.4. The map $D = \operatorname{ad}_A : M_{n,n} \longrightarrow M_{n,n}$ is a derivation. We have to prove that if $D^2B = 0$, then $D^n(B^n) = n!(DB)^n$. For $n = 1$ the statement is obvious. Suppose the statement holds for some n. Then

$$D^{n+1}(B^n) = D[D^n(B^n)] = n!D[(DB)^n] = n!\sum_{i=0}^{n-1}(DB)^i(D^2B)(DB)^{n-1-i} = 0.$$

Clearly,

$$D^{n+1}(B^{n+1}) = D^{n+1}(B \cdot B^n) = \sum_{i=0}^{n+1}\binom{n+1}{i}(D^iB)(D^{n+1-i}(B^n)).$$

Since $D^iB = 0$ for $i \geq 2$, it follows that

$$D^{n+1}(B^{n+1}) = B \cdot D^{n+1}(B^n) + (n+1)(DB)(D^n(B^n)) \\ = (n+1)(DB)(D^n(B^n)) = (n+1)!(DB)^{n+1}.$$

40.5. First, let us prove the required statement for $n = 1$. For $m = 1$ the statement is clear. It is also obvious that if the statement holds for some m, then

$$[A^{m+1}, B] = A(A^mB - BA^m) + (AB - BA)A^m \\ = mA[A, B]A^{m-1} + [A, B]A^m = (m+1)[A, B]A^m.$$

Now, let $m > n > 0$. Multiplying the equality $[A^n, B] = n[A, B]A^{n-1}$ by mA^{m-n} from the right we get

$$m[A^n, B]A^{m-n} = mn[A, B]A^{m-1} = n[A^m, B].$$

40.6. To the operator ad_A in the space $\operatorname{Hom}(V, V)$ there corresponds the operator $L = I \otimes A - A^T \otimes I$ in the space $V^* \otimes V$; cf. 27.5. If A is diagonal with respect to a basis $e_1, \dots, e_n$, then L is diagonal with respect to the basis $e_i \otimes e_j$. Therefore, $\operatorname{Ker} L^n = \operatorname{Ker} L$.

40.7. a) If $\operatorname{tr} Z = 0$, then $Z = [X, Y]$ (see 40.2); hence,

$$\operatorname{tr}(AZ) = \operatorname{tr}(AXY) - \operatorname{tr}(AYX) = 0.$$

Therefore, $A = \lambda I$; cf. Problem 5.1.

b) For any linear function f on the space of matrices there exists a matrix A such that $f(X) = \operatorname{tr}(AX)$. Now, since $f(XY) = f(YX)$, it follows that $\operatorname{tr}(AXY) = \operatorname{tr}(AYX)$ and, therefore, $A = \lambda I$.

41.1. The product of the indicated quaternions is equal to

$$-(x_1x_2 + y_1y_2 + z_1z_2) + (y_1z_2 - z_1y_2)i + (z_1x_2 - z_2x_1)j + (x_1y_2 - x_2y_1)k.$$

41.2. Let $q = a + v$, where a is the real part of the quaternion and v is its imaginary part. Then

$$(a + v)^2 = a^2 + 2av + v^2.$$

By Theorem 41.2.1, $v^2 = -\overline{v}v = -|v|^2 \leq 0$. Therefore, the quaternion $a^2 + 2av + v^2$ is real if and only if av is a real quaternion, i.e., $a = 0$ or $v = 0$.

41.3. It follows from the solution of Problem 41.2 that $q^2 = -1$ if and only if $q = xi + yj + zk$, where $x^2 + y^2 + z^2 = 1$.

41.4. Let the quaternion $q = a + v$, where a is the real part of q, commute with any purely imaginary quaternion w. Then $(a + v)w = w(a + v)$ and $aw = wa$; hence, $vw = wv$. Since $\overline{vw} = \overline{w}\,\overline{v} = wv$, we see that vw is a real quaternion. It remains to notice that if $v \neq 0$ and w is not proportional to $\overline{v}$, then $vw \notin \mathbb{R}$.

41.5. Let $B = W_1 + W_2j$, where W_1 and W_2 are complex matrices. Then

$$AB = Z_1W_1 + Z_2jW_1 + Z_1W_2j + Z_2jW_2j$$

and

$$A_c B_c = \begin{pmatrix} Z_1 W_1 - Z_2 \overline{W}_2 & Z_1 W_2 + Z_2 \overline{W}_1 \\ -\overline{Z}_2 W_1 - \overline{Z}_1 \overline{W}_2 & -\overline{Z}_2 W_2 + \overline{Z}_1 \overline{W}_1 \end{pmatrix}.$$

Therefore, it suffices to prove that $Z_2 j W_1 = Z_2 \overline{W}_1 j$ and $Z_2 j W_2 j = -Z_2 \overline{W}_2$. Since $ji = -ij$, we see that $jW_1 = \overline{W}_1 j$; and since $jj = -1$ and $jij = i$, it follows that $jW_2 j = -\overline{W}_2$.

41.6. a) Since $2(x, y) = \overline{x}y + \overline{y}x$, the equality $(x, y) = 0$ implies that $\overline{x}y + \overline{y}x = 0$. Hence,

$$2(qx, qx) = \overline{qx}qy + \overline{qy}qx = \overline{x}\,\overline{q}qy + \overline{y}\,\overline{q}qx = |q|^2(\overline{x}y + \overline{y}x) = 0.$$

b) The considered map preserves orientation and sends the rectangular parallelepiped formed by the vectors $1, i, j, k$ into the rectangular parallelepiped formed by the vectors q, qi, qj, qk; the ratio of the lengths of the corresponding edges of these parallelepipeds is equal to $|q|$ which implies that the ratio of the volumes of these parallelepipeds is equal to $|q|^4$.

41.7. A tetrahedron can be placed in the space of quaternions. Let a, b, c and d be the quaternions corresponding to its vertices. We may assume that c and d are real quaternions. Then c and d commute with a and b and, therefore,

$$(a - b)(c - d) + (b - c)(a - d) = (b - d)(a - c).$$

It follows that

$$|a - b||c - d| + |b - c||a - d| \geq |b - d||a - c|.$$

41.8. Let $x = a + be$ and $y = u + ve$. By the definition of the double of an algebra,

$$(a + be)(u + ve) = (au - \overline{v}b) + (b\overline{u} + va)e$$

and, therefore,

$$(xy)y = [(au - \overline{v}b)u - \overline{v}(b\overline{u} + va)] + [(b\overline{u} + va)\overline{u} + v(au - \overline{v}b)]e,$$
$$x(yy) = [(a(u^2 - \overline{v}v) - (u\overline{v} + \overline{u}\,\overline{v})b] + [b(\overline{u}^2 - \overline{v}v) + (v\overline{u} + vu)a]e.$$

To prove these equalities it suffices to use the associativity of the quaternion algebra and the facts that $\overline{v}v = v\overline{v}$ and that $u + \overline{u}$ is a real number. The identity $x(xy) = (xx)y$ is similarly proved.

b) Let us consider the trilinear map $f(a, x, y) = (ax)y - a(xy)$. Substituting $b = x + y$ in $(ab)b = a(bb)$ and taking into account that $(ax)x = a(xx)$ and $(ay)y = a(yy)$ we get

$$(ax)y - a(yx) = a(xy) - (ay)x,$$

i.e., $f(a, x, y) = -f(a, y, x)$. Similarly, substituting $b = x + y$ in $b(ba) = (bb)a$ we get $f(x, y, a) = -f(y, x, a)$. Therefore,

$$f(a, x, y) = -f(a, y, x) = f(y, a, x) = -f(y, x, a),$$

i.e., $(ax)y + (yx)a = a(xy) + y(xa)$. For $a = y$ we get $(yx)y = y(xy)$.

43.1. By Theorem 43.2, $R(f,g) = a_0^m \prod g(x_i)$ and $R(f,r) = a_0^k \prod r(x_i)$. Moreover, $f(x_i) = 0$; hence,

$$g(x_i) = f(x_i)q(x_i) + r(x_i) = r(x_i).$$

43.2. Let $c_0, \dots, c_{n+m-1}$ be the columns of the Sylvester matrix $S(f,g)$ and let $y_k = x^{n+m-k-1}$. Then

$$y_0c_0 + \dots + y_{n+m-1}c_{n+m-1} = c,$$

where c is the column $(x^{m-1}f(x), \dots, f(x), x^{n-1}g(x), \dots, g(x))^T$. Clearly, if $k \leq n-1$, then $x^k g(x) = \sum \lambda_i x^i f(x) + r_k(x)$, where λ_i are certain numbers and $i \leq m-1$. It follows that by adding linear combinations of the first m elements to the last n elements of the column c we can reduce this column to the form

$$(x^{m-1}f(x), \dots, f(x), r_{n-1}(x), \dots, r_0(x))^T.$$

Analogous transformations of the rows of $S(f,g)$ reduce this matrix to the form $\begin{pmatrix} A & C \\ 0 & B \end{pmatrix}$, where

$$A = \begin{pmatrix} a_0 & & * \\ & \ddots & \\ 0 & & a_0 \end{pmatrix}, \quad B = \begin{pmatrix} a_{n-1,0} & \dots & a_{n-1,n-1} \\ \vdots & \dots & \vdots \\ a_{00} & \dots & a_{0,n-1} \end{pmatrix}.$$

43.3. To the operator under consideration there corresponds the operator

$$I_m \otimes A - B^T \otimes I_n \text{ in } V^m \otimes V^n,$$

see 27.5. The eigenvalues of this operator are equal to $\alpha_i - \beta_j$, where α_i are the roots of f and β_j are the roots of g; see 27.4. Therefore, the determinant of this operator is equal to $\prod_{i,j}(\alpha_i - \beta_j) = R(f,g)$.

43.4. It is easy to verify that $S = V^TV$, where

$$V = \begin{pmatrix} 1 & \alpha_1 & \dots & \alpha_1^{n-1} \\ \vdots & \vdots & \dots & \vdots \\ 1 & \alpha_n & \dots & \alpha_n^{n-1} \end{pmatrix}.$$

Hence, $\det S = (\det V)^2 = \prod_{i<j}(\alpha_i - \alpha_j)^2$.

44.1. The equations $AX = C$ and $YB = C$ are solvable; therefore, $AA^{\text{“}-1\text{”}}C = C$ and $CB^{\text{“}-1\text{”}}B = C$; see 45.2. It follows that

$$C = AA^{\text{“}-1\text{”}}C = AA^{\text{“}-1\text{”}}CB^{\text{“}-1\text{”}}B = AZB, \text{ where } Z = A^{\text{“}-1\text{”}}CB^{\text{“}-1\text{”}}.$$

44.2. If X is a matrix of size $m \times n$ and $\operatorname{rank} X = r$, then $X = PQ$, where P and Q are matrices of size $m \times r$ and $r \times n$, respectively; cf. 8.2. The spaces spanned by the columns of matrices X and P coincide and, therefore, the equation $AX = 0$ implies $AP = 0$, which means that $P = (I - A^{\text{"}-1\text{"}}A)Y_1$; cf. 44.2. Similarly, the equality $XB = 0$ implies that $Q = Y_2(I - BB^{\text{"}-1\text{"}})$. Hence,

$$X = PQ = (I - A^{\text{"}-1\text{"}}A)Y(I - BB^{\text{"}-1\text{"}}), \text{ where } Y = Y_1Y_2.$$

It is also clear that if $X = (I - A^{\text{"}-1\text{"}}A)Y(I - BB^{\text{"}-1\text{"}})$, then $AX = 0$ and $XB = 0$.

44.3. If $AX = C$ and $XB = D$, then $AD = AXB = CB$. Now, suppose that $AD = CB$ and each of the equations $AX = C$ and $XB = C$ is solvable. In this case $AA^{\text{"}-1\text{"}}C = C$ and $DB^{\text{"}-1\text{"}}B = D$. Therefore,

$$A(A^{\text{"}-1\text{"}}C + DB^{\text{"}-1\text{"}} - A^{\text{"}-1\text{"}}ADB^{\text{"}-1\text{"}}) = C$$

and

$$(A^{\text{"}-1\text{"}}C + DB^{\text{"}-1\text{"}} - A^{\text{"}-1\text{"}}CBB^{\text{"}-1\text{"}})B = D,$$

i.e., $X_0 = A^{\text{"}-1\text{"}}C + DB^{\text{"}-1\text{"}} - A^{\text{"}-1\text{"}}ADB^{\text{"}-1\text{"}}$ is the solution of the system of equations considered.

46.1. Let $J = \begin{pmatrix} 0 & -1 \\ 1 & 0 \end{pmatrix}$. Then $A^2 = -t^2I$, $A^3 = -t^3J$, $A^4 = t^4I$, $A^5 = t^5J$, etc. Therefore,

$$\begin{aligned} e^A &= \left(1 - \frac{t^2}{2!} + \frac{t^4}{4!} - \dots\right)I + \left(t - \frac{t^3}{3!} + \frac{t^5}{5!} - \dots\right)J \\ &= (\cos t)I + (\sin t)J = \begin{pmatrix} \cos t & -\sin t \\ \sin t & \cos t \end{pmatrix}. \end{aligned}$$

46.2. a) Newton's binomial formula holds for the commuting matrices and, therefore,

$$\begin{aligned} e^{A+B} &= \sum_{n=0}^{\infty} \frac{(A+B)^n}{n!} = \sum_{n=0}^{\infty}\sum_{k=0}^{n} \frac{\binom{n}{k}A^kB^{n-k}}{k!} \\ &= \sum_{k=0}^{\infty}\sum_{n=k}^{\infty} \frac{A^k}{k!} \cdot \frac{B^{n-k}}{(n-k)!} = e^Ae^B. \end{aligned}$$

b) Since

$$e^{(A+B)t} = I + (A+B)t + (A^2 + AB + BA + B^2)\frac{t^2}{2} + \dots$$

and

$$e^{At}e^{Bt} = I + (A+B)t + (A^2 + 2AB + B^2)\frac{t^2}{2} + \dots,$$

it follows that

$$A^2 + AB + BA + B^2 = A^2 + 2BA + B^2$$

and, therefore, $AB = BA$.

46.3. There exists a unitary matrix V such that

$$U = VDV^{-1}, \text{ where } D = \operatorname{diag}(\exp(i\alpha_1), \dots, \exp(i\alpha_n)).$$

Let $\Lambda = \operatorname{diag}(\alpha_1, \dots, \alpha_n)$. Then $U = e^{iH}$, where $H = V\Lambda V^{-1} = V\Lambda V^*$ is an Hermitian matrix.

46.4. a) Let $U = e^X$ and $X^T = -X$. Then $UU^T = e^X e^{X^T} = e^X e^{-X} = I$ since the matrices X and $-X$ commute.

b) For such a matrix U there exists an orthogonal matrix V such that

$$U = V \operatorname{diag}(A_1, \dots, A_k, I) V^{-1}, \text{ where } A_i = \begin{pmatrix} \cos\varphi_i & -\sin\varphi_i \\ \sin\varphi_i & \cos\varphi_i \end{pmatrix};$$

cf. Theorem 11.3. It is also clear that the matrix A_i can be represented in the form e^X, where $X = \begin{pmatrix} 0 & -x \\ x & 0 \end{pmatrix}$; cf. Problem 46.1.

46.5. a) It suffices to observe that $\det(e^A) = e^{\operatorname{tr} A}$ (cf. Theorem 46.1.2), and that $\operatorname{tr} A$ is a real number.

b) Let λ_1 and λ_2 be eigenvalues of a real 2×2 matrix A and $\lambda_1 + \lambda_2 = \operatorname{tr} A = 0$. The numbers λ_1 and λ_2 are either both real or $\lambda_1 = \overline{\lambda}_2$, i.e., $\lambda_1 = -\overline{\lambda}_1$. Therefore, the eigenvalues of e^A are equal to either e^α and $e^{-\alpha}$ or $e^{i\alpha}$ and $e^{-i\alpha}$, where in either case α is a real number. It follows that $B = \begin{pmatrix} -2 & 0 \\ 0 & -1/2 \end{pmatrix}$ is not the exponent of a real matrix.

46.6. a) Let A_{ij} be the cofactor of a_{ij}. Then $\operatorname{tr}(\dot{A} \operatorname{adj} A^T) = \sum_{i,j} \dot{a}_{ij} A_{ij}$.

Since $\det A = a_{ij}A_{ij} + \dots$, where dots stand for terms that do not contain a_{ij}, it follows that

$$(\det A)^\cdot = \dot{a}_{ij}A_{ij} + a_{ij}\dot{A}_{ij} + \dots = \dot{a}_{ij}A_{ij} + \dots,$$

where dots stand for terms that do not contain $\dot{a}_{ij}$. Hence, $(\det A)^\cdot = \sum_{i,j} \dot{a}_{ij}A_{ij}$.

b) Since $A \operatorname{adj} A^T = (\det A)I$, then $\operatorname{tr}(A \operatorname{adj} A^T) = n \det A$ and, therefore,

$$n(\det A)^\cdot = \operatorname{tr}(\dot{A} \operatorname{adj} A^T) + \operatorname{tr}(A(\operatorname{adj} A^T)^\cdot).$$

It remains to use a).

46.7. First, suppose that $m > 0$. Then

$$(X^m)_{ij} = \sum_{a,b,\dots,p,q} x_{ia}x_{ab}\dots x_{pq}x_{qj},$$

$$\operatorname{tr} X^m = \sum_{a,b,\dots,p,q,r} x_{ra}x_{ab}\dots x_{pq}x_{qr}.$$

Therefore,

$$\frac{\partial}{\partial x_{ji}}(\operatorname{tr} X^m) = \sum_{a,b,\dots,p,q,r} \left(\frac{\partial x_{ra}}{\partial x_{ji}} x_{ab}\dots x_{pq}x_{qr} + \dots + x_{ra}x_{ab}\dots x_{pq}\frac{\partial x_{qr}}{\partial x_{ji}} \right)$$

$$= \sum_{b,\dots,p,q} x_{ib}\dots x_{pq}x_{qj} + \dots + \sum_{a,b,\dots,p} x_{ia}x_{ab}\dots x_{pj} = m(X^{m-1})_{ij}.$$

Now, suppose that $m < 0$. Let $X^{-1} = \|y_{ij}\|_1^n$. Then $y_{ij} = X_{ji}\Delta^{-1}$, where X_{ji} is the cofactor of x_{ji} in X and $\Delta = \det X$. By Jacobi's Theorem (Theorem 2.5.2) we have

$$\begin{vmatrix} X_{i_1 j_1} & X_{i_1 j_2} \\ X_{i_2 j_1} & X_{i_2 j_2} \end{vmatrix} = (-1)^\sigma \begin{vmatrix} x_{i_3 j_3} & \dots & x_{i_3 j_n} \\ \vdots & \dots & \vdots \\ x_{i_n j_3} & \dots & x_{i_n j_n} \end{vmatrix} \Delta$$

and

$$X_{i_1 j_1} = (-1)^\sigma \begin{vmatrix} x_{i_2 j_2} & \dots & x_{i_2 j_n} \\ \vdots & \dots & \vdots \\ x_{i_n j_2} & \dots & x_{i_n j_n} \end{vmatrix}, \quad \text{where } \sigma = \begin{pmatrix} i_1 & \dots & i_n \\ j_1 & \dots & j_n \end{pmatrix}.$$

Hence, $\begin{vmatrix} X_{i_1 j_1} & X_{i_1 j_2} \\ X_{i_2 j_1} & X_{i_2 j_2} \end{vmatrix} = \Delta \frac{\partial}{\partial x_{i_2 j_2}}(X_{i_1 j_1})$. It follows that

$$\begin{aligned} -X_{j\alpha} X_{\beta i} &= \Delta \frac{\partial}{\partial x_{ji}}(X_{\beta\alpha}) - X_{\beta\alpha} X_{ji} \\ &= \Delta \frac{\partial}{\partial x_{ji}}(X_{\beta\alpha}) - X_{\beta\alpha} \frac{\partial}{\partial x_{ji}}(\Delta) = \Delta^2 \frac{\partial}{\partial x_{ji}} \left(\frac{X_{\beta\alpha}}{\Delta} \right), \end{aligned}$$

i.e., $\frac{\partial}{\partial x_{ji}} y_{\alpha\beta} = -y_{\alpha j} y_{i\beta}$. Since

$$(X^m)_{ij} = \sum_{a,b,\dots,q} y_{ia} y_{ab} \dots y_{qj} \quad \text{and} \quad \operatorname{tr} X^m = \sum_{a,b,\dots,q,r} y_{ra} y_{ab} \dots y_{qr},$$

it follows that

$$\begin{aligned} \frac{\partial}{\partial x_{ji}}(\operatorname{tr} X^m) = &- \sum_{a,b,\dots,q,r} y_{rj} y_{ia} y_{ab} \dots y_{qr} - \dots \\ &- \sum_{a,b,\dots,q,r} y_{ra} y_{ab} \dots y_{qj} y_{ir} = m(X^{m-1})_{ij}. \end{aligned}$$

APPENDIX

A polynomial f with integer coefficients is called *irreducible* over $\mathbb{Z}$ (resp. over $\mathbb{Q}$) if it cannot be represented as the product of two polynomials of lower degree with integer (resp. rational) coefficients.

THEOREM. *A polynomial f with integer coefficients is irreducible over $\mathbb{Z}$ if and only if it is irreducible over $\mathbb{Q}$.*

To prove this, consider the greatest common divisor of the coefficients of the polynomial f and denote it $\operatorname{cont}(f)$, the *content* of f.

LEMMA (Gauss). *If* $\operatorname{cont}(f) = \operatorname{cont}(g) = 1$, *then* $\operatorname{cont}(fg) = 1$.

PROOF. Suppose that $\operatorname{cont}(f) = \operatorname{cont}(g) = 1$ and $\operatorname{cont}(fg) = d \neq \pm 1$. Let p be one of the prime divisors of d; let a_r and b_s be the nondivisible by p coefficients of the polynomials $f = \sum a_i x^i$ and $g = \sum b_i x^i$ with the smallest indices. Let us consider the coefficient at x^{r+s} in the power series expansion of fg. Together with all the coefficients of fg, this one is also divisible by p. On the other hand, it is equal to the sum of numbers $a_i b_j$, where $i + j = r + s$. But only one of these numbers, namely, $a_r b_s$, is not divisible by p, since either $i < r$ or $j < s$. Contradiction. $\square$

Now we are able to prove the theorem.

PROOF. We may assume that $\operatorname{cont}(f) = 1$. Given a factorization $f = \varphi_1 \varphi_2$, where φ_1 and φ_2 are polynomials with rational coefficients, we have to construct a factorization $f = f_1 f_2$, where f_1 and f_2 are polynomials with integer coefficients. Let us represent φ_i in the form $\varphi_i = \dfrac{a_i}{b_i} f_i$, where $a_i, b_i \in \mathbb{Z}$, f_i are polynomials with integer coefficients, and $\operatorname{cont}(f_i) = 1$. Then $b_1 b_2 f = a_1 a_2 f_1 f_2$; hence, $\operatorname{cont}(b_1 b_2 f) = \operatorname{cont}(a_1 a_2 f_1 f_2)$. By the Gauss lemma, $\operatorname{cont}(f_1 f_2) = 1$. Therefore, $a_1 a_2 = \pm b_1 b_2$, i.e., $f = \pm f_1 f_2$, which is the desired factorization. $\square$

A.1. THEOREM. *Let polynomials f and g with integer coefficients have a common root and let f be an irreducible polynomial with the leading coefficient* 1. *Then g/f is a polynomial with integer coefficients.*

PROOF. Let us successively perform the division with a remainder (*Euclid's algorithm*):

$$g = a_1 f + b_1, \quad f = a_2 b_1 + b_2, \quad b_1 = a_3 b_2 + b_3, \quad \ldots, \quad b_{n-2} = a_{n-1} b_n.$$

It is easy to verify that b_n is the greatest common divisor of f and g. All polynomials a_i and b_i have rational coefficients. Therefore, the greatest common divisor of polynomials f and g over $\mathbb{Q}$ coincides with their greatest common divisor over $\mathbb{C}$. But over $\mathbb{C}$ the polynomials f and g have a nontrivial common divisor and, therefore, f and g have a nontrivial common divisor, r, over $\mathbb{Q}$ as well. Since f is an irreducible polynomial with the leading coefficient 1, it follows that $r = \pm f$. $\square$

A.2. THEOREM (Eisenstein's criterion). *Let*

$$f(x) = a_0 + a_1 x + \cdots + a_n x^n$$

be a polynomial with integer coefficients and let p be a prime such that the coefficient a_n is not divisible by p whereas $a_0, \dots, a_{n-1}$ are, and a_0 is not divisible by p^2. Then the polynomial f is irreducible over $\mathbb{Z}$.

PROOF. Suppose that $f = gh = (\sum b_k x^k)(\sum c_l x^l)$, where g and h are not constants. The number $b_0 c_0 = a_0$ is divisible by p and, therefore, one of the numbers b_0 or c_0 is divisible by p. Let, for definiteness sake, b_0 be divisible by p. Then c_0 is not divisible by p because $a_0 = b_0 c_0$ is not divisible by p^2. If all numbers b_i are divisible by p, then a_n is divisible by p. Therefore, b_i is not divisible by p for some i with $0 < i \leq \deg g < n$.

We may assume that i is the least index for which the number b_i is nondivisible by p. On the one hand, by the assumption, the number a_i is divisible by p. On the other hand, $a_i = b_i c_0 + b_{i-1} c_1 + \cdots + b_0 c_i$ and all numbers $b_{i-1} c_1, \dots, b_0 c_i$ are divisible by p whereas $b_i c_0$ is not divisible by p. Contradiction. $\square$

COROLLARY. *If p is a prime, then the polynomial $f(x) = x^{p-1} + \cdots + x + 1$ is irreducible over $\mathbb{Z}$.*

Indeed, we can apply Eisenstein's criterion to the polynomial

$$f(x+1) = \frac{(x+1)^p - 1}{(x+1) - 1} = x^{p-1} + \binom{p}{1} x^{p-2} + \cdots + \binom{p}{p-1}.$$

A.3. THEOREM. *Suppose the numbers*

$$y_1,\ y_1^{(1)},\ \dots,\ y_1^{(\alpha_1 - 1)},\ \dots,\ y_n,\ y_n^{(1)},\ \dots,\ y_n^{(\alpha_n - 1)}$$

are given at points $x_1, \dots, x_n$ and $m = \alpha_1 + \cdots + \alpha_n - 1$. Then there exists a polynomial $H_m(x)$ of degree not greater than m for which $H_m(x_j) = y_j$ and $H_m^{(i)}(x_j) = y_j^{(i)}$.

PROOF. Let $k = \max(\alpha_1, \dots, \alpha_n)$. For $k = 1$ we can make use of the Lagrange interpolation polynomial

$$L_n(x) = \sum_{j=1}^{n} \frac{(x - x_1) \dots (x - x_{j-1})(x - x_{j+1}) \dots (x - x_n)}{(x_j - x_1) \dots (x_j - x_{j-1})(x_j - x_{j+1}) \dots (x_j - x_n)} y_j.$$

Let $\omega_n(x) = (x - x_1) \dots (x - x_n)$. Take an arbitrary polynomial H_{m-n} of degree not greater than $m - n$ and assign to it the polynomial $H_m(x) = L_n(x) + \omega_n(x) H_{m-n}(x)$. It is clear that $H_m(x_j) = y_j$ for any polynomial H_{m-n}. Besides,

$$H_m'(x) = L_n'(x) + \omega_n'(x) H_{m-n}(x) + \omega_n(x) H_{m-n}'(x),$$

i.e., $H'_m(x_j) = L'_n(x_j)+\omega'_n(x_j)H_{m-n}(x_j)$. Since $\omega'_n(x_j) \neq 0$, then at points where the values of $H'_m(x_j)$ are given, we may determine the corresponding values of $H_{m-n}(x_j)$. Furthermore,

$$H''_m(x_j) = L''_n(x_j) + \omega''_n(x_j)H_{m-n}(x_j) + 2\omega'_n(x_j)H'_{m-n}(x_j).$$

Therefore, at points where the values of $H''_m(x_j)$ are given we can determine the corresponding values of $H'_{m-n}(x_j)$, etc. Thus, our problem reduces to the construction of a polynomial $H_{m-n}(x)$ of degree not greater than $m-n$ for which $H^{(i)}_{m-n}(x_j) = z^{(i)}_j$ for $i = 0, \dots, \alpha_j - 2$ (if $\alpha_j = 1$, then there are no restrictions on the values of H_{m-n} and its derivatives at x_j). It is also clear that $m - n = \sum(\alpha_j - 1) - 1$. After $k - 1$ of similar operations it remains to construct Lagrange's interpolation polynomial. □

A.4. Hilbert's Nullstellensatz. We will only need the following special case of *Hilbert's Nullstellensatz.*

THEOREM. *Let $f_1, \dots, f_r$ be polynomials in n indeterminates over $\mathbb{C}$ without common zeros. Then there exist polynomials $g_1, \dots, g_r$ such that $f_1g_1 + \dots + f_rg_r = 1$.*

PROOF. Let $I(f_1, \dots, f_r)$ be the ideal of the polynomial ring $\mathbb{C}[x_1, \dots, x_n] = K$ generated by $f_1, \dots, f_r$. Suppose that there are no polynomials $g_1, \dots, g_r$ such that $f_1g_1 + \dots + f_rg_r = 1$. Then $I(f_1, \dots, f_r) \neq K$. Let I be a nontrivial maximal ideal containing $I(f_1, \dots, f_r)$. As is easy to verify, K/I is a field. Indeed, if $f \notin I$, then $I + Kf$ is the ideal strictly containing I and, therefore, this ideal coincides with K. It follows that there exist polynomials $g \in K$ and $h \in I$ such that $1 = h + fg$. Then the class $\overline{g} \in K/I$ is the inverse of $\overline{f} \in K/I$.

Now, let us prove that the field $A = K/I$ coincides with $\mathbb{C}$.

Let α_i be the image of x_i under the natural projection

$$p : \mathbb{C}[x_1, \dots, x_n] \longrightarrow \mathbb{C}[x_1, \dots, x_n]/I = A.$$

Then

$$A = \left\{\sum z_{i_1 \dots i_n} \alpha_1^{i_1} \dots \alpha_n^{i_n} \mid z_{i_1 \dots i_n} \in \mathbb{C}\right\} = \mathbb{C}[\alpha_1, \dots, \alpha_n].$$

Further, let $A_0 = \mathbb{C}$ and $A_s = \mathbb{C}[\alpha_1, \dots, \alpha_s]$. Then $A_{s+1} = \{\sum a_i \alpha^i_{s+1} | a_i \in A_s\} = A_s[\alpha_{s+1}]$. Let us prove by induction on s that there exists a ring homomorphism $f : A_s \longrightarrow \mathbb{C}$ (which sends 1 to 1). For $s = 0$ the statement is obvious. Now, let us show how to construct a homomorphism $g : A_{s+1} \longrightarrow \mathbb{C}$ from the homomorphism $f : A_s \longrightarrow \mathbb{C}$. For this let us consider two cases.

a) The element $x = \alpha_{s+1}$ is transcendental over A_s. Then for any $\xi \in \mathbb{C}$ we can define a homomorphism g such that $g(a_nx^n + \dots + a_0) = f(a_n)\xi^n + \dots + f(a_0)$. Setting $\xi = 0$ we get a homomorphism g such that $g(1) = 1$.

b) The element $x = \alpha_{s+1}$ is algebraic over A_s, i.e., $b_mx^m + b_{m-1}x^{m-1} + \dots + b_0 = 0$ for certain $b_i \in A_s$. Then for all $\xi \in \mathbb{C}$ such that $f(b_m)\xi^m + \dots + f(b_0) = 0$ there is determined a homomorphism $g(\sum a_kx^k) = \sum f(a_k)\xi^k$ which sends 1 to 1.

As a result we get a homomorphism $h : A \longrightarrow \mathbb{C}$ such that $h(1) = 1$. It is also clear that $h^{-1}(0)$ is an ideal and there are no nontrivial ideals in the field A. Hence, h is a monomorphism. Since $A_0 = \mathbb{C} \subset A$ and the restriction of h to A_0 is the identity map h is an isomorphism.

Thus, we may assume that $\alpha_i \in \mathbb{C}$. The projection p maps the polynomial $f_i(x_1,\dots,x_n) \in K$ to $f_i(\alpha_1,\dots,\alpha_n) \in \mathbb{C}$. Since $f_1,\dots,f_r \in I$, we have $p(f_i) = 0 \in \mathbb{C}$. Therefore, $f_i(\alpha_1,\dots,\alpha_n) = 0$. Contradiction. $\square$

A.5. THEOREM. *Let the polynomials $f_i(x_1,\dots,x_n) = x_i^{m_i} + P_i(x_1,\dots,x_n)$, where $i = 1,\dots,n$, be such that $\deg P_i < m_i$ and let $I(f_1,\dots,f_n)$ be the ideal generated by $f_1,\dots,f_n$.*

a) *Let $P(x_1,\dots,x_n)$ be a nonzero polynomial of the form $\sum a_{i_1\dots i_n} x_1^{i_1}\dots x_n^{i_n}$, where $i_k < m_k$ for all $k = 1,\dots,n$. Then $P \notin I(f_1,\dots,f_n)$.*

b) *The system of equations $x_i^{m_i} + P_i(x_1,\dots,x_n) = 0$ $(i = 1,\dots,n)$ is always solvable over $\mathbb{C}$ and the number of solutions is finite.*

PROOF. Substituting the polynomial $(f_i - P_i)^{t_i} x^{q_i}$ instead of $x_i^{m_i t_i + q_i}$, where $0 \le t_i$ and $0 \le q_i < m_i$, we see that any polynomial $Q(x_1,\dots,x_n)$ can be represented in the form

$$Q(x_1,\dots,x_n) = Q^*(x_1,\dots,x_n,f_1,\dots,f_n) = \sum a_{js} x_1^{j_1}\dots x_n^{j_n} f_1^{s_1}\dots f_n^{s_n},$$

where $j_1 < m_1, \dots, j_n < m_n$. Let us prove that such a representation Q^* is uniquely determined. It suffices to verify that by substituting $f_i = x_i^{m_i} + P_i(x_1,\dots,x_n)$ in any nonzero polynomial $Q^*(x_1,\dots,x_n,f_1,\dots,f_n)$ we get a nonzero polynomial $\tilde{Q}(x_1,\dots,x_n)$. Among the terms of the polynomial Q^*, let us select the one for which the sum $(s_1 m_1 + j_1) + \dots + (s_n m_n + j_n) = m$ is maximal. Clearly, $\deg \tilde{Q} \le m$. Let us compute the coefficient of the monomial $x_1^{s_1 m_1 + j_1}\dots x_n^{s_n m_n + j_n}$ in $\tilde{Q}$. Since the sum

$$(s_1 m_1 + j_1) + \dots + (s_n m_n + j_n)$$

is maximal, this monomial can only come from the monomial $x_1^{j_1}\dots x_n^{j_n} f_1^{s_1}\dots f_n^{s_n}$. Therefore, the coefficients of these two monomials are equal and $\deg \tilde{Q} = m$.

Clearly, $Q(x_1,\dots,x_n) \in I(f_1,\dots,f_n)$ if and only if $Q^*(x_1,\dots,x_n,f_1,\dots,f_n)$ is the sum of monomials for which $s_1 + \dots + s_n \ge 1$. Besides, if $P(x_1,\dots,x_n) = \sum a_{i_1\dots i_n} x_1^{i_1}\dots x_n^{i_n}$, where $i_k < m_k$, then

$$P^*(x_1,\dots,x_n,f_1,\dots,f_n) = P(x_1,\dots,x_n).$$

Hence, $P \notin I(f_1,\dots,f_n)$.

b) If $f_1, \dots, f_n$ have no common zero, then by Hilbert's Nullstellensatz the ideal $I(f_1,\dots,f_n)$ coincides with the entire polynomial ring and, therefore, $P \in I(f_1,\dots,f_n)$; this contradicts part a). Therefore, the given system of equations is solvable. Let $\xi = (\xi_1,\dots,\xi_n)$ be a solution of this system. Then $\xi_i^{m_i} = -P_i(\xi_1,\dots,\xi_n)$, where $\deg P_i < m_i$, and, therefore, any polynomial $Q(\xi_1,\dots\xi_n)$ can be represented in the form $Q(\xi_1,\dots,\xi_n) = \sum a_{i_1\dots i_n} \xi_1^{i_1}\dots\xi_n^{i_n}$, where $i_k < m_k$ and the coefficient $a_{i_1\dots i_n}$ is the same for all solutions. Let $m = m_1\dots m_n$. The polynomials $1, \xi_i, \dots, \xi_i^m$ can be linearly expressed in terms of the basic monomials $\xi_1^{i_1}\dots\xi_n^{i_n}$, where $i_k < m_k$. Therefore, they are linearly dependent, i.e., $b_0 + b_1\xi_i + \dots + b_m\xi_i^m = 0$, not all numbers $b_0,\dots,b_m$ are zero and these numbers are the same for all solutions (do not depend on i). The equation $b_0 + b_1 x + \dots + b_m x^m = 0$ has, clearly, finitely many solutions. $\square$

BIBLIOGRAPHY

Recommended literature

R. Bellman, *Introduction to matrix analysis*, McGraw-Hill, New York, 1960.
M. J. Crowe, *A history of vector analysis*, Notre Dame, London, 1967.
F. R. Gantmakher, *The theory of matrices,* I, II, Chelsea, New York, 1959.
I. M. Gelfand, *Lectures on linear algebra,* Interscience Tracts in Pure and Applied Math., New York, 1961.
W. H. Greub, *Linear algebra*, Springer-Verlag, Berlin, 1967.
———, *Multilinear algebra*, Springer-Verlag, Berlin, 1967.
P. R. Halmos, *Finite-dimensional vector spaces*, Van Nostrand, Princeton, NJ, 1958.
R. A. Horn and Ch. R. Johnson, *Matrix analysis*, Cambridge University Press, Cambridge, 1986.
A. I. Kostrikin and Yu. I. Manin, *Linear algebra and geometry*, Gordon & Breach, New York, 1989.
M. Marcus and H. Minc, *A survey of matrix theory and matrix inequalities*, Allyn and Bacon, Boston, 1964.
T. Muir and W. H. Metzler, *A treatise on the history of determinants*, Dover, New York, 1960.
M. M. Postnikov, *Lectures on geometry. 2nd semester. Linear algebra.*, "Nauka", Moscow, 1986. (Russian)
———, *Lectures on geometry. 5th semester. Lie groups and Lie algebras.*, "Mir", Moscow, 1986.
G. E. Shilov, *Theory of linear spaces*, Prentice-Hall, Englewood Cliffs, NJ, 1961.

References

J. F. Adams, *Vector fields on spheres*, Ann. Math. **75** (1962), 603–632.
S. N. Afriat, *On the latent vectors and characteristic values of products of pairs of symmetric idempotents*, Quart. J. Math. **7** (1956), 76–78.
A. C. Aitken, *A note on trace-differentiation and the Ω-operator*, Proc. Edinburgh Math. Soc. **10** (1953), 1–4.
A. A. Albert, *On the orthogonal equivalence of sets of real symmetric matrices*, J. Math. and Mech. **7** (1958), 219–235.
B. Aupetit, *An improvement of Kaplansky's lemma on locally algebraic operators*, Studia Math. **88** (1988), 275–278.
S. Barnett, *Matrices in control theory*, Van Nostrand Reinhold, London, 1971.
R. Bellman, *Notes on matrix theory.* IV, Amer. Math. Monthly **62** (1955), 172–173.
R. Bellman and A. Hoffman, *On a theorem of Ostrowski and Taussky*, Arch. Math. **5** (1954), 123–127.
M. Berger, *Géometrie.* vol.4 (*Formes quadratiques, quadriques et coniques*), CEDIC/Nathan, Paris, 1977.
O. I. Bogoyavlenskiĭ, *Solitons that flip over*, "Nauka", Moscow, 1991. (Russian)
P.-F. Burgermeister, *Classification des représentations de la double flèche*, L'Enseignement Math. **32** (1986), 199–210.
N. N. Chan and Kim-Hung Li, *Diagonal elements and eigenvalues of a real symmetric matrix*, J. Math. Anal. and Appl. **91** (1983), 562–566.
C. G. Cullen, *A note on convergent matrices*, Amer. Math. Monthly **72** (1965), 1006–1007.
D. Z. Djoković, *On the Hadamard product of matrices*, Math. Z. **86** (1964), 395.
———, *Product of two involutions*, Arch. Math. **18** (1967), 582–584.
———, *A determinantal inequality for projectors in a unitary space*, Proc. Amer. Math. Soc. **27** (1971), 19–23.
M. A. Drazin, J. W. Dungey, and K. W. Gruenberg, *Some theorems on commutative matrices*, J. London Math. Soc. **26** (1951), 221–228.

M. A. Drazin and E. V. Haynsworth, *Criteria for the reality of matrix eigenvalues*, Math. Z. **78** (1962), 449–452.

W. N. Everitt, *A note on positive definite matrices*, Proc. Glasgow Math. Assoc. **3** (1958), 173–175.

H. K. Farahat and W. Lederman, *Matrices with prescribed characteristic polynomials*, Proc. Edinburgh Math. Soc. **11** (1958), 143–146.

H. Flanders, *On spaces of linear transformations with bound rank*, J. London Math. Soc. **37** (1962), 10–16.

H. Flanders and H. K. Wimmer, *On matrix equations $AX - XB = C$ and $AX - YB = C$*, SIAM J. Appl. Math. **32** (1977), 707–710.

P. Franck, *Sur la meilleure approximation d'une matrice donnée par une matrice singulière*, C.R. Acad. Sci. Paris **253** (1961), 1297–1298.

W. M. Frank, *A bound on determinants*, Proc. Amer. Math. Soc. **16** (1965), 360–363.

G. Fregus, *A note on matrices with zero trace*, Amer. Math. Monthly **73** (1966), 630–631.

Sh. Friedland, *Matrices with prescribed off-diagonal elements*, Israel J. Math. **11** (1972), 184–189.

P. M. Gibson, *Matrix commutators over an algebraically closed field*, Proc. Amer. Math. Soc. **52** (1975), 30–32.

C. Green, *A multiple exchange property for bases*, Proc. Amer. Math. Soc. **39** (1973), 45–50.

M. J. Greenberg, *Note on the Cayley–Hamilton theorem*, Amer. Math. Monthly **91** (1984), 193–195.

D. Yu. Grigoriev, *The algebraic computational complexity of a family of bilinear forms*, Zh. Vychisl. Mat. i Mat. Fiz. **19** (1979), no. 3, 563–580; English transl. in J. Comp. Math. and Math. Phys. **19** (1979).

E. V. Haynsworth, *Applications of an inequality for the Schur complement*, Proc. Amer. Math. Soc. **24** (1970), 512–516.

P. L. Hsu, *On symmetric, orthogonal and skew-symmetric matrices*, Proc. Edinburgh Math. Soc. **10** (1953), 37–44.

G. H. Jacob, *Another proof of the rational decomposition theorem*, Amer. Math. Monthly **80** (1973), 1131–1134.

J. Kahane, *Grassmann algebras for proving a theorem on Pfaffians*, Linear Algebra and Appl. **4** (1971), 129–139.

D. C. Kleinecke, *On operator commutators*, Proc. Amer. Math. Soc. **8** (1957), 535–536.

C. Lanczos, *Linear systems in self-adjoint form*, Amer. Math. Monthly **65** (1958), 665–679.

K. N. Majindar, *On simultaneous Hermitian congruence transformations of matrices*, Amer. Math. Monthly **70** (1963), 842–844.

S. V. Manakov, *A remark on integration of the Euler equation for an N-dimensional solid body.*, Funktsional. Analiz i ego Prilozhen. **10** (1976), no. 4, 93–94; English transl. in Functional Anal. Appl. **10** (1976).

M. Marcus and H. Minc, *On two theorems of Frobenius*, Pac. J. Math. **60** (1975), 149–151.

[a] M. Marcus and B. N. Moyls, *Linear transformations on algebras of matrices*, Can. J. Math. **11** (1959), 61–66.

[b] ———, *Transformations on tensor product spaces*, Pac. J. Math. **9** (1959), 1215–1222.

M. Marcus and R. Purves, *Linear transformations on algebras of matrices: the invariance of the elementary symmetric functions*, Can. J. Math. **11** (1959), 383–396.

W. S. Massey, *Cross products of vectors in higher dimensional Euclidean spaces*, Amer. Math. Monthly **90** (1983), 697–701.

R. Merris, *Equality of decomposable symmetrized tensors*, Can. J. Math. **27** (1975), 1022–1024.

L. Mirsky, *An inequality for positive definite matrices*, Amer. Math. Monthly **62** (1955), 428–430.

———, *On a generalization of Hadamard's determinantal inequality due to Szasz*, Arch. Math. **8** (1957), 274–275.

———, *A trace inequality of John von Neumann*, Monatshefte für Math. **79** (1975), 303–306.

E. Mohr, *Einfaher Beweis der verallgemeinerten Determinantensatzes von Sylvester nebst einer Verschärfung*, Math. Nachrichten **10** (1953), 257–260.

E. H. Moore, *General Analysis* Part I, Mem. Amer. Phil. Soc. **1** (1935), 197.

R. W. Newcomb, *On the simultaneous diagonalization of two semi-definite matrices*, Quart. Appl. Math. **19** (1961), 144–146.

L. B. Nisnevich and V. I. Bryzgalov, *On a problem of n-dimensional geometry*, Uspekhi Mat. Nauk **8** (1953), no. 4, 169–172. (Russian)

A. M. Ostrowski, *On Schur's complement*, J. Comb. Theory (A) **14** (1973), 319–323.

R. A. Penrose, *A generalized inverse for matrices*, Proc. Cambridge Phil. Soc. **51** (1955), 406–413.

R. Rado, *Note on generalized inverses of matrices*, Proc. Cambridge Phil.Soc. **52** (1956), 600–601.

A. Ramakrishnan, *A matrix decomposition theorem*, J. Math. Anal. and Appl. **40** (1972), 36–38.

M. Reid, *Undergraduate algebraic geometry*, Cambridge Univ. Press, Cambridge, 1988.

Yu. B. Reshetnyak, *A new proof of a theorem of Chebotarev*, Uspekhi Mat. Nauk **10** (1955), no. 3, 155–157. (Russian)

W. E. Roth, *The equations $AX - YB = C$ and $AX - XB = C$ in matrices*, Proc. Amer. Math. Soc. **3** (1952), 392–396.

H. Schwert, *Direct proof of Lanczos' decomposition theorem*, Amer. Math. Monthly **67** (1960), 855–860.

I. Sedláček, *O incidenčnich maticich orientových grafů*, Časop. Pest. Mat. **84** (1959), 303–316.

Z. Šidak, *O počtu kladných prvků v mochinách nezáporné matice*, Časop. Pest. Mat. **89** (1964), 28–30.

M. F. Smiley, *Matrix commutators*, Can. J. Math. **13** (1961), 353–355.

V. Strassen, *Gaussian elimination is not optimal*, Numerische Math. **13** (1969), 354–356.

H. Väliaho, *An elementary approach to the Jordan form of a matrix*, Amer. Math. Monthly **93** (1986), 711–714.

H. Zassenhaus, *A remark on a paper of O. Taussky*, J. Math. and Mech. **10** (1961), 179–180.

SUBJECT INDEX

Selected Titles in This Series

(Continued from the front of this publication)

109 **Isao Miyadera,** Nonlinear semigroups, 1992
108 **Takeo Yokonuma,** Tensor spaces and exterior algebra, 1992
107 **B. M. Makarov, M. G. Goluzina, A. A. Lodkin, and A. N. Podkorytov,** Selected problems in real analysis, 1992
106 **G.-C. Wen,** Conformal mappings and boundary value problems, 1992
105 **D. R. Yafaev,** Mathematical scattering theory: General theory, 1992
104 **R. L. Dobrushin, R. Kotecký, and S. Shlosman,** Wulff construction: A global shape from local interaction, 1992
103 **A. K. Tsikh,** Multidimensional residues and their applications, 1992
102 **A. M. Il′in,** Matching of asymptotic expansions of solutions of boundary value problems, 1992
101 **Zhang Zhi-fen, Ding Tong-ren, Huang Wen-zao, and Dong Zhen-xi,** Qualitative theory of differential equations, 1992
100 **V. L. Popov,** Groups, generators, syzygies, and orbits in invariant theory, 1992
99 **Norio Shimakura,** Partial differential operators of elliptic type, 1992
98 **V. A. Vassiliev,** Complements of discriminants of smooth maps: Topology and applications, 1992 (revised edition, 1994)
97 **Itiro Tamura,** Topology of foliations: An introduction, 1992
96 **A. I. Markushevich,** Introduction to the classical theory of Abelian functions, 1992
95 **Guangchang Dong,** Nonlinear partial differential equations of second order, 1991
94 **Yu. S. Il′yashenko,** Finiteness theorems for limit cycles, 1991
93 **A. T. Fomenko and A. A. Tuzhilin,** Elements of the geometry and topology of minimal surfaces in three-dimensional space, 1991
92 **E. M. Nikishin and V. N. Sorokin,** Rational approximations and orthogonality, 1991
91 **Mamoru Mimura and Hirosi Toda,** Topology of Lie groups, I and II, 1991
90 **S. L. Sobolev,** Some applications of functional analysis in mathematical physics, third edition, 1991
89 **Valeriĭ V. Kozlov and Dmitriĭ V. Treshchëv,** Billiards: A genetic introduction to the dynamics of systems with impacts, 1991
88 **A. G. Khovanskiĭ,** Fewnomials, 1991
87 **Aleksandr Robertovich Kemer,** Ideals of identities of associative algebras, 1991
86 **V. M. Kadets and M. I. Kadets,** Rearrangements of series in Banach spaces, 1991
85 **Mikio Ise and Masaru Takeuchi,** Lie groups I, II, 1991
84 **Đáo Tròng Thi and A. T. Fomenko,** Minimal surfaces, stratified multivarifolds, and the Plateau problem, 1991
83 **N. I. Portenko,** Generalized diffusion processes, 1990
82 **Yasutaka Sibuya,** Linear differential equations in the complex domain: Problems of analytic continuation, 1990
81 **I. M. Gelfand and S. G. Gindikin, Editors,** Mathematical problems of tomography, 1990
80 **Junjiro Noguchi and Takushiro Ochiai,** Geometric function theory in several complex variables, 1990
79 **N. I. Akhiezer,** Elements of the theory of elliptic functions, 1990
78 **A. V. Skorokhod,** Asymptotic methods of the theory of stochastic differential equations, 1989
77 **V. M. Filippov,** Variational principles for nonpotential operators, 1989
76 **Phillip A. Griffiths,** Introduction to algebraic curves, 1989
75 **B. S. Kashin and A. A. Saakyan,** Orthogonal series, 1989
74 **V. I. Yudovich,** The linearization method in hydrodynamical stability theory, 1989
73 **Yu. G. Reshetnyak,** Space mappings with bounded distortion, 1989
72 **A. V. Pogorelev,** Bendings of surfaces and stability of shells, 1988
71 **A. S. Markus,** Introduction to the spectral theory of polynomial operator pencils, 1988
70 **N. I. Akhiezer,** Lectures on integral transforms, 1988

(See the AMS catalog for earlier titles)